T0261999

Essentials of Artificial Photosynthesis

Essentials of
Artificial Photosynthesis

Edited by **Agatha Wilson**

New York

Published by Callisto Reference,
106 Park Avenue, Suite 200,
New York, NY 10016, USA
www.callistoreference.com

Essentials of Artificial Photosynthesis
Edited by Agatha Wilson

International Standard Book Number: 978-1-63239-327-2 (Hardback)

Printed in the United States of America.

Contents

Preface

Every book is a source of knowledge and this one is no exception. The idea that led to the conceptualization of this book was the fact that the world is advancing rapidly; which makes it crucial to document the progress in every field. I am aware that a lot of data is already available, yet, there is a lot more to learn. Hence, I accepted the responsibility of editing this book and contributing my knowledge to the community.

This book presents an updated account on the technique of artificial photosynthesis. Photosynthesis is one of the most vital processes on Earth, and is a scientific field that is essentially interdisciplinary, with various research groups analyzing it. By carefully studying photosynthesis, we can formulate novel strategies and apply them for use in artificial photosynthesis, a research field attempting to mirror the natural process of photosynthesis. The aim of artificial photosynthesis is to consume the energy of the sun to form various useful products or high-energy chemicals for energy production. This book is targeted at offering important and applicable aspects of artificial photosynthesis through the contributions made by various experts.

While editing this book, I had multiple visions for it. Then I finally narrowed down to make every chapter a sole standing text explaining a particular topic, so that they can be used independently. However, the umbrella subject sinews them into a common theme. This makes the book a unique platform of knowledge.

I would like to give the major credit of this book to the experts from every corner of the world, who took the time to share their expertise with us. Also, I owe the completion of this book to the never-ending support of my family, who supported me throughout the project.

Editor

Part 1

Introduction

Gathering Light:
Artificial Photosynthesis

Mohammad Mahdi Najafpour* and Sara Nayeri
*Chemistry Department, Institute for Advanced Studies in Basic Sciences
(IASBS), Zanjan,
Iran*

1. Introduction

Life is dependent on photosynthesis, directly or indirectly. Photosynthesis is converting sunlight, water, and carbon dioxide into carbohydrates and oxygen (Govindjee et al., 2010). Artificial photosynthesis is a research field that attempts to replicate the natural process of photosynthesis. The goal of artificial photosynthesis is to use the energy of the sun to make different useful material or high - energy chemicals for energy production. In this regard, a good question is that why artificial photosynthesis is important and necessary?
There are some reasons for believing that artificial photosynthesis is necessary in future. Two reasons are very important: the first, oil will become scarcer and more expensive and we should find a better source of energy. The second is causing serious environmental problems by fossil fuels.
Artificial photosynthesis could be divided into a series of approaches:

2. Antenna systems

Absorbing photos by an antenna pigment is the first stage in photosynthesis. Pigments can be a chlorophyll, xanthophylls, phycocyanin, carotenes, xanthophylls, phycoerythrin and fucoxanthin depending on the type of organism and a wide variety of different antenna complexes are found in different photosynthetic systems. Each year, the energy of 10^{24} Joule reaches the planet's surface through solar radiation. It is interesting that this is three orders of magnitude more than what is projected for the future global anthropogenic energy demand of
10^{21} Joule (Moore, 2005). The development of artificial antenna for collecting and harvesting solar energy efficiently is an active and complex field as the distance between the pigments to be used, their respective angle, and electronic coupling, must be engineered carefully. Sakata et al. (2001) have designed and synthesized a well-defined, rigid-sheet-structured oligoporphyrin (Fig. 1) with 21 porphyrin chromophores. The compound is a model for light harvesting compounds in *Nature* and showed promising properties for collecting and harvesting solar energy.

---

* Corresponding Author

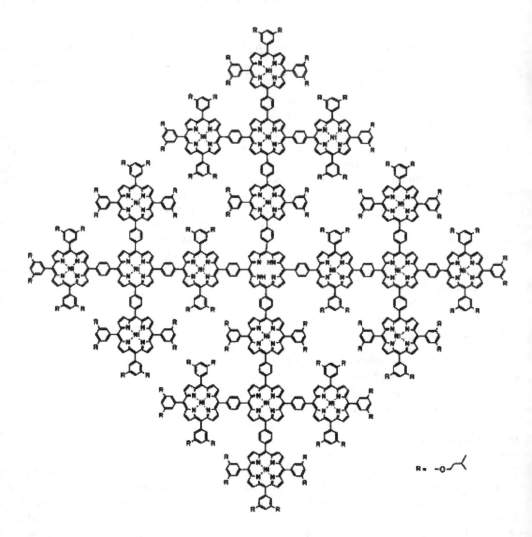

Fig. 1. An artificial antenna for collecting and harvesting solar energy

3. Photoactive chromophore

Synthetic porphyrin derivatives have been widely used to mimic the natural chlorophyll pigments to convert light energy into chemical energy. $[Ru(bpy)_3]^{2+}$ presents an absorption band in the region around 450 nm corresponding to a metal to ligand charge transfer (MLCT) band with an extinction coefficient of about $13,000M^{-1}cm^{-1}$. Upon irradiation in the MLCT band, the input light energy is converted into a $(d^6) \rightarrow (d^5\pi^*)$ excited state, which in turn relaxes to form the lowest triplet state (^{3}MLCT) in less than a picosecond. The oxidation potential of $[Ru(bpy)_3]^{3+}$ is similar to oxidized photosystem II primary donor in natural photosynthesis, therefore making it a suitable candidate to reproduce the oxidation reactions performed by the natural system.

4. Water oxidizing complex

The hydrogen production from water splitting is an appealing solution for the future energy as discussed by Bockris (Bockris, 1977). A strategy is to employ solar, wind, ocean currents, tides or waves energy to water splitting. However, to evolve hydrogen efficiently in a sustainable manner, it is necessary first to synthesize a stable, low cost, and efficient, environmentally friendly and easy to use, synthesis, manufacture catalyst for water oxidation, which is the more challenging half reaction of water splitting. In past few years, there has been a tremendous surge in research on the synthesis of various metal compounds aimed at simulating water oxidizing complex (WOC) of photosystem II. Particular attention has been given to the manganese compounds aimed at simulating the water oxidizing complex of photosystem II (Umena et al., 2011)) not only because it has been used by *Nature* to oxidize water but also because manganese is cheap and environmentally friendly.

5. The dye-sensitized solar cell approach

Photovoltaic is a method of generating electricity by converting light into electricity. Photovoltaic devices work base on the concept of charge separation. A new family of devices, the dye-sensitzed solar cell, is shown in Fig. 2. In the system, there is an oxide layer (for example TiO_2) which to allow for electronic conduction to take place. The material oxide layer is attached to a monolayer of the charge transfer dye. Photo excitation of the dye results in the injection of an electron into the conduction band of the oxide. Usually the iodide/triiodide couple restors the original state of the dye. The dye-sensitized solar cell made of low-cost materials, robust, does not require elaborate apparatus to manufacture, can be engineered into flexible sheets, requiring no protection from minor events like hail or tree strikes. Thus, there are technically attractive. In this devise, light is absorbed by a dye, the sensitizer is grafted onto the TiO_2 surface and then light induced electron injection from the adsorbed dye into the TiO_2 conductive (Grätzel, 2003).

In Fig. 3 comparing between the spectral response of the photocurrent observed with the two sensitizers and TiO_2 is shown (Grätzel, 2003). The incident photon to current conversion efficiency of the dye-sensitized solar cell is plotted as a function of excitation wavelength. Both chromophores show very high incident photon to current conversion efficiency values in the visible range. Some companies work to develop dye-sensitized solar cells for applications in cars and homes.

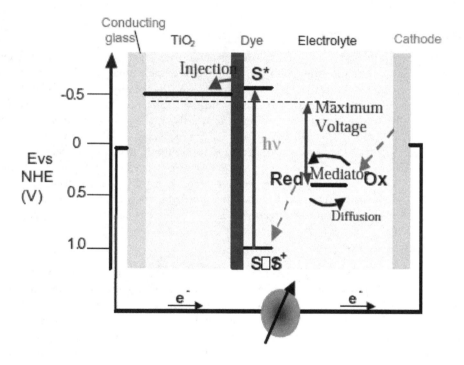

Fig. 2. A schematic presentation of the operating principles of the dye-sensitized solar cell
(The figure was reproduced from Grätzel, 2003).

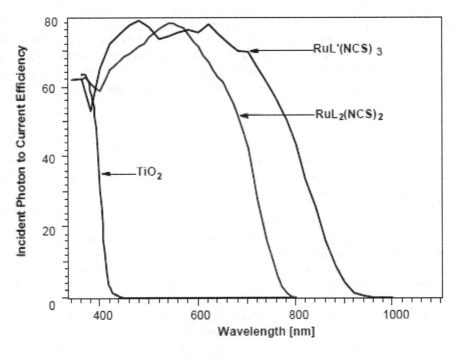

L = 4, 4'-COOH-2,2'-bipyridine
L = 4,4',4" -COOH-2,2':6',2"-terpyridine

Fig. 3. The incident photon to current conversion efficiency of the dye-sensitized solar cell is plotted as a function of excitation wavelength. (The figure was reproduced from (Grätzel, 2003)).

6. Algal systems for hydrogen

Biological hydrogen production is a method of photobiological water splitting which done based on the production of hydrogen by algae. In 1939, it was observed that a green - algae would sometimes switch from the production of oxygen to the production of hydrogen. In the late 1990s, professor Anastasios Melis discovered that if the algae culture medium is deprived of sulfur it will switch from the production of oxygen (normal photosynthesis), to the production of hydrogen. However, under normal conditions where oxygen is a by-product of photosynthesis, sustained algal hydrogen photoproduction cannot be maintained for more than a few minutes. Many research groups are currently trying to find a way to take the part of the hydrogenase enzyme that creates the hydrogen and introduce it into the photosynthesis process. These include molecular engineering of the hydrogenase to remove the oxygen sensitivity and development of physiological means to separate oxygen and hydrogen production. The result would be a large amount of hydrogen, possibly on par with the amount of oxygen created (Federico Rossi & Mirko Filipponi, 2011).

7. Carbon capture and storage

The concentration of carbon dioxide in the atmosphere has risen from 280 to 370 PPM from 1860 to recent years. Industrial emission of CO_2 into the earth's atmosphere presently exceeds 10^{10} tons per year. Storage of the CO_2 either in deep geological formations, in deep ocean masses, or in the form of mineral carbonates is a way to decrease of CO_2 in atmosphere. In the case of deep ocean storage, there is a risk of decreasing pH an issue that also stems from the excess of carbon dioxide already in the atmosphere and oceans. In this regard, several concepts have been proposed. Injection CO_2 by ship or pipeline into the ocean water column at depths of 1000 – 3000 m, forming an upward-plume, and the CO_2 subsequently dissolves in seawater, injecting CO_2 directly into the sea at depths greater than 3000 m, where high-pressure liquefies CO_2, making it denser than water, and is expected to delay dissolution of CO_2 into the ocean and atmosphere, storing CO_2 in solid clathrate hydrates already existing on the ocean floor or using a chemical reaction to combine CO_2 with a carbonate mineral. Geological formations are currently considered the most promising sequestration sites. Geological storage involves injecting carbon dioxide, generally in supercritical form, directly into underground geological formations. Various physical and geochemical trapping mechanisms would prevent the CO_2 from escaping to the atmosphere. Recycling CO_2 is likely to offer the most environmentally and financially sustainable response to the global challenge of significantly reducing greenhouse gas emissions. Using artificial photosynthesis, scientists try to find a way to produce useful organic compounds from CO_2. For example, CO_2 and other captured greenhouse gases could be injected into the membranes containing waste water and select strains of organisms causing an oil rich biomass that doubles in mass every 24 hours or to convert CO_2 into hydrocarbons where it can be stored or reused as fuel or to make plastics (Cook, 2005).

8. Ribulose-1,5-bisphosphate carboxylase oxygenase

Ribulose-1,5-bisphosphate carboxylase oxygenase (RuBisCO) is an enzyme involved in the Calvin cycle that catalyzes a process by which CO_2 are made available to organisms in the form of energy-rich molecules such as glucose. RuBisCO catalyzes either the carboxylation or the oxygenation of ribulose-1,5-bisphosphate.

It is believed that RuBisCO is rate-limiting for photosynthesis in plants and it is proposed that may be possible to improve photosynthetic efficiency by modifying RuBisCO genes in plants to increase its catalytic activity (Spreitzer & Salvucci, 2002). Engineered changes in Rubisco's properties Unpredictable expression of plastid transgenes and assembly requirements of some foreign Rubiscos that are not satisfied in higher-plant plastids provide challenges for future research.

There are the most important titles in artificial photosynthesis but we could increase our list as important titles in artificial photosynthesis and as it is considered by Pace, artificial photosynthesis is an umbrella term. As you see in each title, inspired by natural photosynthesis, in artificial photosynthesis novel approaches used to develop technologies for non-polluting electricity generation, fuel production and carbon sequestration using solar energy. (Pace, 2005). Researchers and scientists are trying to learn a great about the detail of natural photosynthetic systems and have been able to understand at least parts of this process. Therefore, artificial photosynthetic goal and capable of converting sunlight into chemically-bound energy seem to be a realistic scenario in near future.

9. Acknowledgment

Authors are grateful to Institute for Advanced Studies in Basic Sciences for financial support.

10. References

Bockris, J.O.M. (1977) Energy-the solar hydrogen alternative, Wiley&Sons, New York.

Cook; P. J. (2005). Greenhouse gas technologies: A pathway to decreasing carbon intensity. In: Collings, A. F. & Critchley, C. (eds.): Artificial Photosynthesis: From Basic Biology to Industrial Application. Weinheim 2005, pp. 13-34).

Govindjee; Kern, J.F.; Messinger, J.& Whitmarsh, J. (2010) Photosystem II. In: Encyclopedia of Life Sciences (ELS). John Wiley & Sons, Ltd: Chichester

Grätzel, M. (2003) Dye-sensitized solar cells. *J. Photochem. Photobiol., C*, vol.4, pp. 145-153.

Moore, T. A. (2005) Bio-inspired energy security for planet earth. Photochem. Photobiol. Sci, v.4, pp. 927-927).

Pace, 2005 R. J.: An Integrated Artificial Photosynthesis Model. In: Collings, A. F. & Critchley, C. (eds.): Artificial Photosynthesis: From Basic Biology to Industrial Application. Weinheim 2005, pp. 13-34.

Rossi, F. & Filipponi, M. (2011) Hydrogen production from biological systems under different illumination conditions. *Int. J. Hydrogen. Energ.*, v.36, p7479-7486.

Sakata, Y.; Imahori, H. & Sugiura, K. (2001) Molecule-based artificial photosynthesis. *J. Incl. Phenom. Macro.*, v.41, 31-36.

Spreitzer, R.J. & Salvucci, M.E. (2002). Rubisco: structure, regulatory interactions, and possibilities for a better enzyme. *Annu. Rev. Plant. Biol.*, v.53, pp. 449–475.

Umena, Y.; Kawakami, K.; Shen, J.R. & Kamiya, N. (2011) Crystal structure of oxygen-evolving photosystem II at a resolution of 1.9Å. *Nature*, v.473, p.55-60

Part 2

Fundamental Aspects

2

Artificial Photosynthesis from a Chemical Engineering Perspective

Bahar Ipek and Deniz Uner
Middle East Technical University,
Turkey

1. Introduction

Green plants and photosynthetic bacteria are responsible for storing solar energy in chemical bonds via photosynthesis. Photosynthesis is not only the major source of food, fuel and oxygen on earth, but it is also the key player in the global carbon cycle by converting 120 gigatonnes of carbon per year.

Conversion of solar energy into chemical energy through utilization of inorganic materials by photocatalytic CO_2 reduction; which is also known as *'Artificial Photosynthesis'* is the next challenge for a sustainable development. In the present state-of-the art artificial photosynthesis processes, nature is so far mimicked only to the extent that CO_2 is reduced by water to valuable 1- carbon chemicals, not to the multi-carbon equivalents of glucose or cellulose yet. Although mimicking nature is viable by photocatalytic means, enhancing photocatalytic CO_2 reduction rates is vital in order to achieve artificial photosynthesis in industrial scales. To illustrate the gap between photosynthetic and photocatalytic rates, we will compare the turnover frequencies of water oxidation process below.

Water oxidation is the key step both in photocatalysis and photosynthesis for being the carbon free hydrogen source and also for providing oxygen for the oxygen consuming organisms. Completion of an S cycle taking place in a Mn_4 cluster which is responsible for water oxidation was reported to last for 1.59 ms in order to produce one molecule of oxygen at that one particular site (Haumann et al., 2005). In other words, molecular oxygen is produced in photosynthesis, with a turn over frequency of 630 molecule/site/s. On the other hand, typical rates of photocatalytic synthesis of hydrocarbons are of the order of 30-μmoles/g cat/h, (Ozcan et al., 2007; Uner et al., 2011) which amounts to $1.11*10^{-5}$ molecule/site/s if the typical surface areas of 45 m^2/g cat and typical site densities of $10^{15}/cm^2$ are used. Of course the remarkable rates of 9 μmoles of O_2/cm^2/s (Kanan & Nocera, 2008), giving a turn over frequency of 5400 molecule/site/s for an oxygen evolving cobalt- phosphate catalyst operating at neutral water is keenly followed by the academic community. Considering the huge gap between photosynthetic and photocatalytic rates reported above, one can easily claim that there is room for further investigation and development in photocatalytic CO_2 reduction systems.

It is also important to see the thermodynamic energy demand of the some of the reactions between CO_2 and H_2O. For this, a number of products are chosen and the standard Gibbs free energy of formation values are listed in Table 1.1 for comparison. The interesting observation that we make in this table is the following: when compared per mole of hydrocarbon formed, the Gibbs free energies of formation increase with increasing carbon chain length. But when the Gibbs free energy formation values are normalized per mole O_2 formed, one can compare the energy demand of the reactions on a common basis. A close examination of the data in the last column reveals the fact that energetically almost all of the reactions are similar. The second conclusion we can arrive at is that once the water splitting reaction is possible, the formed hydrogen can drive the subsequent reduction reactions, almost spontaneously.

Reaction	ΔG_f (kJ/mol HC product)	ΔG_f (kJ/mol O_2)
$CO_2 + 2H_2O \Rightarrow CH_4 + 2O_2$	801.0	400.5
$CO_2 + 2H_2O \Rightarrow CH_3OH + 1\frac{1}{2}O_2$	689.2	459.5
$2CO_2 + 3H_2O \Rightarrow C_2H_5OH + 3O_2$	1306.6	435.5
$H_2O + CO_2 \Rightarrow 1/6\ C_6H_{12}O_6 + O_2$	2880.0	480.0
$H_2O \Rightarrow H_2 + \frac{1}{2}O_2$	228.6	457.2

Table 1.1. The thermodynamics of the reactions involved in carbon dioxide reduction

2. Photosynthesis

2.1 Overview

Photosynthesis is the world's most abundant process with an approximate carbon turnover number of 300- 500 billion tons of CO_2 per year. In this vital process, green plants, algae and photosynthetic bacteria are converting CO_2 with water into carbohydrates and oxygen (in oxygenic photosynthesis), both of which are essential for sustaining life on earth. Oxygenic photosynthesis is believed to be started 2.5 billion years ago by the ancestors of cyano bacteria. In this remarkable process, energy need for converting stable compounds (CO_2 and H_2O) into comparably less stable arranged molecules (($CH_2O)_n$ and O_2) is supplied from solar energy in which highly sophisticated protein complexes embedded in an internal chloroplast membrane (called thylakoid membrane) are major players.

$$6CO_2 + 12\ H_2O + Light\ Energy \rightarrow C_6H_{12}O_6 + 6O_2 + 6H_2O$$

$$\Delta G^0 = 2880\ kJ\ /mol\ C_6H_{12}O_6$$

Harnessing solar energy into chemical bonds in this process is achieved by light absorption and sequential electron and proton transport processes in which a great deal of number of light harvesting pigments, protein complexes and intermediate charge carriers are involved. CO_2 is being reduced with the indirect products of water oxidation; supplying required energy in the form of redox free energy (from NADPH) and high energy Pi bonds (from ATP).

Overall process can be shown in the reaction scheme below where D: electron donor, A: electron acceptor and T: energy trap (Govindjee, 1975).

Water oxidation: $D^+ \cdot T \cdot A + \frac{1}{2}H_2O \rightarrow D \cdot T \cdot A + \frac{1}{4}O_2 + H^+$

$NADP^+$ reduction: $D^+ \cdot T \cdot A^- + \frac{1}{2}NADP^+ + H^+ \rightarrow D^+ \cdot T \cdot A + \frac{1}{2}NADPH + \frac{1}{2}H^+$

Cyclic Photophosphorylation: $D^+ \cdot T \cdot A^- + ADP + P_i \rightarrow D \cdot T \cdot A + ATP$

CO_2 reduction: $CO_2 + 2\ NADPH + 3\ ATP \rightarrow (CH_2O) + 2NADP^+ + 3\ ADP + 3P_i$

2.2 Reactions

Photosynthesis includes a series of photophysical, photochemical and chemical reactions realized by highly sophisticated protein complexes, energy carriers and enzymes. With all the complexity of their mechanisms, reactions involved in photosynthesis are mainly divided into two stages: (i) light dependent reactions including water oxidation and chemical energy generation through electron and proton transport and (ii) light independent reactions including CO_2 fixation, reduction and regeneration of ribulose 1,5 biphosphate (Calvin Cycle).

2.2.1 Light induced reactions

The light induced reactions occur in a complex membrane system (thylakoid membrane) via electron transfer through light induced generation of cation- anion radical pairs and intermediate charge carriers such as plastoquinone, plastocyanin and ferrodoxin. Light dependent reactions in green plants follow a Z scheme which was first proposed by Hill & Bendall, 1960 (Figure 2.1). In this scheme, light energy is absorbed by light harvesting molecules and funneled to two special reaction center molecules; P680 and P700 which are acting as major electron donors in PS II and PSI respectively. Electron transport from PSII to PS I is realized by intermediate charge carriers and electron need of P680+ (strong oxidant with E^0 = 1.1 eV) in PSII is compensated from water molecules (water oxidation).

Electron transport through thylakoid membrane and water oxidation reactions results in a proton concentration gradient across the thylakoid membrane. Energy created by proton electrochemical potential resulting from this proton gradient is used by ATP synthase to produce ATP from ADP and Pi. The net reaction in light dependent reaction system is the electron transport form a water molecule to a NADP+ molecule with the production of ATP molecules (Figure 2.2).In this complex electron transport system, PS II alone is composed of more than 15 polypeptides and nine different redox components including chlorophylla and b, pheophytin, plastoquinone.

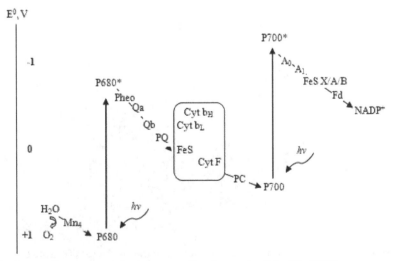

Fig. 2.1. Z scheme electron transfer in terms of redox potentials (Ke, 2001)

Photosystem II is the only protein complex with the capability of oxidizing water into O_2 and protons. In PS II, water is oxidized with an Oxygen Evolving Complex whose components are revealed to be in the form of Mn_4CaO_5 (Umena et al., 2011). This inorganic core oxidizes two water molecules in Kok cycle, comprised of five oxidation states (S states) of PSII donor site. In this model, oxygen formation requires successive four light flashes for four-electron and four-proton release. Recently, presence of an intermediate S_4' state and kinetics of completion of final oxidation cycle responsible for O- O bond formation was revealed with time resolved X ray study of Haumann et al. (Figure 2.3) (Haumann et al., 2005).

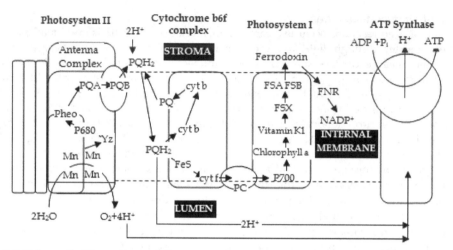

Fig. 2.2. Schematic illustrations of electron and proton transport processes and ATP synthesis in light dependent reactions (Hankamer et al., 1997)

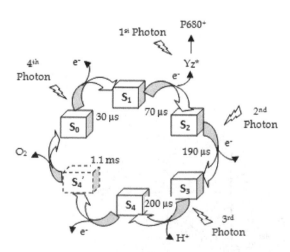

Fig. 2.3. Extension of classical S state cycle of the manganese- calcium complex (Haumann et al., 2005)

This high energy requiring water oxidation reaction with four- proton, four- electron extraction and an oxygen- oxygen bond formation (with a standard free energy requirement of 312 kJ/mol of O_2) necessitates the regeneration of the oxygen evolving complex at every half an hour in order to repair the damage caused by the oxygen production (Meyer, 2008).

In electron transfer from the oxygen evolving complex (OEC) to P680+ molecule, tyrosine (Y_z*) acts as intermediate electron carrier. Protons evolved from OEC are deposited in lumen phase contributing proton concentration gradient (ΔpH) mentioned in ATP synthesis part. Excited electron upon light absorption is transferred to the cytochrome b6f complex through a pheophytin, a tightly bound phylloquinone (Q_A) and a mobile phylloquinone (Q_B). Subsequently reduced phylloquinol (PQH_2) (reduced with electrons from P680* and two protons from stromal phase) releases two additional protons into lumen phase as it binds to cyctochrome b6f complex after diffusion through thylakoid membrane. Electron transfers from cyctochrome b6f complex to PS I (through lumen phase) and from PS I to NADP+ molecule (through stromal phase) are achieved by plastocyanin and ferrodoxin respectively. ATP synthesis reaction in light dependent reactions is driven by the proton electrochemical and charge potential across the membrane resulted from proton concentration difference and charge separation during illumination respectively. NADPH and ATP molecules produced as such are used as energy and proton sources in carbon dioxide reduction reactions in Calvin Cycle.

2.2.2 Dark reactions

The reactions that do not involve solar energy directly are somewhat roughly called the dark reactions. These reactions take place in outer space of thylakoid membrane which is also known as stromal phase. CO_2 enters the leaf structure through small holes called stomata and diffuses into stromal phase in the chloroplast where it is being reduced with reactions in series that are catalyzed by more than ten enzymes. Driving force for the reduction reaction is supplied from NADPH and ATP molecules; hence, the 'catalytic' reaction sequence does not require light as an energy source and called as light independent reactions. However, recent findings indicate light activation of enzymes due to regulatory processes (reductive pentose phosphate).

Melvin Calvin and his collaborators were the first to resolve the photosynthetic CO_2 reduction mechanism with studies involving radioactively labeled CO_2. The Calvin Cycle, also known as reductive pentose phosphate pathway consists of three sections:

1. CO_2 fixation by carboxylation of rubilose 1,5- bisphosphate to two 3-phosphoglycerate molecules,
2. Reduction of 3-phosphoglycerate to triose phosphate, and
3. Regeneration of rubilose 1,5- bisphosphate from triose phosphate molecules (Figure 2.4).

The key reaction in photosynthetic CO_2 reduction is the fixation of a CO_2 molecule to rubilose 1,5- bisphosphate to two phosphoglycerate molecules with a standard free energy of -35 kJ/mol indicating its irreversibility. This reaction is catalyzed with the Ribulose biphosphate Carboxylase/Oxygenase (RubisCO) enzyme which is one of largest enzymes in nature with its 8 large, 8 small subunits (with molecular weights changing from 12 to 58 kDa). This enzyme also catalyzes a side reaction, oxygenation, to give a 3-phospho glycerate and a 2- phosphoglycolate instead of two 3- phosphoglycerates for CO_2 fixation. Although oxygenation occurs with a ratio of 1:4 to 1:2 (oxygenation : carboxylation), oxygenation ratio decreases as CO_2 concentration in the atmosphere is increased. This regulatory measure of photosynthesis is worth appreciation.

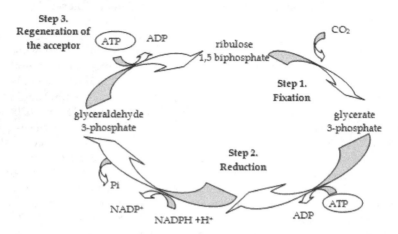

Fig. 2.4. The Calvin Cycle

In carboxylation reaction catalyzed by RubisCO, rubilose 1,5- bisphosphate (RuBP) accepts CO_2 to form a keto intermediate after keto-enol isomerization (Figure 2.5). For the synthesis of glyceraldehydes 3- phosphates, firstly 3- phosphoglyerates are phospholyrated to 1,3-bisphosphoglycerate with phosphoglycerate kinase enzyme. Afterwards, 1,3-biphosphoglycerate is reduced with NADPH to glyceraldehydes 3- phosphate with glyceraldehydes phosphate dehydrogenase enzyme. Redox potential difference between the aldehyde and carboxylate is overcome with the consumption of ATP (Figure 2.6).

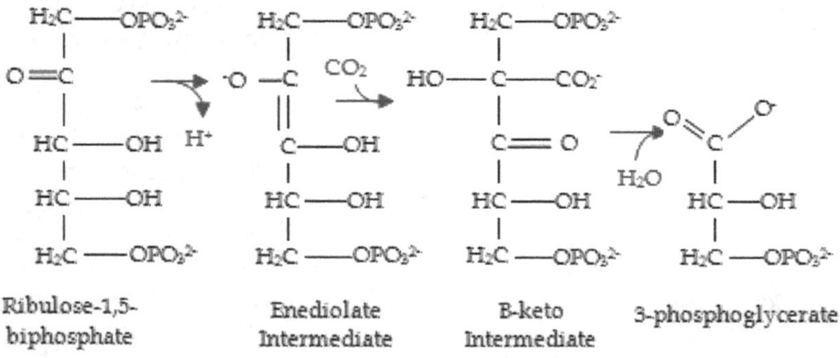

Fig. 2.5. Reaction sequence of carboxylation of RuBP by RubisCO (Diwan, 2009)

After production of glyceraldehyde 3- phosphates, out of six aldehydes produced by fixation of three CO_2 molecules, five of them are used in regeneration of three RuBP molecules together with ATP consumption. Remaining one molecule of glyceraldehyde 3-phosphate is transported into the cytosol for utilization in glucose synthesis.

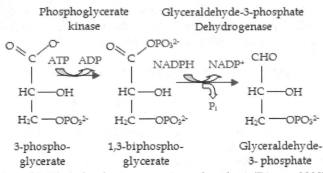

Fig. 2.6. Conversion of 3- Phosphoglycerate to triose phosphate (Diwan, 2009)

2.3 Transport processes

The vesicular thylakoid membrane structure defines a closed space separating outside water phase (stromal phase) and inside water phase (lumen phase). CO_2 fixation reactions occur in the stromal phase while majority of light dependent reactions are realized in the complex membrane system with embedded protein complexes and intermediate charge carriers.

As mentioned in light dependent reactions, electron and proton transport processes through protein complexes and intermediate charge carriers like plastoquinone, plastocynanin and ferrodoxin molecules play an important role in controlling photosynthetic rates. Within a protein complex such as PSII or cytochrome bf complex, electron transfer and pathway is controlled by polypeptide chains of the protein. However between protein complexes, electron transfer via electron carriers is controlled by distance and free energy. Below, electron and proton transport processes taken place in light dependent reactions are illustrated with particle sizes of protein complexes given by Ke, 2001.

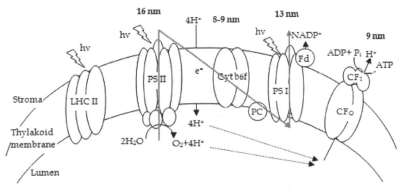

Fig. 2.7. Distribution of photosynthetic complexes in thylakoid membrane and the corresponding Z scheme (Ke, 2001)

Presence of the membrane affects reaction rates in an aspect that it limits electron and proton transport to two dimensions which increases the random encounters. Furthermore, electron transport reactions and special structure and orientation of the membrane and protein complexes contribute to a proton electrochemical potential difference which drives

ATP synthesis reaction; i.e., plays a significant role in energy supply of photosynthesis. The proton electrochemical potential difference across the membrane is created by two main contributions; i. proton concentration gradient (pH difference), and ii. electric potential difference.

The processes contributing *proton concentration difference (ΔpH)* across the membrane can be listed as below:

1. Proton release to the lumen phase as a consequence of water oxidation reaction at PS II.
2. Proton uptake from stromal phase for PQ reduction.
3. Proton release into lumen phase during PQH_2 oxidation at cytochrome b6f complex.
4. $NADP^+$ reduction at stromal phase.

On the other hand, vectoral electron transfer process in PS II and PS I initiated by photon absorption could be accounted as the reason for *electric potential difference (ΔΨ)*. Whitmarsh & Govindjee, 1999 gave the proton electrochemical potential difference with Equation (1).

$$\Delta\mu_{H+} = F\Delta\Psi - 2.3RT\Delta pH \qquad (1)$$

Where F is the Faraday constant, R is the ideal gas constant and T is temperature in Kelvin. They reported that although electric potential difference can be as large as 100 mV, pH difference has a dominating effect in overall electrochemical potential. For a pH difference of 2 (with inner pH 6 and outer pH 8, ΔpH equivalent to 120 mV), the free energy difference across the membrane is about -12 kJ/mol of proton.

In photosynthesis, fastest reactions taking place are the photophysical reactions like light absorption and charge separation in picoseconds orders. They are followed by rapid photochemical processes like electron transfer reactions and with slower biochemical reactions like water splitting and CO_2 reduction.

Since photosynthesis is a series of reactions including photophysical, photochemical and chemical reactions, reaction rates of particular reactions are dependent upon transfer rates of reaction intermediates like electrons or protons. In Figures 2.8 and 2.9, electron transfer times in PS II and PS I are given to illustrate characteristic times of different processes.

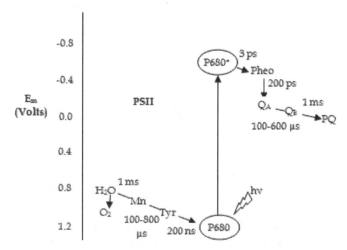

Fig. 2.8. PS II electron transport pathways and transfer times with midpoint potentials of electron carriers (Whitmarsh & Govindjee, 1999)

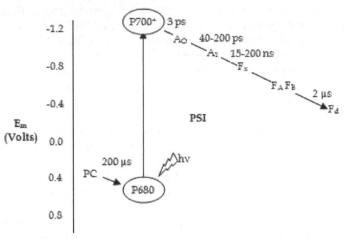

Fig. 2.9. PS I electron transport pathways and transfer times with midpoint potentials of electron carriers (Whitmarsh & Govindjee, 1999)

Water oxidation and CO_2 reduction reactions are the slowest processes in photosynthesis. S-cycle taking place in PS II for water oxidation is completed with a total of 1.59 ms, which is equivalent to production of 630 molecule of O_2/site/s. On the other hand, turnover frequency given for a subunit of RubisCO for CO_2 reduction is given as 3.3 s^{-1} (Heldt, 2010), which is much slower than oxygen evolution. Average photosynthesis rate of a sunflower was given as 13.5 μmol/m^2/s by Whittingham, 1974 and as 12 μmol/m^2/s for Brassica pods with an internal CO_2 concentration of 292 ppm by Singal et al., 1995 where rate of dark CO_2 fixation was given as 400 nmol/ mg protein/h.

In Table 2.1, time characteristics are unraveled for photosynthesis and artificial photosynthesis which indicates similarity in photochemistry but significant difference in time characteristics of chemical reactions.

Photosynthesis		Artificial Photosynthesis (on TiO_2)	
Charge carrier generation	ps	Charge carrier generation	ps
Charge trapping	ps-ns	Charge trapping	10 ns
Electron transport	ns- μs	Interfacial charge transfer	100 ns
Water oxidation	1.59 ms	Water oxidation	670 ms[a]
CO_2 reduction	300 ms	CO_2 reduction	14950 s[b]

Table 2.1. Time characteristics of major processes realized in photosynthesis and artificial photosynthesis

a. Oxygen evolution with cobalt ITO electrode (Kanan & Nocera, 2008)
b. Considering 200 μmole/g_{cat}/h activity and 50 m^2/g_{cat} and 10^{15} sites/cm^2

3. Artificial photosynthesis

Since the pioneering work of Inoue et al., 1979, CO_2 is being reduced with H_2O photocatalytically to mainly one carbon molecules like methane and methanol in the presence of a photo-activated semiconductor. Electrons, generated upon illumination of the semiconductor, are trapped at the electron trap centers and utilized directly in reduction centers without a complex transportation system involving intermediate charge carriers. Similarly holes, generated upon illumination are utilized in oxidation reactions on the catalyst surface. Since photocatalysis lacks a specialized transportation system for generated electrons and holes, majority of the charge carriers recombine at the catalyst surface or in the bulk volume of the catalyst, lowering photocatalytic rates. (Figure 3.1)

Semiconductors, having a band gap, ensure a life-time for generated electrons and holes; however, this lifetime is limited to 10^{-7} s, which is the characteristic time of recombination (for bare TiO_2) (Carp et al., 2004). In order to increase this lifetime of photo-generated electrons and holes, some modifications on materials such as metal addition to semiconductors (Anpo et al., 1997, Tseng et al., 2002, Ozcan et al., 2007) or formation of solid-solid interfaces in composite catalysts (Chen et al., 2009, Woan et al., 2009) were proposed in literature.

Metal addition to semiconductors is suggested to show charge separation effect on photocatalysis by the Schottky Barrier Formation. When metals are brought into contact with semiconductors, electrons populate on metals if Fermi level of the metal is lower than the conduction band of the semiconductor. Hence, metals act like 'charge carrier traps', increasing lifetime of electron hole pairs with charge separation effect.

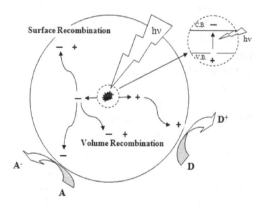

Fig. 3.1. Illustrating scheme of electron/hole pair generation and realization of redox reactions in photocatalysis

The other modification that can hinder recombination of generated electrons and holes is formation of solid-solid interfaces in composite photocatalysts having different band gap energies. To illustrate; commercial TiO_2 catalysts; Degussa P-25, is composed of anatase and rutile crystal phases of TiO_2, having band gap energies of 3.2 eV and 3.0 eV respectively. Mixed phase TiO_2, tends to exhibit higher photocatalytic activity than pure phases, because it allows transfer of the photogenerated electron from rutile to anatase, resulting in charge separation (Carp et al., 2004, Chen et al., 2009).

In photocatalysis, reduction and oxidation reactions occur at similar sites on the same catalyst surface. There are no different reaction centers with certain distances apart.

The half reactions of the photosynthesis; water oxidation and CO_2 reduction, are also realized in photocatalysis.

$$semiconductor \xrightarrow{hv} e^- + h^+$$

$$2H_2O + 4h^+ \rightarrow O_2 + 4H^+$$

$$CO_2 + 6H^+ + 6e^- \rightarrow CH_3OH + H_2O$$

$$CO_2 + 8H^+ + 8e^- \rightarrow CH_4 + 2H_2O$$

In photocatalysis; surface adsorbed species should have appropriate redox potentials regarding flat band positions of the semiconductor for thermodynamic favorability of the reactions, or vice versa. It required that; semiconductors should have conduction bands located at a more negative potential than the reduction potential of CO_2 to hydrocarbons, and valence bands located at a more positive potential than the oxidation potential of H_2O. In Table 3.1 oxidation and reduction reactions taking place in photocatalytic CO_2 reduction is listed with their electro-potentials at pH=7, vs NHE. In order to provide thermodynamic favorability, large band gap semiconductors such as TiO_2 are mostly utilized in photocatalytic CO_2 reduction reactions which render UV light illumination obligatory for photo-activation of the catalysts. Visible light utilization in carbon dioxide reduction is the ultimate goal in photocatalytic studies for a complete carbon free energy generation. There are material modification studies in literature conducted for efficient visible light utilization such as dye sensitization (Ozcan et al., 2007) and anion doping (Asahi et al., 2001).

Reactions	E^0 (V)
$2H_2O + 4h^+ \rightarrow O_2 + 4H^+$	+0.82
$CO_2 + 2H^+ + 2e^- \rightarrow HCOOH$	-0.61
$CO_2 + 2H^+ + 2e^- \rightarrow CO + H_2O$	-0.52
$CO_2 + 4H^+ + 4e^- \rightarrow HCHO + H_2O$	-0.48
$CO_2 + 6H^+ + 6e^- \rightarrow CH_3OH + H_2O$	-0.38
$CO_2 + 8H^+ + 8e^- \rightarrow CH_4 + 2H_2O$	-0.24

Table 3.1. Half cell reactions and their electro-potentials at pH=7 vs NHE (Jitaru, 2007)

One disadvantage of the realization of reduction and oxidation reactions at the same surface is the interaction between reactants on the surface. In one of the gas phase photocatalytic CO_2 reduction experiments realized on Cu/TiO_2 surfaces, when Langmuir- Hinshelwood surface reaction mechanism was selected with competitive adsorption of H_2O and CO_2, it was found that adsorption constant of H_2O dominates that of CO_2 (Wu et al., 2005) indicating that surface is mainly covered with water.

Presence of that much of water on TiO_2 surface would inhibit CO_2 activation ($CO_2+e^- \rightarrow CO_2^{\bullet-}$ E^0_{redox}= -1.9 V vs. NHE at pH 7), which is considered as the essential step in CO_2 reduction (Solymosi & Tombacz, 1994), by oxidizing defect structure of CO_2. Electron affinity of CO_2 molecule is related to the position of lowest unoccupied molecular orbital of CO_2 and conduction band of TiO_2, assuming that electron is transferred from excited state of TiO_2 (Ti^{+3}-O^-) to CO_2. A decrease in lowest unoccupied molecular orbital (LUMO) of CO_2 was reported with lower bond angles that could result from the interaction of the molecule with the surface (Freund & Roberts, 1996). According to Indrakanti et al., 2009, CO_2 gains electrons from oxygen deficient TiO_2 via the formation of bent CO_2 molecules near Ti^{+3} sites, whereas electron transfer is not favorable with defect free TiO_2 due to high LUMO of CO_2.

Initial photocatalytic carbon dioxide reduction rates from literature were summarized in Table 3.2. In those studies, carbon dioxide is reduced with water on the same catalyst surface in a batch reactor system.

Results presented at the table and in photocatalytic carbon dioxide reduction studies in general indicated very low carbon dioxide reduction yields when compared to photosynthetic and even to the catalytic carbon dioxide reduction yields. To illustrate, methanol was reported to be produced with a rate of 220 000 $\mu mol/g_{cat}/h$ with a $Cu/ZnO/Al_2O_3$ catalyst at 45 bar and 250 °C (Sahibzada et al., 1998). For a more proper comparison of catalytic and photocatalytic reduction rates, the results should be evaluated at the same conditions; i.e., at the same temperature and pressure. By this way, one could reveal the 'photocatalytic' effect in carbon dioxide reduction mechanism. Since catalytic carbon dioxide hydrogenation rate data are not available at ambient conditions, the comparison was based on the kinetic models of methanol synthesis on copper surfaces.

BATCH REACTORS					
GAS PHASE			*LIQUID PHASE*		
Photocatalyst	Initial Rates ($\mu mol*gcat^{-1}*h^{-1}$)		Photocatalyst	Initial Rates ($\mu mol*gcat^{-1}*h^{-1}$)	
	CH$_4$	CH$_3$OH		CH$_4$	CH$_3$OH
TiO$_2$ (Anpo & Chiba, 1992)	0,11	0,02	TiO$_2$ (Dey et al., 2004)	5,94	
JRC TiO$_2$ (Anpo et al., 1995)	0,17		TiO$_2$ (Degussa P25) (Tseng et al., 2002)		6,37
Cu/TiO$_2$ (Yamashita et al., 1994)	0,013	0,0015	Cu/TiO$_2$ (Tseng et al., 2002)		19,75
Ti-SBA-15 (Hwang et al., 2005)	63,60	16,62	TiO$_2$/SBA-15 (Yang et al., 2009)		627
Ti-MCM-48, (Anpo et al., 1998)	4,5	1,5	Cu/TiO$_2$/SBA-15 (Yang et al., 2009)		689,7
Pt/ Ti-MCM-48 (Anpo et al., 1998)	7,5	0,48	TiO$_2$ anatase (Koci et al., 2009)	0,38	0,045
TiO$_2$ (Kitano et al., 2007a)	0,2	0,003	Ag/ TiO$_2$ (Koci et al., 2010)	0,38	0,075
Ex-Ti-oxide/ Y-zeolite (Anpo et al., 1997)	4,2	2,4	TiO$_2$ (Kaneco et al., 1998)	0,72	
Pt TiNT (Zhang et al., 2009)	0,07		Rh /TiO$_2$ /WO$_3$ (Solymosi & Tombacz, 1994)		2,7
CdSe/Pt/TiO$_2$ (C.J. Wang et al., 2010)	0,61	0,04	NiO InTaO$_4$ (Z.Y. Wang et al., 2010)		2,8
NT/Cu-600 (Varghese et al., 2009)	2,84		CoPc TiO$_2$ (Zhao et al. 2009)		9,3

Table 3.2. Photocatalytic carbon dioxide reduction rates from literature

3.1 Catalytic vs photocatalytic CO_2 reduction

A microkinetic analysis for catalytic CO_2 hydrogenation on Cu (111) surface was performed in order to reveal catalytic rates at ambient conditions and also to study the effect of water in carbon dioxide reduction mechanism. The microkinetic analysis was expected to reveal rate determining steps in photocatalytic carbon dioxide reduction based on the assumption of similar reaction mechanism with catalytic hydrogenation since both processes involve copper based catalysts for methanol production.

In photocatalytic carbon dioxide reduction mechanism, whether every step in the mechanism is light activated or not is still ambiguous. One consensus could be the carbon dioxide activation to be the essential step for reduction (Solymosi & Tombacz, 1994). But after CO_2 activation, reduction could proceed catalytically.

Presence of catalytic steps in CO_2 hydrogenation in artificial photocatalysis (just like in photosynthesis) was suggested based on the observations from literature such as temperature sensitivity of the photocatalytic CO_2 reduction (Chen et al., 2009, Zhang et al., 2009, Z.Y. Wang et al., 2010) and realization of reduction without light illumination at room conditions (Varghese et al. 2009).

In the microkinetic analysis, the mechanism was selected to include water gas shift reaction together with the carbon dioxide hydrogenation since carbon monoxide and carbon dioxide were simultaneously used in industry for better methanol production rates.

$$CO_2 + 3H_2 \leftrightarrow CH_3OH + H_2O$$

$$CO + H_2O \leftrightarrow CO_2 + H_2$$

The reaction mechanism could be seen in Table 3.3.

Steps	Reactions
1	$5/H_2 + * \leftrightarrow H_2 *$
2	$5/H_2 * + * \leftrightarrow 2H *$
3	$CO_2 + * \leftrightarrow CO_2 *$
4	$CO + * \leftrightarrow CO *$
5	$CO * + O * \leftrightarrow CO_2 * + *$
6	$2/CO_2 * + H * \leftrightarrow HCOO * + *$
7	$2/HCOO * + H * \leftrightarrow H_2CO * + O *$
8	$2/H_2CO * + H * \leftrightarrow H_3CO * + *$
9	$2/H_3CO * + H * \leftrightarrow H_3COH * + *$
10	$2/CH_3OH * \leftrightarrow CH_3OH + *$
11	$O * + H * \leftrightarrow OH * + *$
12	$OH * + H * \leftrightarrow H_2O * + *$
13	$H_2O * \leftrightarrow H_2O + *$
Total	$CO_2 + CO + 5H_2 \leftrightarrow 2CH_3OH + H_2O$

Table 3.3. Elementary reaction steps used in the microkinetic modeling

Enthalpy changes of reaction steps and individual activation energy barriers were calculated using Bond Order Conservation – Morse Potential Method as proposed by Shustorovic & Bell, 1991. In this method, enthalpy changes of the reaction steps were calculated from heats of chemisorptions of surface species on Cu (111) surface and from bond energies (Table 3.4). In the calculation of the pre-exponential factors of the reaction steps, one of the pre-exponential factors is estimated from transition state theory (Dumesic et al., 1993) and the other was calculated accordingly from the entropy change of that elementary step. The entropy change values of the elementary steps were calculated from partition functions of surface intermediates.

$$\frac{Af}{Ar} = \frac{z_C z_D}{z_A z_B} = exp\left(\frac{\Delta S^0}{R}\right) \text{ for the reaction } A + B \rightarrow C + D \tag{2}$$

$k_f = A_f^* exp(-E_{af}/RT)$			$k_r = A_r^* exp(-E_{ar}/RT)$	
A^r (s⁻¹ or bar⁻¹s⁻¹)	E_{af} (kJ/mol)	Reactions	A^r (s⁻¹ or bar⁻¹s⁻¹)	E_{ar} (kJ/mol)
$6.77*10^5$	0	$H_2 +*\leftrightarrow H_2*$	$6*10^{12}$	21
$1*10^{13}$	52.5	$H_2*+ *\leftrightarrow 2H*$	$8.55*10^{12}$	64.5
$1*10^6$	0	$CO_2 +*\leftrightarrow CO_2*$	$1.62*10^{13}$	21
$1*10^6$	0	$CO +*\leftrightarrow CO*$	$8.86*10^{14}$	50
$1*10^{13}$	44.8	$CO*+O*\leftrightarrow CO_2*+*$	$2.38*10^{13}$	115.8
$1*10^{13}$	14	$CO_2*+H*\leftrightarrow HCOO*+*$	$1.79*10^{11}$	6
$1*10^{13}$	79	$HCOO*+H*\leftrightarrow H_2CO*+O*$	$1*10^{13}$	0
$1*10^{13}$	15.5	$H_2CO*+H*\leftrightarrow H_3CO*+*$	$1*10^{13}$	36.5
$1*10^{13}$	41	$H_3CO*+H*\leftrightarrow H_3COH*+*$	$1.06*10^{14}$	75
$9*10^{16}$	63	$CH_3OH*\leftrightarrow CH_3OH+*$	$1*10^6$	0
$1*10^{13}$	86.8	$O*+H*\leftrightarrow OH*+*$	$5.31*10^{11}$	64.8
$1*10^{13}$	6	$OH*+H*\leftrightarrow H_2O*+*$	$9.78*10^{14}$	107
$1.59*10^{14}$	59	$H_2O*\leftrightarrow H_2O+*$	$1*10^6$	0

Table 3.4. Used energy barriers and pre exponential factors in microkinetic modeling

In the microkinetic analysis, coverage trends of the surface intermediates can be easily followed with given individual rate constants. For an initial gas composition of %70 H_2, %25 CO and %5 CO_2, surface coverage trends of CO_2, CO and H at industrial conditions can be seen in Figure 3.1 for a time interval of 0- 10^{-7} s.

From Figure 3.1, it was observed that CO and CO_2 adsorbs and saturates on the surface as soon as 10^{-10}s. However, adsorption of H species controls the vacant site which is dominant in the mechanism.

Effect of water on the catalytic methanol formation rates were studied by changing the initial gas composition to 70 % H_2, 24 % CO, 4 % CO_2, 2% H_2O from 70 % H_2, 25 % CO, 5 % CO_2. The results were shown for a carbon dioxide conversion of 0.00005 at industrial conditions and also at ambient conditions in order to compare with industrial and photocatalytic rates (Table 3.5).

The validity of the microkinetic analysis results was verified with the comparison of the methanol formation rate with literature value (Sahibzada et al., 1998). The inhibitory effect of water on catalytic carbon dioxide hydrogenation mechanism is observed especially at low temperatures. Positive effect of pressure on carbon dioxide reduction rates was also observed with this microkinetic analysis.

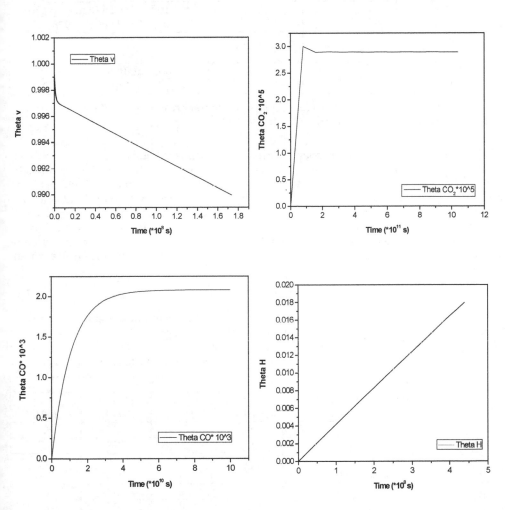

Fig. 3.1. Coverage values of vacant sites, CO, CO_2 and H at 75 bar and 523 K

When the catalytic methanol formation rates were compared with photocatalytic rates, it is seen that at room conditions; i.e., at 300 K and 1 atm, photocatalytic rates (20 µmol/g_{cat}/h ~6*10^{-9} mol/gcat/s (Tseng et al., 2002)) significantly surpass catalytic rates (2.07*10^{-12} mol/gcat/s). Even though estimation of kinetic parameters could contribute to non certainty of the kinetic results, it could be observed from the comparison that photo irradiation results in an obvious improvement in methanol formation rates. This improvement could be attributed to easier activation of molecules such as carbon dioxide through transfer of a photo-generated electron.

Reaction Conditions	Microkinetic Modeling		Literature
	Without initial H_2O	With initial H_2O	
523 K 75 bar	$4.53*10^{-5}$	$3.58*10^{-5}$	$6.1*10^{-5}$ (with Cu/ZnO/Al$_2$O$_3$) (Sahibzada et al., 1998)
300 K 75 bar	$1.07*10^{-9}$	$4.58*10^{-12}$	
300 K 1 bar	$1.9*10^{-11}$	$2.07*10^{-12}$	Photocatalytic: $6*10^{-9}$ (with Cu/TiO$_2$) (Tseng et al., 2002)

Table 3.5. Methanol formation rates (mol/g_{cat}/s) at different reaction conditions with or without initial water concentration

3.2 Rate determining step of methanol formation on Cu (111) surface

Calculation of degree of rate controls for elementary reaction steps in the microkinetic model allows revealing rate limiting steps in methanol formation from CO_2 hydrogenation and water gas shift reaction. Degree of rate control is defined by Campbell such as; the degree of change of the overall rate by a change in rate constant of a single step (Equation 2) (Campbell, 2001). Campbell proposed that steps where degree of rate control is positive be called rate-limiting steps and negative be inhibition steps. The larger the numeric value of degree of rate control, $X_{rc,i}$, the bigger is the influence of its rate constant on the overall reaction rate.

$$X_{rc,i} = (k_i/\delta k_i)*(\delta R/R) \qquad (3)$$

When degree of rate control values were calculated for the microkinetic model of methanol synthesis on Cu (111) surface, the results indicated that H supply (Step 2) to Cu surface as well as formate hydrogenation step (step 7) is essential especially at 300 K, at which artificial photosynthesis occurs (Table 3.6). This study underlines the importance of H supply and

	Elementary reactions	$X_{rc,i}$			
		75 atm 523 K	75 atm 423 K	75 atm 300 K	1 atm 300K
1	$H_2 +*\leftrightarrow H_2 *$	~0	0	0	0
2	$H_2 * + *\leftrightarrow 2H *$	0.36	2.4	3.53	3.7
3	$CO_2 +*\leftrightarrow CO_2 *$	0	0	~0	0
4	$CO +*\leftrightarrow CO *$	0	0	-3.67	-0.21
5	$CO * +O *\leftrightarrow CO_2 * + *$	0	0	0	0
6	$CO_2 * +H *\leftrightarrow HCOO * + *$	0	0	0	0
7	$HCOO * +H *\leftrightarrow H_2CO * +O *$	0.997	~1	~1	1
8	$H_2CO * +H *\leftrightarrow H_3CO * + *$	~0	~0	0.71	0.57
9	$H_3CO * +H *\leftrightarrow H_3COH * + *$	0	0.56	~1	1
10	$CH_3OH *\leftrightarrow CH_3OH +*$	-	-	-	-
11	$O * +H *\leftrightarrow OH * + *$	0	0	0	0
12	$OH * +H *\leftrightarrow H_2O * + *$	0	0	0	0
13	$H_2O *\leftrightarrow H_2O +*$	0	0	0	0

Table 3.6. The degree of rate control values with respect to r_f10 ($CH_3OH *\leftrightarrow CH_3OH +*$) found by finite difference method at t= $5.18*10^{-7}$ s

concentration on the surface for methanol formation rates. Since water splitting reaction is the only source of H in photocatalytic CO_2 reduction mechanism, it could be said that water oxidation rates are rate limiting in artificial photosynthesis systems whereas water oxidation rate surpass carbon dioxide reduction rates at photosynthesis (Table 2.1).

4. Similarities and differences between photosynthesis and artificial photosynthesis

Analogy between photosynthesis and artificial photosynthesis is in the similar tools and methods utilized in both systems. Collecting solar energy for triggering chemical reactions by chlorophyll pigments packed in thylakoid membrane or by semiconductors; oxidizing water into molecular oxygen and protons and reducing CO_2 with transported electrons and H+s are among the similarities of the two systems. However, the gap between the design of the systems and number of reaction sites and intermediate molecules result in more sophisticated and simpler products in photosynthesis $((CH_2O)_6)$ and in photocatalysis (CH_4 or CH_3OH) respectively.

In photosynthesis, there are three major reaction centers in light dependent reactions, regulating electron and proton transport together with the intermediate charge carriers (redox components). In photocatalysis, on the other hand, design of the system is limited to the presence of a pool of charges wandering on the semiconductor/metal surface in an unregulated fashion, increasing the chance of recombination of charge carriers. In addition, realization of oxidation and reduction reactions on the same catalyst surface results in interactions between the surface adsorbates which is proven to be inhibitory on reaction rates as studied with the microkinetic model in Section 3.1.

In photosynthesis, CO_2 diffusion from atmosphere to stromal phase in chloroplasts is controlled by stomata activities and permeability of chloroplast membranes. Photosynthetic rate is limited with the CO_2 concentration in stromal phase for values lower than a saturation value; i.e., the photosynthetic rate is linearly increasing with CO_2 concentration. For CO_2 concentrations above the saturation value, photosynthetic rate stays constant, limited by the rate of the enzyme system. Since CO_2 concentration in the stromal phase is related to CO_2 diffusion, photosynthetic rate is dependent upon diffusion rates.

In photocatalysis, diffusion of dissolved CO_2 and other reactants/products to/from the catalyst surface is largely dependent upon the reactor types, reaction media and stirring rates. The photocatalytic experiment parameters are not standardized in literature, resulting in confusion about the proper comparison of the real kinetic data. For photocatalytic tests performed in liquid media, which constitute the majority of the tests reported in literature, presence of gas-liquid-solid interfaces imposes non negligible mass transfer limitations in observed rates. A study performed to reveal the effect of stirring rates on photocatalytic hydrogen evolution rates indicated the importance of boundary layer and gas-liquid equilibrium in liquid phase photocatalytic experiments (Figure 4.1) (Ipek, 2011).

Increase in photo catalytic rates with increasing stirring rates (from 350 rpm to 900 rpm) up to a certain hydrogen concentration could be interpreted as decreased mass transfer limitation effects due to thinning of the boundary layer surrounding the catalyst particles whereas after that concentration, hydrogen seems to accumulate in the gas phase with the same limiting liquid-gas transfer rate. The limitation at the gas-liquid interface could also be inferred from the overlapping hydrogen amounts accumulated in the gas phase regardless of the catalyst amount or the reaction mixture volume (Figure 4.2).

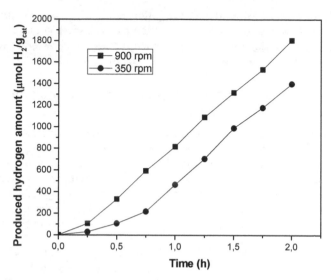

Fig. 4.1. Effect of stirring rates on photocatalytic hydrogen evolution with methanol as sacrificial agent, with 0.5 wt % Pt/TiO$_2$, 250 ml deionized water, 2 ml methanol (■) 900 rpm, (●) 350 rpm

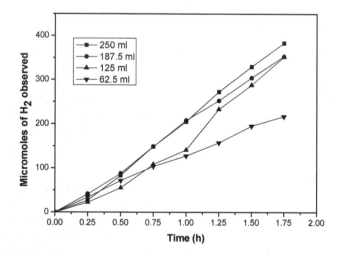

Fig. 4.2. Observed hydrogen amount in the gas phase with changing reaction mixture, (■) 62.5 ml (●) 125 ml (▲) 187.5 ml (▼) 250 ml, CH$_3$OH/ H$_2$O: 1/125 (v/v) and C$_{cat}$: 1 g/L for each case

Furthermore, temperature sensitivity of the photocatalytic hydrogen evolution reactions indicate presence of diffusion limitations with found 12 and 19.5 kJ/mol apparent activation energies for Pt/TiO$_2$ and Cu/TiO$_2$ reactions respectively (Ipek, 2011).

In PSII of photosynthesis, there are over 15 polypeptides and 9 different redox components responsible for water oxidation and electron transport. Even the oxygen evolving complex is regenerating itself at every 30 minutes in order to sustain its stability. Along with the sophistication of light dependent reactions including numerous intermediate charge carriers, difference in the CO_2 reduction mechanism (activating CO_2 by fixing it into another chemical) with 13 specific enzymes result in higher photosynthetic rates and more complicated products (such as glucose) in photosynthesis. On the other hand, C-C bond making is still remaining as a challenge in artificial photosynthesis systems. Even with one carbon chemical synthesis, photocatalytic rates are well below photosynthetic rates. To illustrate, CO_2 reduction using titanium nanotubes resulted in nearly 1 nmol/m²/s CH_4 production rate (Schulte et al., 2010) whereas an avarage photosynthetic rate is 12 μmol/m²/s.

Apart from the sophisticated numerous enzymes taking part in carbon dioxide reduction mechanism, the major gap between carbon dioxide reduction rates are suggested to result from undeveloped charge and H transport systems in artificial photosynthesis systems. Recently, regulated electron and hole transport is reported with zeolites increasing the charge separation and water oxidation activity (Dutta & Severance, 2011). Furthermore, H^+ transport through oxidation center to the reduction center is reported to be realized with H^+ permeable electron conducting membrane (Hou et al., 2011). Design of a reaction system as well as the photocatalyst is of uttermost importance in artificial photosynthesis systems for better activities. As indicated in the microkinetic analysis part, carbon dioxide reduction rates mainly suffer from insufficient H supply to the reduction centers at artificial photosynthesis. In photosynthesis, H transport is highly regulated via the intermediate charge carriers such as NADP+ and via the electrochemical potential difference which may be attributed to the presence of an interstitial membrane. Utilization of such a membrane in photocatalytic systems was first suggested by Kitano et al., 2007b who used the membrane for H transport in water splitting reaction. Similarly, enhanced electron and H transport system should be implemented for carbon dioxide reduction also which would carry the artificial photosynthesis systems to industrial levels.

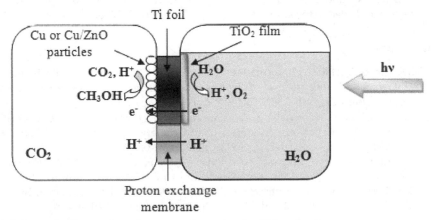

Fig. 4.3. Schematic illustration of suggested reactor for CO_2 reduction

Design of a photocatalytic reactor would work in such a way that in one compartment light is harvested by the semiconductor in order to split water to form H+s and by the transfer of produced protons and electrons, CO_2 could be reduced with another catalyst, such as copper, in the other compartment. In this way, with the help of proton exchange membranes and separate reaction centers, interaction between reactants or products and therefore reverse reactions could be prevented by supplying Hs to catalytic compartment at the same time. On the other hand, a high conductance electron membrane would prevent charge recombination as illustrated in Figure 4.3.

Such a system has already been proposed by Kitano et al. (2007b) for water splitting reaction. Introducing CO_2 in the picture will be the next generation modification of the artificial photosynthesis systems. Further refinement of the design will be possible with more understanding about the rates of chemical conversions and the rates of transport.

5. Conclusions

Photosynthesis and artificial photosynthesis system were compared in this study with an emphasis on charge and H transport which is indicated to be the main reason for the difference in resulting carbon dioxide reduction yields and rates. The gap in artificial systems was found to be in the design of the photocatalytic systems which could be developed with a membrane which would enhance charge separation and H transport. H supply to carbon dioxide reduction centers were found to be limiting the existing photocatalytic carbon dioxide reduction rates, indicating the important role of water splitting in artificial photosynthesis systems. Water being in contact with carbon dioxide on photocatalyst surface was found to act negatively on the methanol formation rates by inhibiting carbon dioxide activation. For a better carbon dioxide reduction with existing photocatalysts, separation of the reaction centers was proposed which would enhance the charge and H transport at the same time.

6. Acknowledgements

This study was funded through The Scientific and Technological Research Council of Turkey (TUBITAK) under research grant numbers: 106Y075 and 107M447. Bahar Ipek would like to acknowledge the support from TUBITAK BIDEB 2228.

7. References

Anpo, M. & Chiba, K. (1992). Photocatalytic Reduction of CO_2 on Anchored Titanium-Oxide Catalysts, *7th International Symposium on Relations between Homogeneous and Heterogeneous Catalysis*, ISBN 0304-5102, Tokyo, Japan, May 1992

Anpo, M.; Yamashita, H.; Ichihashi, Y. & Ehara, S. (1995). Photocatalytic Reduction of CO_2 with H_2O on Various Titanium-Oxide Catalysts. *Journal of Electroanalytical Chemistry*, Vol. 396, No. 1-2, pp. 21-26, ISSN 0022-0728

Anpo, M.; Yamashita, H.; Ichihashi, Y.; Fujii, Y. & Honda, M. (1997). Photocatalytic Reduction of CO_2 with H_2O on Titanium Oxides Anchored within Micropores of Zeolites: Effects of the Structure of the Active Sites and the Addition of Pt. *Journal of Physical Chemistry B*, Vol. 101, No. 14, pp. 2632-2636, ISSN 1089-5647

Anpo, M.; Yamashita, H.; Ikeue, K.; Fujii, Y.; Zhang, S. G.; Ichihashi, Y.; Park, D. R.; Suzuki, Y.; Koyano, K. & Tatsumi, T. (1998). Photocatalytic Reduction of CO_2 with H_2O on Ti-MCM-41 and Ti-MCM-48 Mesoporous Zeolite Catalysts. *Catalysis Today*, Vol. 44, No. 1-4, pp. 327-332, ISSN 0920-5861

Asahi, R.; Morikawa, T.; Ohwaki, T.; Aoki, K. & Taga, Y. (2001). Visible-Light Photocatalysis in Nitrogen-Doped Titanium Oxides. *Science*, Vol. 293, No. 5528, pp. 269-271 ISSN 0036-8075

Campbell, C. T. (2001). Finding the Rate-Determining Step in a Mechanism - Comparing Dedonder Relations with the "Degree of Rate Control". *Journal of Catalysis*, Vol. 204, No. 2, pp. 520-524, ISSN 0021-9517

Carp, O.; Huisman, C. L. & Reller, A. (2004). Photoinduced Reactivity of Titanium Dioxide. *Progress in Solid State Chemistry*, Vol. 32, No. 1-2, pp. 33-177, ISSN 0079-6786

Chen, L.; Graham, M. E.; Li, G. H.; Gentner, D. R.; Dimitrijevic, N. M. & Gray, K. A. (2009). Photoreduction of CO_2 by TiO_2 Nanocomposites Synthesized through Reactive Direct Current Magnetron Sputter Deposition. *Thin Solid Films*, Vol. 517, No. 19, pp. 5641-5645, ISSN 0040-6090

Dey, G. R.; Belapurkar, A. D. & Kishore, K. (2004). Photo-Catalytic Reduction of Carbon Dioxide to Methane using TiO_2 as Suspension in Water. *Journal of Photochemistry and Photobiology a-Chemistry*, Vol. 163, No. 3, pp. 503-508, ISSN 1010-6030

Diwan, J. J. (January, 2009). Calvin Cycle- Photosynthetic Dark Reactions, In: *Rensselaer Polytechnic Institute*, 27.03.2011, Available from http://www.rpi.edu/dept/bcbp/molbiochem/ MBWeb/mb2/part1/dark.htm

Dutta, P. K. & Severance, M. (2011). Photoelectron Transfer in Zeolite Cages and Its Relevance to Solar Energy Conversion. *Journal of Physical Chemistry Letters*, Vol. 2, No. 5, (March 2011), pp. 467-476, ISSN 1948- 7185

Freund, H. J. & Roberts, M. W. (1996). Surface Chemistry of Carbon Dioxide. *Surface Science Reports*, Vol. 25, No. 8, pp. 225-273, ISSN 0167-5729

Govindjee, (Ed.). (1975). *Bioenergetics of Photosynthesis*, Academic Press, ISBN, New York

Hankamer, B.; Barber, J. & Boekema, E. J. (1997). Structure and Membrane Organization of Photosystem II in Green Plants. *Annual Review of Plant Physiology and Plant Molecular Biology*, Vol. 48, pp. 641-671, ISSN 1040- 2519

Haumann, M.; Liebisch, P.; Muller, C.; Barra, M.; Grabolle, M. & Dau, H. (2005). Photosynthetic O-2 Formation Tracked by Time-Resolved X-ray Experiments. *Science*, Vol. 310, No. 5750, (November 2005), pp. 1019-1021, ISSN 0036- 8075

Heldt, H. W.; Piechulla B. & Heldt, F. (Eds.). (2010). *Plant Biochemistry*, Elsevier Science & Technology, ISBN 9780123849861, London, 4th Edition.

Hill, R. & Bendall, F. (1960). Function of the Two Cytochrome Components in Chloroplasts: A Working Hypothesis. *Nature*, Vol. 186, No. 4719, pp. 136-137, ISSN 0028-0836

Hou, Y.; Abrams, B.L.; Vesborg, P.C.K.; Bjorketan, H.E.; Herbst, K.; Bech, L.; Setti, A.M.; Damsgaard, C.D.; Redersen, T.; Hansen, O.; Rossmeisl, J.; Dahl, S.; Norskov, J.K. & Chorkendorff, I. (2011). Bioinspired Molecular co-Catalysts Bonded to a Silicon Photocathode for Solar Hydrogen Evolution. *Nature Materials*, Vol. 10, pp. 434-438, ISSN 1476-1122

Hwang, J. S.; Chang, J. S.; Park, S. E.; Ikeue, K. & Anpo, M. (2005). Photoreduction of Carbondioxide on Surface Functionalized Nanoporous Catalysts. *Topics in Catalysis*, Vol. 35, No. 3-4, (July 2005), pp. 311-319, ISSN 1022-5528

Indrakanti, V. P.; Schobert, H. H. & Kubicki, J. D. (2009). Quantum Mechanical Modeling of CO_2 Interactions with Irradiated Stoichiometric and Oxygen-Deficient Anatase TiO_2 Surfaces: Implications for the Photocatalytic Reduction of CO_2. *Energy & Fuels*, Vol. 23, (October 2009), pp. 5247-5256, ISSN 0887- 0624

Inoue, T.; Fujishima, A.; Konishi, S. & Honda, K. (1979). Photoelectrocatalytic Reduction of Carbon-Dioxide in Aqueous Suspensions of Semiconductor Powders. *Nature*, Vol. 277, No. 5698, pp. 637-638, ISSN 0028-0836

Ipek, B. (April 2011). Photocatalytic Carbon Dioxide Reduction in Liquid Media. M.Sc. Thesis, Chemical Engineering Department, Middle East Technical University, Ankara, Turkey

Jitaru, M. (2007). Electrochemical Carbon Dioxide Reduction- Fundamental and Applied Topics (Review). *Journal of the University of Chemical Technology and Metallurgy*, Vol. 42, No. 4, pp. 333-344, ISSN 1311-7629

Kaneco, S.; Shimizu, Y.; Ohta, K. & Mizuno, T. (1998). Photocatalytic Reduction of High Pressure Carbon Dioxide using TiO_2 Powders with a Positive Hole Scavenger. *Journal of Photochemistry and Photobiology a-Chemistry*, Vol. 115, No. 3, pp. 223-226, ISSN 1010-6030

Ke, B. (2001). *Photosynthesis: Photobiochemistry and Photobiophysics*, Kluwer Academic Publishers, ISBN 0-7923-6334-5, Dordrecht

Kitano, M.; Matsuoka, M.; Ueshima, M. & Anpo, M. (2007a). Recent Developments in Titanium Oxide-Based Photocatalysts. *Applied Catalysis a-General*, Vol. 325, No. 1, pp. 1-14, ISSN 0926- 860X

Kitano, M.; Takeuchi, M.; Matsuoka, M.; Thomas, J. A. & Anpo, M. (2007b). Photocatalytic Water Splitting using Pt-Loaded Visible Light-Responsive TiO_2 Thin Film Photocatalysts. *Catalysis Today*, Vol. 120, No. 2, pp. 133-138, ISSN 0920-5861

Koci, K.; Mateju, K.; Obalova, L.; Krejcikova, S.; Lacny, Z.; Placha, D.; Capek, L.; Hospodkova, A.; & Solcova, O. (2010). Effect Of Silver Doping on The TiO_2 for Photocatalytic Reduction of CO_2. *Applied Catalysis B-Environmental*, Vol. 96, No. 3-4, pp. 239-244, ISSN 0926-3373

Koci, K.; Obalova, L.; Matejova, L.; Placha, D.; Lacny, Z.; Jirkovsky, J. & Solcova, O. (2009). Effect of TiO_2 Particle Size on the Photocatalytic Reduction of CO_2. *Applied Catalysis B-Environmental*, Vol. 89, No. 3-4, pp. 494-502, ISSN 0926-3373

Meyer, T. J. (2008). Catalysis - The Art of Splitting Water. *Nature*, Vol. 451, No. 7180, pp. 778-779, ISSN 0028-0836

Ovesen, C. V.; Clausen, B. S.; Schiotz, J.; Stoltze, P.; Topsoe, H. & Norskov, J. K. (1997). Kinetic Implications of Dynamical Changes in Catalyst Morphology During Methanol Synthesis over Cu/ZnO Catalysts. *Journal of Catalysis*, Vol. 168, No. 2, pp. 133-142, ISSN 0021- 9517

Ozcan, O.; Yukruk, F.; Akkaya, E. U. & Uner, D. (2007). Dye Sensitized Artificial Photosynthesis in the Gas Phase over Thin and Thick TiO_2 Films under UV and Visible Light Irradiation. *Applied Catalysis B-Environmental*, Vol. 71, No. 3-4, pp. 291-297, ISSN 0926-3373

Sahibzada, M.; Metcalfe, I. S. & Chadwick, D. (1998). Methanol Synthesis from $CO/CO_2/H_2$ over $Cu/ZnO/Al_2O_3$ at Differential and Finite Conversions. *Journal of Catalysis*, Vol. 174, No. 2, pp. 111-118, ISSN 0021- 9517

Schulte, K. L.; DeSario, P. A. & Gray, K. A. (2010). Effect of Crystal Phase Composition on the Reductive and Oxidative Abilities of TiO_2 Nanotubes under UV and Visible Light. *Applied Catalysis B-Environmental*, Vol. 97, No. 3-4, pp. 354-360, ISSN 0926-3373

Shustorovich, E. & Bell, A. T. (1991). An Analysis of Methanol Synthesis from CO and CO_2 on Cu and Pd Surfaces by the Bond-Order-Conservation Morse-Potential Approach. *Surface Science*, Vol. 253, No. 1-3, pp. 386-394, ISSN 0039-6028

Singal, H. R.; Talwar, G.; Dua, A. & Singh, R. (1995). Pod Photosynthesis and Seed Dark CO_2 Fixation Support Oil Synthesis in Developing Brassica Seeds. *Journal of Biosciences*, Vol. 20, No. 1, pp. 49-58, ISSN 0250-5991

Solymosi, F. & Tombacz, I. (1994). Photocatalytic Reaction of H_2O+CO_2 over Pure and Doped Rh/TiO_2. *Catalysis Letters*, Vol. 27, No. 1-2, pp. 61-65, ISSN 1011-372X

Tseng, I. H.; Chang, W. C. & Wu, J. C. S. (2002). Photoreduction of CO_2 using Sol-Gel Derived Titania and Titania-Supported Copper Catalysts. *Applied Catalysis B-Environmental*, Vol. 37, No. 1, pp. 37-48, ISSN 0926-3373

Umena, Y.; Kawakami, K.; Shen, J. R. & Kamiya, N. (2011). Crystal Structure of Oxygen-Evolving Photosystem II at a Resolution of 1.9 A. *Nature*, Vol. 473, No. 7345, pp. 55-60, ISSN 0028-0836

Uner, D.; Oymak, M. M. & Ipek, B. (2011). CO_2 Utilization by Photocatalytic Conversion to Methane and Methanol. *International Journal of Global Warming*, Vol. 3, No. 1-2, pp. 142-162, ISSN 1758-2083

VandenBussche, K. M. & Froment, G. F. (1996). A Steady-State Kinetic Model for Methanol Synthesis and the Water Gas Shift Reaction on a Commercial $Cu/ZnO/Al_2O_3$ Catalyst. *Journal of Catalysis*, Vol. 161, No. 1, pp. 1-10, ISSN 0021-9517

Varghese, O. K.; Paulose, M.; LaTempa, T. J. & Grimes, C. A. (2009). High-Rate Solar Photocatalytic Conversion of CO_2 and Water Vapor to Hydrocarbon Fuels. *Nano Letters*, Vol. 9, No. 2, pp. 731-737, ISSN 1530-6984

Wang, C. J.; Thompson, R. L.; Baltrus, J. & Matranga, C. (2010). Visible Light Photoreduction of CO_2 Using $CdSe/Pt/TiO_2$ Heterostructured Catalysts. *Journal of Physical Chemistry Letters*, Vol. 1, No. 1, pp. 48-53, ISSN 1948-7185

Wang, Z. Y.; Chou, H. C.; Wu, J. C. S.; Tsai, D. P. & Mul, G. (2010). CO_2 Photoreduction using $NiO/InTaO_4$ in Optical-Fiber Reactor for Renewable Energy. *Applied Catalysis a-General*, Vol. 380, No. 1-2, pp. 172-177, ISSN 0926-860X

Whitmarsh, J.; Govindjee (1999). The Photosynthetic Process, In: *Concepts in Photobiology: Photosynthesis and Photomorphogenesis*, Singhal, G.S.; Renger, G.; Sopory, S. K.; Irrgang, K. D. & Govindjee, Narosa Publishing House, ISBN 0-7923-5519-9, New Delhi

Whittingham, C. P. (1974). *The Mechanism of Photosynthesis*, American Elsevier Pub. Co, ISBN 9780444195524 , New York

Woan, K.; Pyrgiotakis, G. & Sigmund, W. (2009). Photocatalytic Carbon-Nanotube-TiO_2 Composites. *Advanced Materials*, Vol. 21, No. 21, pp. 2233-2239, ISSN 0935-9648

Wu, J. C. S.; Lin, H. M. & Lai, C. L. (2005). Photo Reduction of CO_2 to Methanol using Optical-Fiber Photoreactor. *Applied Catalysis a-General*, Vol. 296, No. 2, pp. 194-200, 0926-860X

Yamashita, H.; Nishiguchi, H.; Kamada, N.; Anpo, M.; Teraoka, Y.; Hatano, H.; Ehara, S.; Kikui, K.; Palmisano, L.; Sclafani, A.; Schiavello, M. & Fox, M. A. (1994).

Photocatalytic Reduction of CO_2 with H_2O on TiO_2 and Cu/TiO_2 Catalysts. *Research on Chemical Intermediates*, Vol. 20, No. 8, pp. 815-823, ISSN 0922-6168

Yang, H. C.; Lin, H. Y.; Chien, Y. S.; Wu, J. C. S. & Wu, H. H. (2009), Mesoporous TiO_2/SBA-15 and $Cu/TiO_2/SBA$-15 Composite Photocatalysts for Photoreduction of CO_2 to Methanol. *Catalysis Letters*, Vol. 131, No. 3-4, pp. 381-387, ISSN 1011-372X

Zhang, Q. H.; Han, W. D.; Hong, Y. J. & Yu, J. G. (2009). Photocatalytic Reduction of CO_2 with H_2O on Pt-loaded TiO_2 Catalyst. *Catalysis Today*, Vol. 148, No. 3-4, pp. 335-340, ISSN 0920-5861

Zhao, Z. H.; Fan, J. M.; Xie, M. M. & Wang, Z. Z. (2009). Photo-Catalytic Reduction of Carbon Dioxide with in-situ Synthesized $CoPc/TiO_2$ under Visible Light Irradiation. *Journal of Cleaner Production*, Vol. 17, No. 11, pp. 1025-1029, ISSN 0959-6526

3

Mutations in the CP43 Protein of Photosystem II Affect PSII Function and Cytochrome C$_{550}$ Binding

Brandon D. Burch[1], Terry M. Bricker[2] and Cindy Putnam-Evans[1]

[1]*East Carolina University/Biology Department*
[2]*Louisiana State University/Department of Biological Sciences*
USA

1. Introduction

During the process of photosynthesis, light energy is converted to chemical energy that is utilized in the biosynthesis of carbohydrate. The initial events of photosynthesis in cyanobacteria, green algae and higher plants are the photoinduced transfer of electrons from water to plastoquinone. These reactions are catalyzed by Photosystem II (PSII), a large pigment-protein complex embedded within the thylakoid membrane (Renger, et al., 2008). The reactions catalyzed by PSII proceed as follows. First, light is trapped by pigments, predominately chlorophyll, associated with the thylakoid membrane (Glazer, 1983). Excitation energy is then transferred to the photochemically active chlorophyll species P$_{680}$. Excited P$_{680}$ then donates an electron (charge separation) to the primary PSII electron acceptor, a protein-bound pheophytin molecule (Klimov et al., 1980). Pheophytin is then oxidized by a tightly bound plastoquinone, Q$_A$, which in turn reduces a loosely bound plastoquinone, Q$_B$. A second light-induced charge separation results in the formation of plastoquinol (Q$_B$H$_2$). Photooxidized P$_{680}$ is reduced by the primary donor of PSII, a tyrosyl radical (Y$_Z$•) that is Tyr161 of the D1 protein of PSII (Debus et al., 1988a; Debus et al., 1988b). Y$_Z$ then is reduced by an electron from the oxygen-evolving complex of PSII, located on the lumenal face of the thylakoid. The catalytic site of the oxygen-evolving complex consists of a metal ion cluster of four manganese atoms and one calcium atom bound via five μ-oxo-bridges (Mn$_4$CaO$_5$ cluster; Umena et al., 2011). Two chloride ions bound nearby are also required for activity. The oxygen-evolving complex functions to extract electrons and protons from water, ultimately resulting in the release of molecular oxygen. The manganese atoms cycle through a series of redox states, or S states (Joliot et al., 1969; Kok et al., 1970), from S$_0$ to S$_4$, with each S state representing a successively more oxidized form of the cluster. One molecule of O$_2$ is evolved when four electrons and four protons are extracted by the Mn$_4$CaO$_5$ cluster.

PSII consists of both intrinsic thylakoid polypeptides and extrinsic polypeptides located within the thylakoid lumen. Together, these proteins ligate the Mn$_4$CaO$_5$ cluster, chlorophylls and other pigments, and the electron transport chain components. The intrinsic polypeptides necessary to form a PSII complex capable of evolving oxygen are CP47, CP43, D1, D2, the α and β subunits of cytochrome b$_{559}$, and the 4kDa *psbI* gene product (Bricker et al., 2011). In higher plants, three extrinsic proteins, PsbO (previously termed the manganese-stabilizing protein), PsbP and PsbQ, are required for maximal rates of oxygen evolution under physiological conditions (Bricker, 1992). In cyanobacteria, cytochrome c_{550} and a 12 kDa protein (PsbU) perform similar functions to PsbP and PsbQ. However, cyanobacteria also contain homologues of the higher plant PsbP and PsbQ proteins, though their roles are not well defined (Roose et al., 2007). In the absence of these extrinsic proteins, PSII complexes retain the ability to evolve oxygen, but at significantly reduced rates. This ability to evolve oxygen is dependent upon the presence of high, non-physiological, concentrations of calcium and chloride. The extrinsic proteins appear to act as a diffusional barrier that sequesters chloride and calcium in the vicinity of the oxygen-evolving complex (Bricker et al., 2011). They also act to protect the Mn_4CaO_5 cluster from exogenous reductants (Ghanotakis et al., 1984).

CP43 is an integral thylakoid protein. CP43 binds 13 chlorophyll molecules and is a component of the proximal antenna of PSII. X-ray analysis of PS II from *Thermosynechococcus elongatus* and *Thermosynechococcus vulcanus* shows that CP43 contains six transmembrane alpha helices and five hydrophilic loops that connect the membrane-spanning domains (Zouni et al., 2001; Kamiya & Shen, 2003; Ferreira et al., 2004; Umena et al., 2011). One of these loops, the large extrinsic loop E, spans amino acid residues [278]Asn-[410]Trp and is located between the fifth and sixth membrane-spanning helices. The large extrinsic loop is exposed to the lumenal side of the thylakoid membrane and lies close to all three of the extrinsic PSII proteins as well as the Mn_4CaO_5 cluster. We previously constructed and extensively characterized the R320S (R305S in our original numbering scheme[1]) mutant in the large extrinsic loop E of CP43 in the cyanobacterium *Synechocystis* 6803, and showed that it has impaired PSII activity under chloride-limiting conditions (Knoepfle et al., 1999; Young et al., 2002). Isolated PSII from this mutant failed to bind the extrinsic PSII protein, cytochrome c_{550} (Bricker et al., 2002).

In this chapter, we will briefly review the role of both chloride and the CP43 large extrinsic loop E in the water-oxidizing process, and present new data on the characterization of additional mutations within the large extrinsic loop E that provide information about the nature of interaction of CP43 with the extrinsic PSII proteins. These data will be discussed within the context of the most current x-ray structure of PSII.

2. Structure of the Mn_4CaO_5 cluster

The structure of the manganese cluster has been the focus of intense study. Early x-ray structures obtained at low to medium resolution were known to be affected by radiation damage, which alters the valence state of the manganese and also potentially alters the ligand field (Grabolle et al., 2006). Nonetheless, these structures provided the first data on the positions of the protein subunits, chlorophylls and other cofactors. However, they did not allow for a highly refined, detailed structure of the manganese cluster. The manganese cluster models were proposed relying on interatomic distances derived from EXAFS experiments (Dau et al., 2008). The recent x-ray structure at 1.9 Å from *Thermosynechococcus*

---

[1] Due to differences in initiation sites, the CP43 proteins from various cyanobacteria are of varying lengths. In this article, unless otherwise specified, we will follow the common practice of numbering the amino acids using the corresponding numbering in *Thermosynechococcus vulcanus*, the strain from which the latest x-ray structure was derived.

vulcanus provided a well-resolved, detailed structure of the 4 manganese, calcium, and two chloride atoms, as well as revealing the positions of potential substrate water molecules, thus paving the way for an understanding of the mechanistic aspects of water oxidation (Umena et al., 2011).

The Mn_4CaO_5 cluster has a distorted, chair-like structure in which three of the manganese (Mn1, Mn2, Mn3) and the calcium atom form a distorted cubane-like structure (see Fig. 1). The fourth manganese atom (Mn4), the so-called "dangler" manganese, lies outside of the cubane. Mn1, Mn2 and Mn3 and the calcium atom are linked via four oxygen atoms (μ–oxo bridges). Mn4 is linked by a fifth oxygen to Mn1 and Mn3. In fact, each two adjacent manganese atoms are linked by a di-μ-oxo bridge and the calcium atom is linked to all four manganese via oxo bridges. Four water molecules were observed in the x-ray structure in close proximity to the Mn_4CaO_5 cluster. Two of these appear to be associated with Mn4 and the other two with the calcium atom. It has been speculated that least one of these is a substrate for the water oxidizing reaction. Two chloride ions were also revealed in the structure. These flank the cluster and are not bound directly to the manganese or calcium atoms, but rather are ligated via two amino acids and two water molecules each. For each of these chloride ions, bound indirectly to the Mn_4CaO_5 cluster, one ligand is contributed by the backbone nitrogen of a glutamate residue.

In addition to the metal atoms and water, the protein ligands for the Mn_4CaO_5 cluster were identified. While most of the ligands are contributed by the D1 protein, CP43 contributes one bi-dentate ligand to the cluster, Glu354. Glu354 ligates both Mn2 and Mn3. What is additionally interesting is that this glutamate residue binds one of the chloride ions flanking the cluster. The other chloride ion is bound by Glu333 of the D1 protein. Glu333 is a bi-dentate ligand to Mn3 and Mn4. It has been proposed that the chloride ions help to maintain the structural integrity of the Mn_4CaO_5 cluster by maintaining the stable coordination of these glutamate residues to manganese (Umena et al., 2011; Kawakami et al., 2011). Also, the chloride ions lie at the beginning of two putative proton and/or water exit channels. Chloride may also play a role in stabilizing these channels.

3. Role of chloride in oxygen evolution

Chloride is required for both the assembly and the stability of the oxygen-evolving complex and is sequestered at the active site by the PSII extrinsic proteins. Chloride depletion of isolated PSII preparations has profound effects on PSII function. Such preparations show large decreases in steady state oxygen evolution rates, and S-state defects including stabilization of the S_2 state lifetimes, and slowing of the $S_2 \rightarrow S_3$ transition. Chloride is necessary for S-state advancement in both the $S_2 \rightarrow S_3$, and $S_3 \rightarrow S_4 \rightarrow S_0$ transitions (Wincencjusz, et al., 1997). Removal of the extrinsic proteins from PSII results in a requirement for high, non-physiological concentrations of both calcium and chloride to maintain some oxygen-evolving activity (Bricker, et al., 2011). In PSII membranes depleted of PsbO, and also in *Synechocystis* mutants lacking the gene encoding PsbO, two manganese atoms were lost at chloride concentrations less than 100 mM (Kuwabara et al., 1985; Burnap et al., 1994). In both the higher plant and cyanobacterial systems, removal of the PsbO protein resulted in increases in the S_2 and/or S_3 lifetimes (Miyao et al., 1987; Burnap et al., 1992; Liu et al., 2007) and also in a slowing of the $S_3 \rightarrow S_4 \rightarrow S_0$ transition (Ono & Inoue, 1985; Burnap et al., 1992; Liu et al., 2007). Removal of the PsbP and PsbQ extrinsic proteins in higher plant PSII also results in a large decrease in oxygen-evolving activity, which can be partially restored by high concentrations of calcium and chloride (Popelkova & Yocum, 2007). The absence of these proteins also makes the Mn_4CaO_5 cluster vulnerable to the effects of exogenous reductants (Ghanotakis et al., 1984).

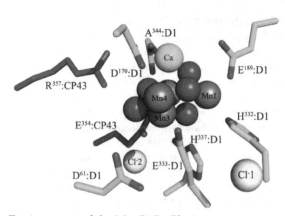

Fig. 1. Coordination Environment of the Mn_4CaO_5 Cluster.
The amino acid residues in the first and second coordination sphere are labeled. Please note that D342:D1 is obscured in this view by the metal cluster. The manganese ions are shown in purple and are labeled. Calcium is shown in cyan and the oxygens are shown in red. The proximal and distal chlorides are shown in yellow and labeled Cl-1 and Cl-2, respectively. This figure was generated in Pymol from the crystal structure of *Thermosynechococcus vulcanus* (PDB: 3ARC; Umena et al., 2011).

In cyanobacteria, deletion of the PsbU protein results in a mutant capable of growth in normal media, but which shows a reduced growth rate in media depleted of chloride (Shen et al., 1997). This mutant has reduced rates of oxygen evolution, a more stable S_2 state, and is more susceptible to damage to its PSII centers by both light and heat stress (Shen et al., 1997; Shen et al., 1998; Inoue-Kashino et al., 2005). Mutants constructed by deletion of the extrinsic cytochrome c_{550} protein, encoded by the *psbV* gene, do not grow well photoautotrophically, and exhibit no capacity for photoautotrophic growth in media depleted of either calcium or chloride (Shen et al., 1998; Katoh et al., 2001). The *psbV* deletion mutants also show decreased ability to evolve oxygen and reduced numbers of PSII centers in the thylakoids (Shen et al., 1995; Katoh et al., 2001). They also exhibit S-state cycling defects, including an increase in miss and double hit factors and a slowing of the $S_3 \rightarrow S_4 \rightarrow S_0$ transition (Shen et al., 1998).

These observations are all supportive of a role of the extrinsic proteins in maintaining the proper ionic environment around the Mn_4CaO_5 cluster. How chloride is actually bound and transported to the active site is currently unknown. Interestingly, the current crystal structure identified a third chloride ion bound near the C-terminal amino acid of PsbU, which is located between PsbU and cytochrome c_{550} (Umena et al., 2011; Kawakami et al., 2011). This chloride is ligated by water molecules and lies close to the exit of a proposed hydrogen-bonding network leading from the Mn_4CaO_5 cluster to the lumen that could possibly serve to transport anions, water or protons (Gabdulkhakov et al., 2009; Vassiliev et al., 2010; Kawakami et al., 2011).

4. Structure and function of CP43

The product of the *psbC* gene, the CP43 mature protein from *Synechocystis* 6803 consists of 460 amino acid residues and is well conserved. The nucleotide sequence of *psbC* has been determined from more than 50 species. The predicted plant apoproteins show approximately 95% homology with each other, and 85% and 77% homology with *Chlamydomonas* and cyanobacterial CP43, respectively. Early crystallographic studies confirmed that CP43 contains six transmembrane alpha helices (Zouni et al., 2001; Kamiya and Shen, 2003; Ferreira et al., 2004). The transmembrane helices of CP43 contain a number of conserved histidyl residues that function as chlorophyll-*a* ligands. Replacement of these histidyl residues with either tyrosyl or arginyl residues (Shen et al., 1993) results in mutants showing decreased PSII stability and defects in light-harvesting efficiency. Thirteen chlorophyll-*a* molecules associated with CP43 were assigned in the most current crystal structure (Umena et. al., 2011).

In addition to the six transmembrane alpha helices, CP43 contains five hydrophilic loops that connect the membrane-spanning domains. The positions of these loops are as follows (*Synechocystis* 6803 numbering): Loop A, Glu58-Phe98; Loop B, Gly123-Thr145; Loop C, Gly170-Asp219; Loop D, Lys242-Tyr258 and Loop E, Asn280-Arg410. Loop E, with 130 amino acids, is quite large and contains approximately 30% of the amino acids in this protein. Loops A, C and the large extrinsic Loop E are exposed to the lumenal side of the thylakoid membrane. The N- (Val1-Lys36) and C-termini (Arg448-Asp460), in addition to Loops B and D, face the stromal surface of the membrane (von Heijne and Gavel, 1988; Sayre and Wrobelboerner, 1994;). The CP47 protein has a similar structure and also possesses a large extrinsic loop.

All crystal structures of dimeric PSII show that CP43 is positioned on the D1 side of the D1-D2 heterodimer and CP47 on the D2 side. The two PSII monomers are related by a local-C2 rotation axis oriented perpendicular to the plane of the membrane. CP47 lies at the interface of the two monomers while CP43 is located at the periphery. This location is speculated to be necessary for the removal and replacement of damaged D1 during photoinhibition. In both CP43 and CP47, the six transmembrane helices are arranged as a "trimer of dimers". The large extrinsic loop of CP47 lies adjacent to the PsbO protein, consistent with a large body of biochemical evidence (Bricker & Frankel, 2002), and is also close to the extrinsic PsbU protein. The large extrinsic loop of CP43 is located adjacent to cytochrome c$_{550}$ and PsbU, as well as the manganese cluster.

Early mutagenesis and biochemical studies clearly outlined a role for CP43 in oxygen evolution and the stable assembly of the PSII complex. CP43 is required for optimal oxygen-evolving activity from isolated PSII preparations (Bricker, 1990). *Chlamydomonas* strains incorporating mutations that affected either the synthesis or stability of CP43 were deficient in PSII activity (Rochaix et al., 1989). In these mutants, levels of the other PSII core proteins were severely reduced. *Synechocystis* mutants lacking the *psbC* gene as a result of deletion mutagenesis accumulated PSII core complexes (minus CP43) to only 10% of wild-type levels and could not grow photoautotrophically or evolve oxygen (Rogner et al., 1991; Carpenter et al., 1990). Interruption of the *psbC* gene by insertional mutagenesis also produced a mutant incapable of evolving oxygen, but which exhibited primary charge separation (Vermaas et al., 1988). A *Synechocystis* mutant recovered by nitrosoguanidine mutagenesis was unable to evolve oxygen or support PSII electron transport from water to either dichlorobenzoquinone or methyl viologen (Dzelzkalns & Bogorad, 1988). Further characterization of this mutant showed that it contained a short deletion within the *psbC* gene. Thylakoid membranes isolated from this mutant had decreased levels of the reaction center protein D2.

In studies specifically targeting the large extrinsic loop E of CP43, eight short deletions were produced in *Synechocystis* CP43 (Kuhn & Vermaas, 1993). Significantly, all resulting mutants showed complete loss of photoautotrophic growth and the ability to evolve oxygen. These mutants also contained decreased levels of the PSII reaction center proteins D1, D2 and CP47. This work clearly established a role for the large extrinsic loop in PSII function.

Site-directed mutagenesis of the large extrinsic loop of CP43 in *Synechocystis* has revealed a number of amino acids required for normal PSII function. In general, these mutants segregate into two groups. One group contains mutations at residues that appear to be involved in manganese ligation and catalysis. These sites are Glu354 and Arg357. Glu354 and Arg357 lie, respectively, within the first and second coordination spheres of the Mn_4CaO_5 cluster. Glu354, as stated above, provides bidentate ligation of Mn3, and also binds one chloride ion. The effects of mutation of this residue have been examined in three independently constructed mutants in which the glutamate was replaced with glutamine (Rosenburg et al., 1999; Strickler et al., 2008; Shimada et al., 2009; Service et al., 2011). The E354Q mutant grows slowly and exhibits only 20% of control rates of oxygen-evolving activity. The PSII centers of this mutant can advance to the S_2 and S_3 states, but do not advance past S_3 (Shimada et al., 2009; Service et al, 2011). Glu354 appears to ligate manganese bound to a substrate water molecule (Shimada et al., 2009; Service et al, 2011) that undergoes rapid exchange in the S_3 state (Service et al., 2011). Arg357 lies in the second coordination sphere of the Mn_4CaO_5 cluster. In the crystal structure, one guanidinium nitrogen of this residue is hydrogen-bonded to two of the μ-oxo bridges (O2 and O4) and the other is hydrogen-bonded to two residues of the D1 protein, Asp170, which ligates Mn4 and the calcium ion, and Ala344, which ligates Mn2. Mutation of this arginine to serine abolished oxygen-evolving activity and photoautotrophic growth (Knoepfle et al., 1999; Ananyev et al., 2005). A more conservative replacement of the arginine with lysine (Hwang et al., 2007), also produced a mutant that failed to grow photoautotrophically but retained a low rate of oxygen-evolving activity. S-state cycling defects were observed in this mutant. Specifically, the strain exhibited a high miss factor and a decreased yield of the S_2 state. Using isotopic labeling coupled with FTIR spectroscopy, structural coupling between an arginine and the Mn_4CaO_5 cluster was observed (Shimada, et al., 2011). The arginine was presumed to be Arg357 of CP43, since this is the only arginine residue in close proximity to the Mn_4CaO_5 cluster. The data are consistent with a role for Arg357 in the structural stabilization of the cluster and/or possibly in abstracting protons during water oxidation, as has been proposed previously (McEvoy & Brudvig, 2006). These experimental data are consistent with the crystallographic data and with a direct role of both Glu354 and Arg357 of CP43 in the water splitting mechanism.

Of particular interest is a second group of CP43 mutants showing defects associated with the chloride requirement of PSII. These are mutations at Arg320 and Asp321. We have extensively characterized a *Synechocystis* mutant in which Arg320 was replaced by serine (Knoepfle et al., 1999; Young, et al., 2002). When grown in complete BG-11 media (480 μM chloride), the R320S mutant strain exhibited photoautotrophic growth rates comparable to the control strain and assembled approximately 70% of the PSII centers found in the control strain (Knoepfle et al., 1999). These centers are capable of evolving oxygen to 60-70% of control rates at saturating light intensities. However, they show an enhanced susceptibility to photoinactivation. Additionally, the fluorescence rise time for this mutant was increased by a factor of two over the control. Thus, the PSII centers in this mutant do not function normally, even under normal growth conditions. Significantly, when the R320S mutant was grown in media depleted of chloride (30 μM chloride), it exhibited a severely reduced photoautotrophic growth rate. The effect of chloride depletion on the growth rate of the mutant was reversed by the addition of 480 μM bromide to the chloride-depleted BG-11 media. It is well known that bromide will functionally replace chloride at the oxygen-evolving site in cyanobacteria grown on bromide-containing media (Yachandra et al., 1993). Additionally, it is well known that bromide can replace chloride in supporting oxygen evolution in isolated thylakoids (Kelly & Izawa, 1978) and PSII membranes (Sandusky & Yocum, 1983). Oxygen evolution rates for the mutant were further depressed to about 22% of the rate observed in control cells under chloride-limiting conditions. Addition of bromide restored these rates to those observed under chloride-sufficient conditions. The mutant exhibited a significantly lower relative quantum yield for oxygen evolution than did the control strain and this was exacerbated under chloride-limiting conditions. Fluorescence yield measurements indicated that both the mutant and control strains assembled fewer PSII reaction centers under chloride-limiting conditions. The reaction centers assembled by the mutant exhibited an enhanced sensitivity to photoinactivation under chloride-limiting conditions, with a $t_{1/2}$ of photoinactivation of 2.6 min under chloride-limiting conditions compared to a $t_{1/2}$ of 4.7 min under normal growth conditions. The mutant also exhibited an enhanced stability of its S_2 state and an increased number of centers in the S_1 state following dark incubation. These results indicate that the mutant R320S exhibited a defect in its ability to utilize chloride in support of efficient oxygen evolution in PSII. This was the first mutant of this type described in the CP43 protein. In terms of this chloride effect, R320S has a similar phenotype to the R448G, K321G, and F363R mutants constructed in the CP47 large extrinsic loop (Putnam-Evans & Bricker, 1994, 1997; Clarke & Eaton-Rye, 1999). These results were significant because the R320S mutation represented the first site in the CP43 protein that alters the chloride requirement of PSII and inferred that the large extrinsic loop of CP43 is involved in chloride binding/sequestration at the PSII active site.

The phenotype of R320S closely resembles that of mutants in which the extrinsic cytochrome c$_{550}$ protein has been deleted. Cytochrome c$_{550}$ appears to regulate the efficiency of the S_1 → S_2 and/or the S_2 → S_3 state transitions. The *psbV* deletion strains exhibit a stabilization of the S_2 state and cannot grow photoautotrophically under chloride- or calcium-limiting conditions (Shen et al., 1998). Cytochrome c$_{550}$ functions in a manner similar to that of the PsbP protein in higher plant PSII. We recently have shown that R320S also does not grow under conditions of calcium depletion (Putnam-Evans, unpublished). In order to assess whether or not cytochrome c$_{550}$ is associated with PSII in the mutant, a histidine-tagged version of the R320S mutant was produced to facilitate the isolation of PSII particles. These particles were analyzed for the presence of cytochrome c$_{550}$ (Bricker et al., 2002). Reduced minus oxidized difference spectroscopy and chemiluminescence staining on western blots indicated that cytochrome c$_{550}$ was absent in these PSII particles. Whole cell extracts from the R320S mutant, however, contained a similar amount of cytochrome c$_{550}$ to that observed in the control strain. These results indicate that the mutation R320S in CP43 prevents the strong association of cytochrome c$_{550}$ with the PSII core complex. Thus, the Arg320 residue may be involved in the formation of the binding domain for the cytochrome. This is the first residue identified in any PSII protein that potentially provides a binding site for cytochrome c_{550}. We have also begun preliminary investigation of mutants produced at the residue adjacent to Arg320, Asp321. Replacement of this aspartate by asparagine produced a mutant with a phenotype very similar to that of the R320S mutant (Putnam-Evans, unpublished). That is, this mutant exhibits almost normal PSII function when grown in complete BG-11 media, but fails to grow photoautotrophically under chloride-limiting conditions and exhibits a marked decrease in oxygen evolution rates under chloride depletion.

Examination of the newest crystal structure reveals that Arg320 lies at the interface between cytochrome c_{550} and PsbU, within 2.9 angstroms of Asn49 of cytochrome c_{550}, close enough to form a hydrogen bond. Another potential hydrogen bonding partner in *Synechocystis* is Asn51. However, in *Thermosynechococcus*, the organism from which the x-ray structure is derived, this residue is a serine residue. Nevertheless, it is located only 3.6 angstroms from Arg320. Asp321 lies within 2.7 angstroms of Asn99. Additionally, Arg320 and Asp321 of CP43 lie within 2.8 angstroms of each other. Thus, these sites are candidates for potential ligands to cytochrome c_{550} and PsbU. It is tempting to speculate that Arg320 of CP43 is a central player in these interactions, perhaps via a hydrogen bonding network involving several of these and perhaps other residues. Here we report the construction and preliminary characterization of two additional mutations at site 320, R320D and R320K, towards the goal of better understanding the role of this amino acid in these potential protein-protein interactions in PSII.

5. Experimental methods

5.1 Growth conditions and growth measurements

Control and mutant *Synechocystis* 6803 were grown in liquid BG-11 media (Williams, 1988) at 30°C and a light intensity of 25 µmol photons*m^{-2}*s^{-1}, with shaking at 200 rpm on a rotary shaker. Glucose was added, when appropriate, to a final concentration of 5 mM (photomixotrophic growth). For fluorescence measurements, the BG-11 media was supplemented with 5 mM glucose and 10 µM N'-(3,4-dichlorophenyl)-N,N-dimethylurea (DCMU). Antibiotics were added to the medium to a final concentration of 10 µg/ml. For growth on solid media, BG-11 media was supplemented with 1.5 % agar, 0.3 % sodium thiosulfate, and 10 mM N-tris-(hydroxymethyl)methyl-2-aminoethanesulfonic acid/potassium hydroxide (TES/KOH), pH 8.2, and 10 µM DCMU.

For growth under chloride-limiting conditions, chloride was excluded from the media by the addition of calcium nitrate, cobalt nitrate, and manganese sulfate in place of their respective chloride salts. To prevent the leaching of chloride from glass, polycarbonate flasks and carboys were used for growing these cultures. BG-11 prepared in this manner contains approximately 30 µM chloride as compared to normal media, which has a chloride concentration of 480 µM (Young et al., 2002). For growth assessment, cells were grown as above and the OD_{730} was determined at the same time each day for 9 days.

5.2 PCR and DNA sequencing

Genomic DNA was isolated as follows: A small ball of cells was scraped from a culture grown on solid media. The cells were washed with liquid BG-11 media and pelleted by centrifugation at 3326 x *g* for 1 minute. The pelleted cells were resuspended in 400 µl 5 mM Tris-HCl, pH 8.0, 50 mM NaCl, 5 mM EDTA. Next, 100 µl of lysozyme (50 mg/ml in dH_2O)

was added and the mixture was incubated at 37°C for 15 minutes. Then, 20 µl of 500 mM EDTA, 50 µl of proteinase K (10 mg/ml in 50% glycerol), and 55 µl of 10% sarkosyl were added. The mixture was then incubated at 55°C for 15 minutes. After 5 minutes at room temperature, 600 µl of TE (5 mM Tris-HCl, pH 8.0, 5 mM EDTA) saturated phenol was added and the mixture incubated at room temperature for 10 additional minutes with gentle inversion. The mixture was then centrifuged at 18000 x g for 3 minutes and the aqueous phase was transferred to a new tube. Next, 100 µl of 5 M NaCl, 100 µl of 10% CTAB, and 600 µl of chloroform were added to the sample. The tubes were then shaken on a rotary shaker for 15 minutes, centrifuged at 8300 x g for 3 minutes, and the aqueous phase was transferred to a new tube. Next, an equal volume of cold 100% isopropanol was added to the sample, and the tubes were gently inverted. The mixture was allowed to incubate at room temperature for 20 minutes to precipitate the DNA before centrifuging the sample at 8200 x g for 10 minutes at 4°C to pellet the DNA. After centrifugation, the supernatant was removed and discarded, and 1 ml of cold 70% ethanol was added to the tube to wash the DNA pellet. The tubes were then centrifuged again for 10 minutes at 4°C. Following this, the supernatant was quickly removed and the DNA pellet was allowed to air dry. Then, 100 µl of TE buffer was added to the tube and the DNA was allowed to resuspend at room temperature overnight. Afterward, the DNA was stored at 4°C.

The *psbC* gene was amplified by PCR. A typical reaction consisted of 79.7 µl of sterile dH$_2$O, 10 µl of 10X PCR buffer (Invitrogen, Inc.), 4 µl of 50 mM MgCl$_2$, 0.8 µl of 10 mM dNTPs, 2 µl of forward primer (2 pmol/µl) (Invitrogen, Inc.), 2 µl of reverse primer (2 pmol/µl) (Invitrogen, Inc.), 30-60 ng of genomic DNA, and 0.5 µl of Taq DNA polymerase (2.5 units, recombinant). The thermal cycling routine consisted of the following steps: Step 1 – 1 cycle of 94°C for 3 minutes, 60°C for 45 seconds, and 72°C for 2 minutes, Step 2 – 25 cycles of 94°C for 1 minute, 60°C for 45 seconds, and 72°C for 2 minutes, Step 3 – 5 cycles of 94°C for 1 minute, 60°C for 45 seconds, and 72°C for 2 minutes with a 5 second extension added to the elongation step each cycle, and Step 4 – 72°C for 7 minutes.

Samples were cleaned using a QIAquick PCR Purification Kit (Qiagen, Inc.) and resuspended in 40 µl of sterile dH$_2$O. Following the cleanup, the samples were subjected to sequencing reactions in a PTC-100™ Programmable Thermal Cycler (MJ Research, Inc.). The components of these sequencing reactions were 5 µl of sterile dH$_2$O, 2 µl of Big Dye (Applied Biosystems, Inc.), 3 µl of Big Dye reaction buffer, 2 µl of primer (2 pmol/µl), and 30-60 ng of purified PCR product DNA. The cycling routine was as follows: 26 cycles of 94°C for 1 minute, 55°C for 1 minute, and 72°C for 2 minutes, followed by a final elongation at 72°C for 10 minutes.

After the completion of the sequencing reactions, the DNA was alcohol precipitated with ethanol and was resuspended in a 5:1 formamide:50 mM EDTA/50 mg/ml Blue Dextran mixture. Then, samples were loaded in an ABI Prism 377 DNA Sequencer (Applied Biosystems, Inc.) and the sequencer was run for 7 hours. Sequences were analyzed using the Auto Assembler DNA sequence analysis software (Applied Biosystems v.1.4.0).

5.3 Oxygen evolution assays and photoinactivation assays

Oxygen evolution activity was measured by O$_2$ polarography with a Hansatech oxygen electrode (Knoepfle et al., 1998). Assays were performed at 25°C on whole cells in complete or chloride-depleted BG-11 media with 1 mM 2,6-dichlorobenzoquinone (DCBQ) added as an artificial electron acceptor. Cells were allowed to grow for 5 days under the appropriate growth conditions, as above, and were harvested on the fifth day by centrifuging at 12000 x g for 5 minutes at 4°C. The cells were then resuspended in 1-2 ml of complete BG-11 media. Next, the chlorophyll concentration of each sample was determined by taking absorbance readings at 678 nm and 710 nm and calculating the concentration of chlorophyll (in µg/ml) by the equation: $(OD_{678}-OD_{710})$ * 14.96 * dilution factor (modified from Williams, 1988). The light intensity used during these assays was 2500 µmol photons*m^{-2}*s^{-1} of white light, verified with a spectroradiometer equipped with a quantum probe (Li-Cor, Inc.).

For photoinactivation experiments, cells were incubated at 25°C in BG-11 media, at a chlorophyll concentration of 10 µg/ml, and exposed to a light intensity of 5000 µmol photons*m^{-2}*s^{-1}. Cells were removed from exposure to the photoinactivating light in increments of 2 minutes and assayed for oxygen-evolving activity, as above. The $t_{0.5}$ for photoinactivation was calculated by fitting the data to a single exponential decay.

5.4 Fluorescence measurements

Fluorescence yield measurements were performed on a Waltz PAM 101 fluorometer as described previously (Nixon & Diner; 1992, Chu et al., 1994). Samples were incubated in the dark for 5 minutes in the presence of 1 mM potassium ferricyanide and 330 µM DCBQ. Next, DCMU was added to a final concentration of 40 µM followed one minute later by the addition of hydroxylamine hydrochloride, pH 6.5, to a concentration of 20 mM. After 20 s, the weak monitoring flashes were turned on, followed 1 s later by continuous actinic illumination (1000 µmol photons*m^{-2}*s^{-1}). The variable fluorescence was measured and a second set of assays was performed, this time omitting the hydroxylamine, in order to measure the variable fluorescence with water as the electron donor. For both assays, F_{max} was measured 5 s after the onset of actinic illumination.

5.5 Introduction of poly-histidine tags and purification of photosystem II

A recombinant pTZ18 plasmid was created previously that contained the *psbB* gene encoding CP47 with six histidyl codons at the 3' end of the gene (Young et al., 2002). An antibiotic resistance gene encoding resistance to spectinomycin was cloned into the 3' non-coding portion of the *psbB* gene downstream of the His-tag. This plasmid was transformed into *Synechocystis* as previously described (Williams, 1988). Briefly, 5 x 10^8 cells in 100 µl of BG-11 media were added to a sterile 15 ml centrifuge tube. To these cells, 2.5 µg of plasmid DNA was added. The tubes were incubated for 4-6 hours at room temperature and exposure to a light intensity of 25 µmol photons*m^{-2}*s^{-1}. After incubation, the entire volume of cells was plated onto solid BG-11/glucose/DCMU to which no antibiotic had been added. After 48 hours, the plates were underlaid with antibiotic solution (196 µl of BG-11 media + 8 µl spectinomycin (50 mg/ml stock)). Transformants were allowed to sort out by serial streaking on BG-11 plates containing glucose, DCMU, and spectinomycin. The insertion of the histidyl codons was confirmed by DNA sequencing, as above.

Cells were grown in either complete BG-11 media or in BG-11 media deficient in chloride. Cultures were supplemented with 5 mM glucose and 5 µg/ml spectinomycin. The cells were grown at 25°C, with aeration, at a light intensity of 25 µmol photons*m^{-2}*s^{-1}, for 7-9 days. Cells from 15 l cultures were centrifuged 10000 x g for 10 minutes at 4°C. The cells were resuspended in an adequate amount of Buffer A (50 mM 2-(N-morpholino)ethanesulfonic acid (MES)-NaOH, pH 6.0, 10 mM MgCl$_2$, 5 mM CaCl$_2$, 25% glycerol) (~10 ml per pellet), combined into a single container, and centrifuged again at 10000 x g for 10 min at 4°C. Excess Buffer A was decanted and the pellet was frozen at –70 °C.

To purify the His-tagged PSII, the procedure of Bricker et al. (1998) was used. Frozen cells were thawed and resuspended in 5 ml of Buffer A. The cells were then assayed for chlorophyll concentration and brought to 1 mg/ml of chlorophyll. Then, the cell suspension was brought to 1.0 mM phenylmethylsulfonyl fluoride (PMSF), 1.0 mM benzamidine, 1 mM ε-amino caproic acid, and 50 µg/ml DNAase. The sample was added to an ice water cooled bead beater apparatus (Bio-Spec Products, Inc.) and glass beads (1.0 mm) were added at a 1:1 ratio to the cell suspension. The cells were then broken at 4°C over 12-14 break cycles, each consisting of 15 seconds of homogenization followed by 4 minutes of cooling. The cell homogenate and glass beads were then transferred to a beaker and the glass beads allowed to settle. The cell homogenate was decanted and the beads were washed several times with Buffer A to recover additional homogenate. The cell homogenate was brought to 1% n-dodecyl-β-D-maltoside by the addition of a 10% n-dodecyl-β-D-maltoside stock solution (freshly prepared) and was immediately centrifuged at 36000 x g for 10 minutes to remove unbroken cells and residual glass beads. The solubilized cell homogenate was then loaded onto a 25 ml cobalt metal affinity column (Clontech, Inc.) which had been pre-equilibrated with Buffer A + 0.04% n-dodecyl-β-D-maltoside at 4°C. The column was washed with several bed volumes of Buffer A + 0.04% n-dodecyl-β-D-maltoside and the bound His-tagged PSII particles were eluted with Buffer A + 0.04% n-dodecyl-β-D-maltoside + 50 mM L-histidine. The eluted fractions were then pooled and the PSII complex was precipitated by addition of an equal volume of 25% PEG-8000 in 50 mM MES-NaOH, pH 6.0, and incubation at 4°C for 30 minutes. The precipitated PSII particles were harvested by centrifugation at 36000 x g for 30 minutes. The precipitated PSII particles were resuspended in 1.0 ml of Buffer A + 0.04% n-dodecyl-β-D-maltoside and stored immediately at –70 °C.

5.6 Spectrophotometric assays

For the determination of absorbance spectra, the procedure of Bricker and coworkers (2002) was employed. First, purified PSII particles were assayed for chlorophyll concentration by methanol extraction, and samples were diluted to 20 µg/ml chlorophyll in a solution of 50 mM MES-NaOH, pH 6.5, 10 mM MgCl$_2$, 5 mM CaCl$_2$, 50 mM NaCl, 0.04% β-D-dodecyl maltoside, and 25% glycerol. After performing a baseline correction on a Cary 100 dual-beam spectrophotometer (Varian, Inc.), the samples were either oxidized with a few crystals of potassium ferricyanide, for the reference cuvette, or reduced with a few crystals of sodium dithionite, for the sample cuvette. After a 10-minute dark incubation, reduced minus oxidized difference spectra were collected by scanning the samples from 540-580 nm for the cytochrome-specific spectra, and from 400-800 nm for the overall absorbance spectra.

6. Results

6.1 Verification of genomic DNA sequence

The charge-switching mutants constructed for this study were created by site-directed mutagenesis of Arg320 of CP43 to either lysine or aspartate. DNA sequencing of the portion of the *psbC* gene surrounding the mutation site was used to verify each point mutation. Additionally, because the mutant sequence was engineered within a plasmid insert containing the AccI/AccI fragment of *psbC*, the DNA sequence included between these restriction sites in the genomic DNA was also sequenced. This segment consists of 1221 base pairs, along with an inserted kanamycin resistance cartridge. DNA sequencing confirmed the presence of the desired point mutations, as well as the integrity of the sequence within the AccI/AccI fragment.

Additional mutants were created for the insertion of a poly-histidine tag on the CP47 protein of control and mutant strains. This was done by inserting a series of six consecutive histidine-encoding codons, along with a spectinomycin resistance cartridge, into *psb*B. The poly-histidine encoding sequence was verified by DNA sequencing for each strain.

6.2 Growth measurements

The mutant and control strains were grown in the absence of glucose to determine the effect of each mutation on the photoautotrophic growth characteristics of the organism. Each strain was grown in complete BG-11 media, in BG-11 media deficient in chloride, or in chloride-deficient BG-11 media supplemented with 480 μM sodium bromide, in order to determine any effect chloride had on the growth of the strains. In complete media, both R320K and R320D grew at rates comparable to the control strain (Fig. 2A). In chloride-deficient media, the control strain exhibited a normal growth rate. (Fig. 2B). The mutant strains, however, were affected under these conditions, with R320K growing at a lower rate than it did in complete media and R320D showing virtually no photoautotrophic growth. When the strains were grown in chloride-deficient media supplemented with bromide (Fig. 2C), growth rates for each mutant strain were restored to near those observed for the strains grown in complete media. It is well known that bromide can substitute for chloride at the PSII active site (Sandusky & Yocum, 1983; Yachandra et al., 1993).

6.3 Oxygen evolution

Oxygen evolution rates were measured for control and mutant strains grown either photoautotrophically or in the presence of glucose. When grown photoautrophically (Fig. 3), the control strain evolved oxygen at similar rates regardless of the media conditions. In complete media, the R320K strain evolved oxygen at a rate about 80% that of the control. When chloride was lacking in the growth media, this rate fell to 35% of the control. The R320D strain was the least successful of these strains for evolving oxygen. When grown photoautotrophically in complete media, the rate of oxygen evolution for this strain was depressed to 60% of the control. The effect on oxygen evolution was greatly exacerbated when the strain was grown under chloride-limiting conditions, where the observed rate of oxygen evolution was only 5% of the control rate. For both mutants, the addition of bromide restored this rate to levels near those observed for cells grown in complete media. These same results were observed when the strains were grown photomixotrophically (not shown).

6.4 Variable fluorescence yields

Variable fluorescence yield assays are used to make qualitative estimates of the number of PSII centers present in the thylakoid membranes of a particular strain, and also allow for a determination of the functionality of these centers. The results of these experiments for cells grown in complete media are shown in Table 1. Both mutant strains assemble relatively

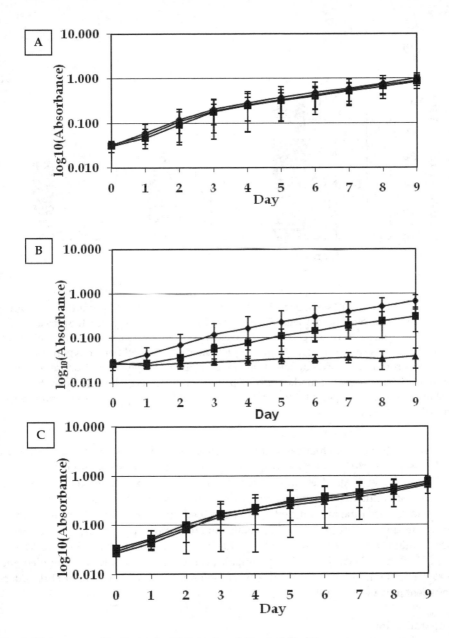

Fig. 2. Photoautrophic Growth of Mutant and Control Strains.
Growth curves showing control (diamonds) and mutant strains (R320K squares; R320D triangles) grown in either **A.** complete media, **B.** chloride-deficient media, or **C.** chloride-deficient media supplemented with bromide. These results are the average of at least 3 independent experiments. The error bars represent plus and minus one standard deviation.

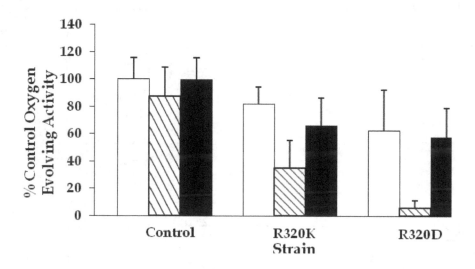

Fig. 3. Oxygen Evolution Rates of Control and Mutant Strains Grown Photoautotrophically. The control rate in complete media was 391 μmol O_2*(mg chlorophyll)$^{-1}$ *hr^{-1}. Rates are expressed as the percentage of the control rate in cells grown in complete media. White bars, oxygen evolution rates from cells grown in complete media; Striped bars, oxygen evolution rates from cells grown in chloride deficient media; Black bars, oxygen evolution rates from cells grown in chloride depleted media with 480μM bromide added. The results are the average of at least three independent experiments. The error bars represent plus and minus one standard deviation.

large numbers of PS II centers that can transfer electrons from hydroxylamine to Q_A. R320K and R320D assembled 80% and 83%, respectively, of the PS II centers assembled in the control strain. If water is used as an electron donor, the variable fluorescence yield is a measure of the amount of PSII centers that can oxidize water. The data from the water experiment indicate that the mutant centers remain mostly functional under normal growth conditions, with the fluorescence yields for both mutants being approximately 80% that of the control strain. Data is also included for the psbC deletion strain, which lacks any functional PSII centers. Data for the R320S mutant (Young et al., 2002) is also included and shows that this mutant, like R320D and R320K, contains substantial numbers of PSII centers that are efficient in water oxidation.

6.5 Photoinactivation
Photoinactivation assays are used to determine the relative stability of the PSII complex of a strain. As summarized in Table 2, when grown photoautotrophically, regardless of the media, the control strain had a $t_{0.5}$ for photoinactivation of approximately 5.6 min. In complete media, R320K exhibited a $t_{0.5}$ close to the control of 5.3 min and R320D a $t_{0.5}$ slightly lower than the control of 3.6 min. When grown in media deficient in chloride, however, the

mutant strains exhibited large increases in photoinactivation demonstrated by $t_{0.5}$ values of 2.9 and 0.9 min, respectively, for R320K and R320D. These data indicate that the PSII complexes of the mutants are less stable, and are much more susceptible to the effects of photoinactivating light intensities. It should be noted that the $t_{0.5}$ values for the R320K closely mirror those of the R320S mutant (Young et al., 2002).

Strain	Hydroxylamine	Water
Control	0.46 ± 0.069	0.45 ± 0.056
psbC Deletion	0.02 ± 0.000	-0.02 ± 0.029
R320K	0.37 ± 0.034	0.35 ± 0.022
R320D	0.38 ± 0.029	0.33 ± 0.020
R320S*	0.41 ± 0.030	0.46 ± 0.050

Table 1. Variable Fluorescence Yields.
Results of variable fluorescence yield assays performed on control, psbC deletion, R320K, and R320D, and R320S* strains.
(* Results for the R320S (R320S) strain obtained from Young et al., 2002. The control values for these experiments were hydroxylamine = 0.58 and water = 0.67.)

Strain	Complete Media	Chloride-Deficient Media
Control	5.4 min.	5.7 min.
R320K	5.3 min.	2.9 min.
R320D	3.6 min.	0.9 min.
R320S*	4.7 min.	2.6 min.

Table 2. $t_{0.5}$ Values for Photoinactivation.
$t_{0.5}$ values for photoinactivation for each strain after growth in either complete or chloride-deficient media and exposure to photoinactivating light (5000 μmol photons*m^{-2}*s^{-1}). The results are the average of at least two independent experiments.
(* Results from Young et al., 2002)

6.6 Visible absorption spectra

Absorbance spectra in the range of 400-800 nm were obtained for PSII particles purified from His-tagged control and mutant strains grown photomixotrophically in complete media. As can be seen in Fig. 4, there were no appreciable differences in these spectra, regardless of the strain, which indicates that no major disruption of the pigment-containing portion of PSII occurred within the mutant strains. The major peaks located at 410, 430, and 680 nm represent absorption by chlorophyll-a molecules, with the major peak at 680 nm representing the reaction center chlorophyll of PSII, P680. The minor peaks arise from accessory pigment molecules associated with the photosystem. These spectra are similar to spectra obtained from PSII preparations of non-His-tagged control strains (Noren et al., 1991; Tang & Diner, 1994) and to spectra obtained from other His-tagged PSII preparations (Bricker et al., 1998).

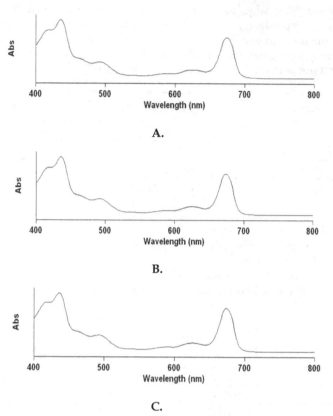

Fig. 4. Visible Absorption Spectra
Visible absorbance spectra for **A.** control, **B.** R320K, and **C.** R320D PSII histidine-tagged
particles purified from cells grown photomixotrophically in complete BG-11 media.

6.7 Reduced-minus oxidized difference spectra

Reduced minus oxidized difference spectra were collected in the wavelength range of 540-
580 nm for PSII particles isolated from the His-tagged control and mutant strains grown
photomixotrophically in complete media (Fig. 5) Plots of the data reveal a non-symmetrical
absorbance curve for the control strain. This non-symmetrical curve is also present in control
PSII particles derived from cells grown under chloride depletion (not shown). This curve
reveals a maximum absorbance at 559 nm that tapers off into a wide shoulder toward 550
nm, while toward 570 nm, there is no indication of a shoulder. The signal at 559 nm
indicates the presence of cytochrome b_{559}, while the broadened shoulder arises due to the
absorbance properties of cytochrome c_{550} (Bricker et al., 2002). The plots of data derived
from the mutant strains, like the control, exhibit an absorbance maximum at 559 nm.
However, unlike the control strain, the plots for the mutant strains are symmetrical, giving
no indication of a broad shoulder on either side of the peak, regardless of the media,
suggesting the absence of cytochrome c_{550} within these preparations. The same symmetrical
peaks are observed for both mutants grown under chloride-limited conditions (not shown).

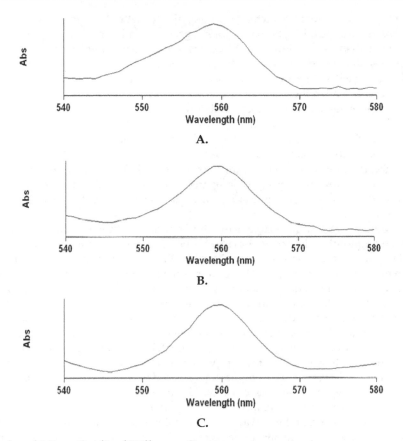

Fig. 5. Reduced Minus Oxidized Difference Spectra

Reduced minus oxidized absorbance difference spectra for **A.** control, **B.** R320K, and **C.** R320D His-tagged PSII particles purified from cells grown photomixotrophically in complete media. Samples were diluted to 20 μg/ml chlorophyll in 50 mM MES-NaOH, pH 6.5, 10 mM MgCl$_2$, 5 mM CaCl$_2$, 50 mM NaCl, 0.04 % β-D-dodecyl maltoside, and 25 % glycerol. After performing a baseline correction on a Varian Cary 100 dual-beam spectrophotometer, the samples were either oxidized with a few crystals of potassium ferricyanide, for the reference cuvette, or reduced with a few crystals of sodium dithionite, for the sample cuvette. After a 10-minute dark incubation, reduced minus oxidized difference spectra were collected by scanning the samples from 580-540 nm. The peak absorbance at 560 nm measured approximately 0.10 absorbance units.

7. Discussion

The purpose of the present study was to elucidate further the interaction of CP43 with the extrinsic cytochrome c$_{550}$ protein of PSII by studying the effect of additional alterations of the arginine residue at position 320 of the CP43 large extrinsic loop. Two additional mutations, R320K and R320D, were introduced into the large extrinsic loop and the effects of each on PSII activity were characterized. First, the growth characteristics of the mutant strains were assessed. Under photoautotrophic growth conditions (Fig. 2A), when chloride levels in the media were not limiting, the control and mutant strains grew at the same rate. This indicates that under these growth conditions, there appear to be no discernible anomalies associated with PSII. When grown in chloride-deficient media, however, major differences arose. Under these conditions, the control strain grew at rates comparable to those in complete media, but R320D exhibited very little capacity for photoautotrophic growth (Fig. 2B). The R320K strain exhibited intermediate growth between that of the control and R320D. When grown in chloride-deficient media that had been supplemented with bromide, all strains grew at very similar rates, with the rates of the mutant strains being restored to those observed in complete media (Fig. 2C). As stated previously, bromide is well known to substitute for chloride at the active site of PSII, and these data clearly indicate that the effect on growth arises as a consequence of a defect associated with the chloride requirement of PSII.

Oxygen evolution rates were obtained for the control and mutant strains (Fig. 3). When grown photoautotrophically, the control strain evolves oxygen at similar rates, regardless of the media in which the cells are grown. The R320K strain exhibits a modest reduction in oxygen evolution rate when grown in complete media, which is greatly magnified when chloride is limiting in the media. Under these conditions, the R320K strain appears to evolve oxygen at less than 40% of the control strain. This rate is restored when bromide is included in the chloride-deficient media. The R320D strain reveals the most dramatic phenotype. When grown in complete media or in chloride-deficient media supplemented with bromide, this strain evolves oxygen at a rate of approximately 60% of the control strain. When chloride is limiting, however, this rate drops to only about 5% of that of the control strain. Taken together, these results indicate that there is a defect in both mutant strains in their ability to evolve oxygen, regardless of the media in which they are grown. This defect is exacerbated by a deficiency of chloride in the media, which again points to a physical and/or mechanistic defect at the PSII active site involving some interaction with chloride.

The data from growth experiments and oxygen evolution assays indicated that both mutant strains, which are relatively functional under normal growth conditions, must assemble a fairly normal content of PSII centers. Confirmation of this was obtained from variable fluorescence yield assays. These assays allow for a qualitative estimate of the number of PSII centers present when using hydroxylamine as an artificial electron donor to measure the reduction of Q_A (Nixon & Diner, 1992). The results presented in Table 1 show that the mutant strains assemble a near normal complement of PSII centers within their thylakoid membranes compared to the control strain. Thus, the mutations do not prevent the assembly and incorporation of PSII centers into the thylakoids of these strains.

When water is used as the electron donor, the variable fluorescence yield assay provides information on the functionality of the PSII centers present within the membrane, with respect to their ability to conduct electron transport from water to Q_A. As shown in Table 1, the variable fluorescence yields for both mutant strains approximated those of the control strain, indicating that the mutant PSII centers are fully capable of oxidizing water. These results mirrored those previously obtained for the R320S strain (Table 1; Young, et al., 2002). It should be noted that the psbC deletion strain, which contains no functional PSII centers, did not give a fluorescence signal.

Although the R320K and R320D strains appear to assemble near control amounts of PSII centers that one might assume from the above data are functioning relatively normally, this is not the case, particularly under chloride-limiting conditions. The susceptibility of the PSII centers to photoinactivation was assessed by exposing the cells to a high light intensity. The process of photoinactivation leads to a number of structural changes within the PSII complex. These events include the rapid turnover of the D1 protein and proteolysis of the D2 protein and CP43 (Nedbal et al., 1990; Mori and Yamamoto, 1992; Aro et al., 1993), with a resultant drop in oxygen-evolving activity. The results for the control strain (Table 2) showed no appreciable differences between the $t_{0.5}$ values for samples grown in complete or chloride-deficient media. When grown in complete media and subjected to photoinactivating light, the R320K strain exhibited a $t_{0.5}$ value similar to the control value. The calculated $t_{0.5}$ for photoinactivation for the R320D mutant was 1.4 times less than that of control cells, indicating that this mutant is more sensitive to photoinactivation than the control. This suggests that, in complete media, there is a defect in the stability of the R320D PSII complex, while no defect is apparent for the R320K strain under these conditions.

Photoinactivation experiments were also performed on strains grown under conditions of chloride depletion (Table 2). The results of these experiments indicate that, while $t_{0.5}$ value for the control strain remained essentially the same compared to its value in complete media, the PSII stability of both mutant strains was altered under these conditions. For the R320K strain grown in chloride-deficient media, the $t_{0.5}$ value was 2.9 min compared to 5.7 min for the control, indicating that this mutant is photoinactivating at twice the rate of the control under these conditions. Photoinactivation was difficult to measure in the R320D mutant due to its low rate of oxygen evolution. Given this, R320D cells demonstrated an extreme sensitivity to photoinhibition, with a calculated $t_{0.5}$ for photoinactivation 6 times less than that of the control. Thus, the PSII centers in all the mutants become destabilized under chloride-limiting conditions.

Comparison of the results of this photoinactivation study to those obtained for the R320S mutant (Table 2; Young et al., 2002) highlights some notable trends. The PSII stability of these strains seems to follow the following pattern in complete media: Control ≈ R320K > R320S > R320D. This trend is only slightly modified when the strains are grown in chloride-deficient media, and follows the pattern: Control > R320K ≈ R320S > R320D. When chloride is not limiting in the media, it appears that various amino acid substitutions can be made at this site with limited effect on the stability of PSII. However, the results also indicate that when chloride is limiting, even substitution of an amino acid with similar properties to arginine, such as lysine, is not sufficient to preserve the integrity of the interactions in which this site is involved.

In order to gain a better understanding of the effect of the mutations on possible interactions with cytochrome c$_{550}$, PSII core particles were isolated from His-tagged control and mutant strains using cobalt metal affinity chromatography. The effectiveness of this method for isolating active PSII particles has been documented (Bricker et al., 1998). Fig. 4 shows the optical absorption spectra obtained from the PSII particles derived from the control and two mutant strains. There are no apparent differences in the spectra, each of which shows a characteristic absorption maximum at 674 nm. Reduced minus oxidized difference spectra were obtained from the isolated control, R320K, and R320D His-tagged particles (Fig. 5). These indicated that while the control particles were able to retain cytochrome c$_{550}$ within the particle, the mutant strains were not (Fig. 5). This was true whether the cells were grown in complete (Fig. 5) or chloride-deficient media (not shown). These results are similar to those obtained previously with the R320S mutant, which was unable to retain the association between PSII and cytochrome c_{550} after purification of these particles, regardless of the media in which the cells were grown. Thus, even a conservative substitution (lysine) at this site, which results in a fairly functional mutant under normal growth conditions, prevents the strong association of cytochrome c_{550}. It is possible that some cytochrome c_{550} associates with these mutants in vivo, but that the relatively harsh conditions used to isolate the particles disrupt this interaction. We are currently probing the kinetics of binding of cytochrome c_{550} to His-tagged PSII from these strains in order to examine this.

8. Conclusion

Site-directed mutagenesis has been employed extensively to probe the function of the CP43 protein of PSII. In particular, substitutions at Arg320 have revealed a potential role of this amino acid in the interaction of CP43 with the extrinsic PSII protein, cytochrome c_{550}. In the present study, we have further defined the requirement at position 320 by constructing and characterizing additional mutant strains R320K and R320D. The results indicate that both the R320K and R320D strains are defective in their ability to utilize chloride to carry out efficient oxygen evolution and the effect is magnified under low chloride conditions. Spectrophotometrically detectable cytochrome c_{550} is lacking in these mutants. Additionally, we have preliminary evidence that these mutants also lack PsbU (Bricker and Putnam-Evans, unpublished). The phenotype of the R320S mutant previously described is intermediate between that of the K and D mutants. Thus, the order of effectiveness of amino acids at this site in supporting normal PSII function is R>K>S>D. A positively charged residue appears to be required at this site for proper function. The 1.9 Å crystal structure reveals a number of amino acid residues within both cytochrome c_{550} and PsbU that are potential hydrogen bonding partners to Arg320. Our operating hypothesis is that CP43, cytochrome c_{550} and PsbU interact via a hydrogen-bonding network that is involved in promoting a stable environment for chloride incorporation into the active site. While the position and ligands of chloride around the manganese cluster are now known, what other proteins bind chloride, how chloride is funneled to the active site, and where potential chloride channels occur remain unclear. Continuing studies to address these questions will no doubt expand our knowledge of the protein-protein and protein-ligand interactions in PSII that contribute to the proper functioning of the oxygen-evolving complex.

9. Acknowledgment

This work was supported by a U.S. National Science Foundation grant (MCB9982981) to CP-E, and a U.S. Department of Energy grant (DE-FG02-98ER20310) to TMB.

10. References

Ananyev G., Nguyen T., Putnam-Evans C., & Dismukes, C. (2005) Mutagenesis of CP43-arginine-357 to serine reveals new evidence for (bi)carbonate functioning in the water oxidizing complex of Photosystem II. *Photochemical and Photobiological Sciences*, Vol. 4, pp. 991–998.

Aro, E.-M., Virgin, I., & Andersson, B. (1993) Photoinhibition of photosystem II. Inactivation, protein damage and turnover. *Biochimica et Biophysica Acta*, Vol. 1143, pp. 113-134.

Bricker, T. (1990) The structure and function of CPa-1 and CPa-2 in photosystem II. *Photosynthesis. Research*, Vol. 24, pp. 1-13.

Bricker, T. (1992) Oxygen evolution in the absence of the 33 kDa manganese-stabilizing protein. *Biochemistry*, Vol. 31, pp. 4623-4628.

Bricker, T., Morvant, J., Masri, N., Sutton, H., & Frankel, L. (1998) Isolation of a highly active photosystem II preparation from Synechocystis 6803 using a histidine-tagged mutant of CP 47. *Biochimica et Biophysica Acta*, Vol. 1409, pp. 50-7.

Bricker, T., A. Young, L.K. Frankel, and C. Putnam-Evans. (2002) Introduction of the [305]Arg→ [305]Ser mutation in the large extrinsic loop E of the CP43 protein of Synechocystis sp. PCC 6803 leads to the loss of cytochrome c_{550} binding to photosystem II. *Biochimica et Biophysica Acta*, Vol. 1556, pp. 92-96.

Bricker, T. & Frankel, L. (2002) The structure and function of CP47 and CP43 in Photosystem II. *Photosynthesis Research*, Vol. 72, pp. 131-146.

Bricker, T., Roose, J., Fagerlund, R., Frankel, L., & Eaton-Rye, J. (2011) The extrinsinc proteins of photosystem II. *Biochimica et Biophysica Acta*, doi:10.1016/j.bbabio.2011.07.006 [Epub ahead of print].

Burnap, R., Qian, M., Shen, J.-R., Inoue, Y., & Sherman, L. (1994) Role of disulfide linkage and putative intermolecular binding residues in the stability and binding of the extrinsic manganese-stabilizing protein to the photosystem II reaction center. *Biochemistry*, Vol. 33, pp. 13712-13718.

Burnap, R., Shen, J.-R., Jursinic, P., Inoue, Y., & Sherman, L. (1992) Oxygen yield and thermoluminescence characteristics of a cyanobacterium lacking the manganese-stabilizing protein of Photosystem II. *Biochemistry*, Vol. 31, pp. 7404–7410.

Carpenter, S., Charite, J., Eggars, B., & Vermaas, W. (1990) The psbC start codon in *Synechocystis* sp. PCC 6803. *FEBS Letters*, Vol. 1, pp. 135-137.

Chu, H.-A., Nguyen, A. & Debus, R. (1994) Site-directed photosystem II mutants with perturbed oxygen-evolving properties. 1. Instability or inefficient assembly of the manganese cluster in vivo. *Biochemistry*, Vol. 33, pp. 6137-6149.

Clarke, S., & Eaton-Rye, J. (1999) Mutation of Phe-363 in the Photosystem II protein CP 47 impairs photoautotrophic growth, alters the chloride requirement, and prevents photosynthesis in the absence of either PSII-O or PSII-V in *Synechocystis* sp. PCC 6803. *Biochemistry*, Vol. 38, pp. 2707-2715.

Dau, H., & Haumann, M. (2008) The manganese complex of photosystem II in its reactioncycle—basic framework and possible realization at the atomic level. *Coordination Chemistry Reviews*, Vol. 252, pp. 273–295.

Debus, R., Barry, B., Sithole, I., Babcock, G., & McIntosh, L. (1988a) Directed mutagenesis indicates that the donor to P680+ in photosystem II is tyr-161 of the D1 polypeptide. *Biochemistry*, Vol. 27, pp. 907-9074.

Debus, R., Barry, B., Babcock, G., & McIntosh, L. (1988b) Site-specific mutagenesis identifies a tyrosine radical involved in the photosynthetic oxygen-evolving complex. *Proceedings of the National Academy of Science* (USA), Vol. 85, pp. 427- 430.

Dzelzkalns, V., & Bogorad, L. (1988) Molecular analysis of a mutant defective in photosynthetic oxygen evolution and isolation of a complementing clone by a novel screening procedure. *EMBO Journal*, Vol. 7, pp. 333-338.

Ferreira, K., Iverson, T., Maghlaoui, K., Barber, J., & Iwata, S. (2004) Architecture of the photosynthetic oxygen-evolving center. *Science*, Vol. 303, pp.1831-8.

Gabdulkhakov, A., Guskov, A., Broser, M., Kern, J., Müh, F., Saenger, W., & Zouni, A. (2009) Probing the accessibility of the Mn4Ca cluster in Photosystem II; channels calculation, noble gas derivatization, and cocrystallization with DMSO. *Structure*, Vol. 17, pp. 1223–1234.

Ghanotakis, D., Topper, J., & Yocum, C. (1984) Structural organization of the oxidizing side of photosystem II: exogenous reductants reduce and destroy the Mn complex in photosystem II membranes depleted of the 17 and 23 kDa polypeptides. *Biochimica et Biophysica Acta*, Vol. 767, pp. 524–531.

Ghanotakis, D., Babcock, G., & Yocum, C. (1985) On the role of water-soluble polypeptides (17, 23 kDa) calcium and chloride in photosynthetic oxygen evolution. *FEBS Letters*, Vol. 192, pp. 1–3.

Glazer, A. (1983) Comparative biochemistry of photosynthetic light-harvesting systems. *Annual Review of Biochemistry*, Vol. 52, pp.125-157.

Grabolle, M., Haumann, M., Müller, C., Liebisch, P. & Dau, H. (2006) Rapid loss of structural motifs in the manganese complex of oxygenic photosynthesis by Xray irradiation at 10–300 K. *Journal of Biological Chemistry*, Vol. 281, pp. 4580–4588.

Hwang, H., Dilbeck, P., Debus, R., & Burnap, R. (2007) Mutation of arginine 357 of the CP43 protein of photosystem II severely impairs the catalytic S-state cycle of the H2O oxidation complex. *Biochemistry*, Vol. 46, pp. 11987–11997.

Inoue-Kashino, N., Kashino, Y., Satoh, K., Terashima, I., & Pakrasi, H. (2005) PsbU provides a stable architecture for the oxygen-evolving system in cyanobacterial Photosystem II. *Biochemistry*, Vol. 44, pp. 12214–12228.

Joliot, P., Barbieri, G. & Cjabaud, R. (1969) Un nouveau modele des centres photochemiques du systeme II. *Photochemistry and Photobiology*, Vol.10, pp. 309-329.

Kamiya, N. & Shen, J.-R. (2003) Crystal structure of oxygen-evolving photosystem II from Thermosynechococcus vulcanus at 3.7-Å resolution. *Proceedings of the National Academy of Science* (USA), Vol. 100, pp. 98-103.

Katoh, H., Itoh, S., Shen, J.-R., & Ikeuchi, M. (2001) Functional analysis of psbVand a novel c-type cytochrome gene psbV2 of the thermophilic cyanobacterium Thermosynechococcus elongatus Strain BP-1. *Plant Cell Physiology*, Vol. 42, pp. 599–607.

Kawakami, K., Umena, Y., Kamiya, N., & Shen, J-R. (2011) Structure of the catalytic, inorganic core of oxygen-evolving photosystem II at 1.9 Å resolution. *Journal of Photochemistry and Photobiology B*, Vol. 104, pp. 9-18.

Kelley, P., & Izawa, S. (1978) The role of chloride ion in photosystem II. I. Effects of chloride ion on photosystem II electron transport and on hydroxylamine inhibition. *Biochimica et Biophysica Acta*, Vol. 502, pp. 198-210.

Klimov, V., Dolan, E. & Ke, B. (1980) EPR properties of an intermediary electron acceptor (pheophytin) in photosystem II reaction centers at cryogenic temperatures. *FEBS Letters*, Vol. 112, pp. 97-100.

Knoepfle, N., Bricker, T., & Putnam-Evans, C. (1999) Site-directed mutagenesis of basic arginine residues 305 and 342 in the CP 43 protein of photosystem II affects oxygen-evolving activity in Synechocystis 6803. *Biochemistry*, Vol. 38, pp. 1582-1588.

Kok, B., Forbush, B. & McGloin, M. (1970) Cooperation of charges in photosystem II O$_2$ evolution I: A linear 4-step model. *Photochemistry and Photobiology*, Vol. 11, pp. 457-475.

Kuhn, M., & Vermaas, W. (1993) Deletion mutations in a long hydrophilic loop in the photosystem II chlorophyll-binding protein CP43 in the cyanobacterium *Synechocystis* sp. PCC 6803. *Plant Molecular Biology*, Vol. 23, pp. 123-133.

Kuwabara, T., Miyao, M., Murata, T., & Murata, N. (1985) The function of the 33 kDa protein in the oxygen evolution system studied by reconstitution experiments. *Biochimica et Biophysica Acta*, Vol. 806, pp. 283-289.

Liu, H., Frankel, L., & Bricker, T. (2007) Functional analysis of Photosystem II in a PsbO-1 deficient mutant in Arabidopsis thaliana. *Biochemistry*, Vol. 46, pp. 7607-7613.

McEvoy, J., & Brudvig, G. (2006) Water-splitting chemistry of photosystem II. *Chemical Reviews*, Vol. 106, pp. 4455-4483.

Miyao, M., Murata, N., Lavorel, J., Maison-Peteri, B., Boussac, A., & Etienne, A-L. (1987) Effect of the 33-kDa protein on the S-state transitions in photosynthetic oxygen evolution. *Biochimica et Biophysica Acta*, Vol. 890, pp. 151-159.

Mori, H., & Yamamoto, Y. (1992). Deletion of antenna chlorophyll-a-binding proteins CP43 and CP47 by tris-treatment of PS II membranes in weak light: Evidence for a photo-degradative effect on the PS II components other than the eaction center-binding proteins. *Biochimica et Biophysica Acta*, Vol. 1100, pp. 293-298.

Nedbal, L., Masojídek, J., Komenda, J., Prášil, O., &. Šetlík, I. (1990). Three types of PSII photoinactivation. 2. Slow processes. *Photosynthesis. Research*, Vol. 24, pp. 89-97.

Nixon, P., & Diner, B. (1992) Aspartate 170 of the photosystem II reaction center polypeptide D1 is involved in the assembly of the oxygen-evolving manganese cluster. *Biochemistry*, Vol. 31, pp. 942-948.

Noren, G., Boerner, R., & Barry, B. (1991) EPR characterization of an oxygen-evolving photosystem II preparation from the transformable cyanobacterium Synechocystis 6803. *Biochemistry*, Vol. 30, pp. 3943-3950.

Ono, T., & Inoue, Y. (1984) Reconstitution of photosynthetic oxygen-evolving activity by rebinding of 33 kDa protein to CaCl$_2$-extracted PS II particles. *FEBS Letters*, Vol. 166, pp. 381-384.

Popelkova, H., & Yocum, C. (2007) Current status of the role of Cl$^-$ ion in the oxygen-evolving complex. *Photosynthesis Research*, Vol. 93, pp. 111-121.

Putnam-Evans, C. & Bricker, T. (1994) Site-directed mutagenesis of the Cpa-1 protein of Photosystem II:alteration of the basic residue [448]Arg to [448]Gly prevents the stable assembly of functional Photosystem II centers under chloride-limiting conditions. *Biochemistry*, Vol. 33, pp. 10770-10776.

Putnam-Evans , C., & Bricker T. (1997) Site-directed mutagenesis of the basic residues 321K
    to 321G in the CP 47 protein of photosystem II alters the chloride requirement for
    growth and oxygen-evolving activity in Synechocystis 6803. *Plant Molecular Biology*,
    Vol. 34, pp. 455-63.
Renger, G., & Renger, T. (2008) Photosystem II: the machinery of photosynthetic water
    splitting. *Photosynthesis Research*, Vol. 98, pp. 53–812.
Rochaix, J.-D., Kuchka, M., Mayfield, S., Schirmer-Rahire, M., Girard-Bascou, J., & Bennoun,
    P. (1989) Nuclear and choroplast mutations affect the synthesis or stability of the
    chloroplast psbC gene product in *Chlamydomonas reinhardtii*. *EMBO Journal*, Vol. 8,
    pp. 1013-1021.
Rogner, M., Chisholm, D. & Diner, B. (1991) Site-directed mutagenesis of the *psbC* gene of
    photosystem II: isolation and functional characterization of CP43-less photosystem
    II core complexes. *Biochemistry*, Vol. 30, pp. 5387-5395.
Roose, J., Kashino, Y., & Pakrasi, H. (2007) The PsbQ protein defines cyanobacterial
    Photosystem II complexes with highest activity and stability. *Proceedings of the
    National Academy of Science* (USA), Vol. 104, pp. 2548-2553.
Rosenberg, C., Christian, J., Bricker, T., & Putnam-Evans, C. (1999) Site-directed mutagenesis
    of glutamate residues in the large extrinsic loop of the Photosystem II protein CP 43
    affects oxygen-evolving activity and PSII assembly. *Biochemistry*, Vol. 38, pp. 15994-
    16000.
Sandusky, P., & Yocum, C. (1983) The mechanism of amine inhibition of the photosynthetic
    oxygen-evolving complex: amines displace functional chloride from a ligand site on
    manganese. *FEBS Letters*, Vol 162, pp. 339–343.
Sayre, R., & Wrobelboerner, E. (1994) Molecular topology of the Photosystem II chlorophyll
    alpha binding protein, CP43:topology of a thylakoid membrane protein.
    Photosynthesis Research, Vol. 40, pp. 11-19.
Service, R., Yano, J., McConnell, I., Hwang, H., Niks, D., Hille, R., Wydrzynski, T., Burnap,
    R., Hillier, W., & Debus, R. (2011) Participation of glutamate-354 of the CP43
    polypeptide in the ligation of manganese and the binding of substrate water in
    photosystem II. *Biochemistry*, Vol. 50, pp. 63-81.
Shen, G., Eaton-Rye, J., & Vermaas, W. (1993) Mutation of histidine residues in CP47 leads to
    a destabilization of the photosystem II complex and to impairment of light energy
    transfer. *Biochemistry*, Vol. 32, pp. 5109-5115.
Shen, J.-R., Burnap, R., & Inoue, Y. (1995) An independent role of cytochrome c-550 in
    cyanobacterial Photosystem II as revealed by double-deletion mutagenesis of the
    psbO and psbV genes in Synechocystis sp. PCC 6803. *Biochemistry*, Vol. 34, pp.
    12661–12668.
Shen, J.-R., Ikeuchi, M., & Inoue, Y. (1997) Analysis of the psbU gene encoding the 12-kDa
    extrinsic protein of Photosystem II and studies on its role by deletion mutagenesis
    in Synechocystis sp. PCC 6803. *Journal of Biological Chemistry*, Vol. 272 , pp. 17821–
    17826.
Shen, J.-R. Qian, M., Inoue, Y., & Burnap, R. (1998) Functional characterization of
    Synechocystis sp. PCC 6803 ΔpsbU and ΔpsbV mutants reveals important roles of

cytochrome c-550 in cyanobacterial oxygen evolution. *Biochemistry*, Vol. 37, pp. 1551-1558.

Shimada, Y., Suzuki, H., Tsuchiya, T., Tomo, T., Noguchi, T., & Mimuro, M. (2009) Effect of a single-amino acid substitution of the 43 kDa chlorophyll protein on the oxygen-evolving reaction of the cyanobacterium Synechocystis sp. PCC 6803: analysis of the Glu354Gln mutation. *Biochemistry*, Vol. 48, pp. 6095-103.

Shimada, Y., Suzuki, H., Tsuchiya, T., Mimuro, M., & Noguchi, T. (2011) Structural coupling of an arginine side chain with the oxygen-evolving Mn4Ca cluster in photosystem II as revealed by isotope-edited Fourier transform infrared spectroscopy. *Journal of the American Chemical Society*, Vol. 133, pp. 3808-11.

Strickler, M., Hwang, H., Burnap, R., Junko, Y., Walker, L., Service, R., Britt, D., Hillier, W., & Debus, R. (2008) Glutamate- 354 of the CP43 polypeptide interacts with the oxygen-evolving Mn4Ca cluster of photosystem II: a preliminary characterization of the Glu354Gln mutant. *Philosophical Transactions of the Royal Society B*, Vol. 363, pp. 1179- 1188.

Tang, X.-S., & Diner, B. (1994) Biochemical and spectroscopic characterization of a new oxygen-evolving photosystem II core complex from the cyanobacterium Synechocystis PCC 6803. *Biochemistry*, Vol. 33, pp 4594–4603.

Umena, Y., Kawakami, K., Shen, J-R., & Kamiya, N. (2011) Crystal structure of oxygen-evolving photosystem II at a resolution of 1.9 Å. *Nature*, Vol. 473, pp. 55-60.

Vassiliev, S., Comte, P., Mahboob, A., & Bruce, D. (2010) Tracking the flow of water through photosystem II using molecular dynamics and streamline tracing. *Biochemistry*, Vol. 49, pp. 1873–1881.

Vermaas, W., Rutherford, A., & Hansson, O. (1988) Site directed mutagenesis in photosystem II of the cyanobacterium Synechocystis sp. PCC 6803: Donor D is a tyrosine residue in the D2 protein. *Proceedings of the National Academy of Science (USA)*, Vol. 85, pp. 8477-8481.

von Heijne, G., & Gavel, Y. (1988) Topologenic signals in integral membrane proteins. *European Journal of Biochemistry*, Vol. 174, pp. 671-678.

Williams, J. (1988) Construction of specific mutations in photosystem II reaction center by genetic engineering methods in Synechocystis 6803. *Methods in Enzymology*, Vol. 167, pp. 766-778.

Wincencjusz, H., van Gorkom, H., & Yocum, C. (1997) The photosynthetic oxygen-evolving complex requires chloride for its redox state S2→S3 and S3→S0 transitions but not for S0→S1 or S1→S2 transitions. *Biochemistry*, Vol. 36, pp. 3663- 3670.

Yachandra, V., DeRose, V., Latimer, M., Mukerji, I., Sauer, K., & Klein, M. (1993) A structural model for the photosynthetic oxygen-evolving manganese complex. *Japanese Journal of Applied Physics*, Vol. 32s, 523-526.

Young, A., McChargue, M., Frankel, L., Bricker, T. & Putnam-Evans, C. (2002) Alterations of the oxygen-evolving apparatus induced by a 305Arg → 305Ser mutation in the CP42 protein of photosystem II from Synechocystis sp. PCC 6803 under chloride-limiting conditions. *Biochemistry*, Vol. 41, pp. 15747-15753.

Zouni, A., Witt, H., Kern, J., Fromme, P., Krauss, N., Saenger, W. & Orth., P. (2001) Crystal
    structure of photosystem II from *Synechococcus elongatus* at 3.8 Å resolution. *Nature*,
    Vol. 409, pp. 739-743.

4

Semiconductors in Organic Photosynthesis

Cristian Gambarotti, Lucio Melone
and Carlo Punta
Department of Chemistry, Materials and Chemical Engineering "Giulio Natta",
Politecnico di Milano
Italy

1. Introduction

It is commonly accepted that "Organic Synthesis" was born in 1828, when the german chemist Friedrich Wöhler succeeded to make urea from simple materials (Friedrich, 1828). After only few years, Becquerel reported the first example of photo-induced electrochemical reaction (Becquerel 1839). In his work, he found that a voltage and an electric current were obtained by illuminating a silver chloride electrode immersed in an electrolyte solution and connected to a counter electrode. But the world had to wait until 1955, when Brattain and Garret, working on germanium semiconductor electrodes, well understood the origin of that strange photovoltaic phenomenon (Brattain & Garret, 1955). The modern "Photo-electrochemistry" era was born.

However, the year was 1900 when it appeared the first article onto the use of light to promote an organic reaction (Albini & Fagnoni, 2008; Ciamician & Silber 1900). Ciamician and his colleague Silber carried out the first systematic studies on the behaviour of organic substances in the presence of sunlight, thus, nowadays they are regarded as the fathers of modern Organic Photochemistry.

Nowadays, if we open SciFinder Scholar client and we write the word "Synthesis" in the "Research topic" field, almost 10 millions of references will be shown. This means that in the last two centuries more than 50000 papers per year onto the argument have been published. Moreover, we are considering only an average value, whereas it is well known that the number of publications has known an exponential-type growth in the last decades. However, only 70000 references are reported for "Photocatalysis", indicating how relatively young is this research field, although the first paper was published more than a century ago.

The question is: why in the last two-three decades the use of light in chemistry has aroused a such worldwide attention? The answer could be summarized mainly in one word: environment.

A great impetus to the development of photocatalysis is derived from the growing demand to reduce the environmental pollution (air and ground). This has led to the development of several photo-induced protocols for the oxidative degradation of organic pollutants, in which semiconductors play a key role in the reaction mechanism, focusing the attention onto the photodegradation of water and air pollutants.

2. Semiconductors

It was 1782 when Alessandro Volta used for the first time the term "semiconducting". One century later, in 1883, Michael Faraday documented the first observation of a semiconductor effect. Faraday noted that the electrical resistance of silver sulphide decreased with increasing temperature, showing a behaviour opposite to that of metals (Łukasiak & Jakubowski, 2010).

Nowadays, everywhere we look, we can certainly see articles that contain semiconductors. The modern microelectronics applications are based on semiconductor technologies. Moreover, building materials, health care products, materials for special applications and much more, often employ semiconductors due to their chemical and physical properties. It is sufficient to write "semiconductor applications" into any internet search engine to get many tens of millions of responses.

2.1 Electrical behavior of semiconductors

By definition, a semiconductor is characterized by the absence of a continuum between the states (as for metals) but shows a band structure. The filled levels, called "valence band" (VB), are an energetically closely spaced array of orbitals composed by the valence electrons of the material. A similar, higher energetic, spaced array is formed by the unoccupied orbitals and it is called "conduction band" (CB). The gap existing between the top edge of VB (E_c) and the lower edge of CB (E_c) is called "band-gap" (Chattopadhyay & Rakshit, 2010).

2.1.1 Intrinsic semiconductors

Pristine semiconductors are generally called "intrinsic semiconductors". In an intrinsic semiconductor (IS) the Fermi-level (E_F), which can be the highest occupied energy level at T = 0 K, lies near the middle of the band-gap and the corresponding Fermi-Dirac distribution $f(E)$ is plotted in Fig. 1. At T = 0 K the probability of an electron to occupy a state in the CB is zero and the VB is totally full (see the thick black line). At room temperature (blue sigmoid curve) some electrons can jump into the CB and fill the states closed to the bottom of the CB. In this case the tail of $f(E)$ is extended into the CB, thus, there is a probability to have electrons there. In this condition an equal number of holes exists close to the top of VB. E_F is now the energy level at which the probability of occupancy is half and it lies near the middle of the band-gap.

2.1.2 Extrinsic semiconductors

The *ad hoc* introduction (doping) of impurities into the semiconductor lattice allow to modulate its electrical properties: these doped semiconductors are generally called "extrinsic semiconductors". Extrinsic semiconductors with a larger electron concentration than hole concentration are known as "*n*-type" semiconductors, whereas those with a larger hole concentration than electron concentration are known as "*p*-type" semiconductors. The energy band diagrams for *n*-type and *p*-type semiconductors are given in Fig. 1. In the figure, E_A and E_D represent respectively the energy level of the acceptor and the donor specie. Assuming a complete ionization of the donor atom at a given temperature, we see that the free electrons coming from the donor atoms fill the state close to the bottom of the CB. Thus, it is more difficult for the electrons in the VB to cross the band-gap only by thermal agitation. Therefore, the number of holes in the VB decreases. This means that, if we consider E_F as the energy for which the occupancy probability is 0.5, E_F must move closer to the CB for a n-type semiconductor (Fig. 1 scheme on the right). Similarly, E_F must move closer to the VB for a p-type semiconductor (Fig. 1 scheme on the left). In the extreme cases, in which the doping level is very high, the Fermi levels move into the CB for n-type semiconductor and into the VB for p-type semiconductor.

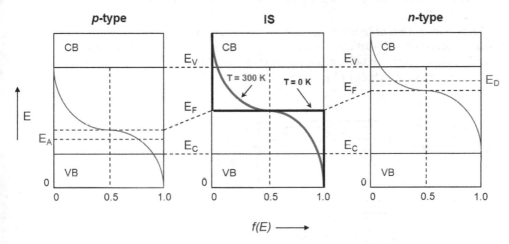

Fig. 1. Energy diagram and Fermi distribution function $f(E)$ for semiconductors

2.2 Electrical properties of illuminated n-type semiconductors

The application of several n-type semiconductors on photocatalysis has experienced an exponential growth in the last 20 years (Fox & Dulay, 1993; Hashimoto et al., 2005; Hoffmann et al., 1995; Maldotti et al., 2002a). Let we consider now the influence of photo-irradiation onto n-type semiconductor properties (Fox, 2001).

When a semiconductor surface is brought into contact with an electrolyte containing a redox couple, interfacial electron transfer can transpire, to equilibrate the solution phase potential and the energy levels of the bulk semiconductor. As electron exchange takes place across the interface, equilibration with the solution occurs and the bulk E_F moves to the solution-phase equilibrium potential, whereas the band-edge positions at the surface remain fixed at their original values (Fig. 2). Thus, in moving inward from the semiconductor-electrolyte

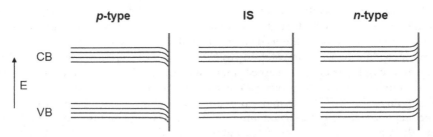

Fig. 2. Band bending into n-type and p-type semiconductors. The vertical lines represent semiconductor / solution interface.

interface, the bands bend. Charge injection in this region of bent bands forces electrons to move in the opposite direction from holes: in a *n*-type semiconductor, electrons move away from the surface and toward the bulk of the semiconductor while the holes move toward the surface. Thus, band bending assists in improving the efficiency of photoinduced charge separation. As a result of band bending, the surface of an irradiated *n*-type semiconductor becomes electron-deficient and acts as a photoanode toward an oxidizable adsorbate.

2.3 Redox reactions onto irradiated *n*-type semiconductors

When a semiconductor surface is irradiated by light of an energy higher than the band-gap, a band to band transition takes place. An electron (e⁻) moves from the filled VB to the empty CB, leaving an electron hole (h⁺) in the VB (Fig. 3).

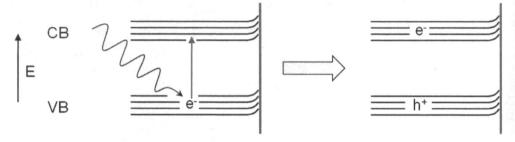

Fig. 3. Photo-irradiation of a semiconductor particle

Depending on their own band-gap, for each semiconductor exists a minimum light frequency to promote the electron jump. Thus, the absorption of a photon induces a charge separation, generating a highly energetic electron and creating a lower-energy vacancy in the valence band. This intermediate species is called "electron-hole pair". In this excited state, the recombination of the electron-hole pair, due to back electron transfer, is of course strongly thermodynamically favored. But, because of the absence of states in the semiconductor band-gap, quantum restrictions slow recombination by back electron transfer and the lifetime of the photogenerated electron-hole pair is sufficiently long to allow charge trapping. Thus, the life time of VB hole, in principle, can be enough long to act as an anode, whereas, similarly, the CB can act as cathode. As a consequence, the excited semiconductor is able to promote redox reactions on substrates that may be present on its surface or in solution. In *n*-type semiconductors, the surface becomes electron-deficient and acts as an oxidative site toward oxidizable organic substrates. If an adsorbed substrate has an appropriate redox potential, it can trap one of the charge carriers faster than the electron-hole recombination and photoinduced chemistry will be observed. As represented in Fig. 4, the e⁻ promoted in the CB is transferred to the adsorbed acceptor (A) and the VB hole is filled by an e⁻ coming from the adsorbed donor species (D). The rate of the electron-transfer processes compete with spontaneous back electron transfer of the electron-hole pair.

If the trapping of both VB hole and CB e⁻ are faster than the spontaneous back electron transfer, singly oxidized and reduced intermediates are formed on the photocatalyst surface. Because these adsorbed species are free to move along the surface, novel chemistry is likely to ensue. This is a unique characteristic of photochemical catalysts, that are able to promote oxidations and reductions on the same surface.

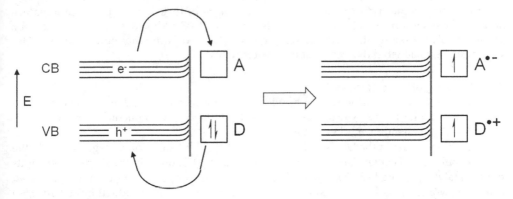

Fig. 4. Interfacial electron transfer on semiconductor surface

For practical applications, because the simultaneous presence of cathodic and anodic site onto the same particle, the semiconductor photocatalyst can be considered as a "short circuit" electrochemical cell, when prepared by deposition island of an inert metal onto its surface (Fig. 5) (Bard, 1979).

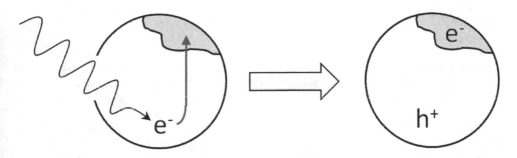

Fig. 5. Irradiation of metalized semiconductor particle

These metalized particles bear regions for both photoactivated oxidation of organic substrates adsorbed at the semiconductor-electrolyte interface and collection of the photogenerated electrons on the metal island. Thus, as in a electrochemical cell, these metalized powders include sites for photoinduced oxidation and reduction, analogous to the working anode and cathode. On platinized TiO_2 powders, however, no external current flow accompanies these transformations.

Anyway, even pristine wide band-gap semiconductors often show photoactivity if the two redox sites can be separated; metalized semiconductors are essential to promote some redox processes, such as selective hydrogenations of alkenes and alkynes (Fox, 2001).

3. Applications of n-type semiconductors

In the last decades photocatalytic processes induced by semiconductors have attracted the great interest, due to their low environmental impact. Nowadays, many n-type semiconductors are studied and applied in several application fields, such as energy production, smart-materials technology, environment depollution, chemical synthesis, whereas p-type semiconductors are rarely used because of their limited presence in nature and their usually too small band-gap (Palmisano et al., 2007a; Mills & Lee, 2002; Mills & Le Hunte, 1997). Certainly, the most studied and widely used is the titanium dioxide (TiO_2), which can be considered the "king" of the photocatalysts.

3.1 TiO_2: King of the photocatalysts

Between the various semiconductors applied in photocatalysis, certainly TiO_2 is the most used. The reasons for its widespread are due to its high environmental tolerance, large commercially availability and low price.

Its history began just in the late 60th dacades when Fujishima and Honda began to study the photoelectrolysis of water. The first paper, published in 1972, brought the world aware of its photo-catalytic potential (Fujishima & Honda, 1972). This important observation promoted extensive works focused on the solar energy conversion for the production of hydrogen from water. Moreover, it soon became apparent that novel redox reactions of organic and inorganic substrates could also be induced by band-gap irradiation of a variety of semiconductor particles, of sizes ranging from clusters and colloids to powders and large single crystals.

TiO_2 is present in the three common allotropic forms: anatase, rutile and brookite: anatase and rutile belong to thetragonal crystal system, while brookite belongs to orthorhombic system (Linsebigler et al., 1995). Anatase and rutile are the two principal catalytic forms. Commercially available anatase is typically less than 50nm in size. These particles have a band-gap of 3.2 eV, corresponding to a UV wavelength of 385 nm. The adsorptive affinity of anatase for organic compounds is higher than that of rutile, and anatase exhibits lower rates of recombination in comparison to rutile due to its 10-fold greater rate of hole trapping. In contrast, though some exceptions exist, the thermodynamically stable rutile phase generally exists as particles larger than 200 nm. Rutile has a smaller band-gap of 3.0 eV with excitation wavelengths that extend into the visible at 410 nm. Nevertheless, anatase is generally regarded as the more photochemically active phase of titania, presumably due to the combined effect of lower rates of recombination and higher surface adsorptive capacity.

In the last decade, Gray and co-workers reported the enhanced photoactivity in the mixed-phase (anatase and rutile) Degussa P25 TiO_2 (Deanna et al., 2003). This fact is explained by the presence of small rutile crystallites, which creates a structure where rapid electron transfer from rutile to lower energy anatase lattice trapping sites under visible illumination, leads to a more stable charge separation. Transfer of the photogenerated electron to anatase lattice trapping sites allows holes that would have been lost to recombination to reach the surface. Subsequent electron-transfer moves the electron from anatase trapping sites to surface trapping sites, further separating the electron/hole pair. By competing with recombination, the stabilization of charge separation activates the catalyst and the rutile-originating hole can participate in oxidative chemistry. Three main factors are employed in this increase in the photoactivity: (1) the smaller band-gap of rutile extends the useful range of photoactivity into the visible region; (2) the stabilization of charge separation by electron transfer from rutile to anatase slows recombination; (3) the small size of the rutile crystallites facilitates this transfer, making catalytic hot spots at the rutile/anatase interface. This process depends critically on the interface between the TiO_2 phases and particle size. The atypically small size of the rutile particles in this formulation, and the intimate contact with anatase that the comparable size allows, are crucial to enhancing the catalyst activity.

3.2 Applications of TiO_2

Nowadays, due to its most photoactivity, highest stability, low cost and non-toxicity, TiO_2 is most probably the only photocatalyst suitable for heavy industrial applications (Hashimoto et al., 2005; Ravelli et al., 2011). In the last 30 years various research groups have developed several photocatalityc processes in which it plays a key role as catalyst. Generally those applications regard the photo-induced redox reactions of adsorbed substances, the photo-induced hydrophilic and/or hydrophobic behaviour of TiO_2 itself and the use as white pigment from ancient times (Anpo & Kamat, 2010; Ohama & Gemert, 2011).

3.2.1 Water splitting

Water reduction to H_2 and oxidation to O_2 requires that the bottom of the conduction band lies at a more negative potential than $E°_{red}(H^+/H_2)$, 0 V, and the top of the valence band at a more positive value than $E°_{ox}$ (H_2O/O_2), 1.23 V. The two-electron process from one mol of H_2O to give one mol of H_2 and half mol of O_2 is termodinamically unfavored, $DG° = 237$ kJ/mol = 2.46 eV. Thus, the minimum energy required to drive the reaction corresponds to that of two photons of 1.23 eV, corresponding to l ca. 1000 nm, in the near infrared region. However, to overpass the activation barrier, higher photon energy is required in order to promote water splitting at a reasonable rate. Water splitting for hydrogen production aroused a great interest during the second oil crisis, in the 1970s. In this period Fujishima and Honda reported the possibility to easily obtain H_2 and O_2 using sunlight as energy source (Fig. 6) (Hashimoto et al., 2005).

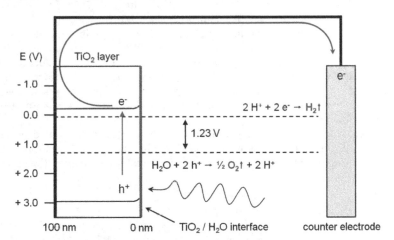

Fig. 6. Schematic representation of water splitting process at pH = 1

Then, photocatalysis drew the attention of many researchers as one of the promising methods for hydrogen production. However, despite the high reaction efficiency, TiO_2 can adsorb only the UV light present in a solar spectrum. This means that only about 3% of solar light was effective on the reaction and, in few years, from the productive technological point view, TiO_2 photocatalysis lost its initial actractive. Thus, in the last 20 years, many efforts have been made in order to improve the photocatalytic performance of TiO_2 by different approaches, such as doping/mixing with several other metals and semiconductors, like Pt, Cr, Ce, Ag, CdS, CdSe, use of zeolites and nanostructured TiO_2 nano-arrays, and more (Balzani et al., 2008; Esswein & Nocera, 2007; Li et al., 2010; Maeda & Domen, 2007; Zhang et al., 2010).

3.2.2 Photodegradation of pollutants

During the same period, Frank and Bard applied TiO_2 photocatalysis in the destruction of pollutants (Frank & Bard, 1997). Their preliminary studies on the oxidation of cyanide to cyanate, by using four different forms of TiO_2 under xenon lamp, opened the way for the photodegradation of environmental pollutants.

In the last three decades, TiO_2 powders have known wide applications into the recovery of water of industrial, agricultural or civil origin, as well as the decontamination of atmosphere and soil, through the mineralization of the pollutants, or at least their transformation into non-toxic compounds (Fox & Dulay, 1993; Gaya & Abdullaha, 2008; Hoffmann et al., 1995; Ravelli et al., 2011). Of course, the great advantage of these photocatalytic systems depends on the fact that they do not need the use of stoichiometric quantities of chemicals, potentially themselves polluting, but they act under light (often sunlight represents the best choice) in the presence of oxygen (Malato et al., 2007).

Moreover, the high band-gap value of TiO_2 allow to oxidize various organic substrates, such as hydrocarbons and their derivatives to lower molecular weight ozidized species and CO_2, volatile organic compounds (VOCs) and nitrogen oxides (NO_x) present in urban atmosphere (Carp et al., 2004).

3.2.3 Anti-bacterial materials

Photocatalytic decomposition reactions can be applied to the destruction of bacteria. Escherichia coli (*E. Coli*) cells are completely destroyed in one week under 1 mW/cm^2 UV irradiation on TiO_2 (Evans & Sheel, 2007; Fu et al., 2005; Liu et al., 2008; Sunada et al., 1998). Usually, an analogous anti-bacteria effect can be achieved in nearly 1 h under outdoor UV light intensity. However, the typical indoor UV light intensity is in the order of mW/cm^2, thus the photocatalytic disinfection under indoor conditions require too long time to be considered useful from the practical point of view. The antibacterial function of TiO_2 is strongly enhanced, even with weak UV intensity, using a fluorescent lamp assisted by the presence of Ag or Cu (Sunada et al., 2003). TiO_2 photoactivity reaction assists the intrusion of the copper ions into the cell, which is probably the cause of the destruction of the *E. Coli* colonies even under very weak UV light.

Actually several nanostructured TiO_2 derivatives are studied and applied as germicidal species, especially on the treatment of water contaminated with pathogenic micro-organisms presenting a potential hazard to animals and human beings (Mccullagh et al., 2007; Skorb et al., 2008).

3.2.4 Wettability of TiO_2 and self-cleaning materials

During '90s, Fujishima and Heller independently reported the potential utilization of TiO_2 in the development of self-cleaning ceramic materials (Heller, 1995; Watanabe et al., 1992). In these works, thin films of organic contaminants were photocatalytically oxidized on TiO_2-coated surfaces. The typical oxidation-rate, evaluated in 1-5 picometers per day was sufficient to maintain clean the surface then in the case of low flux of contaminants. Typically several hundred mW/cm^2 of UV light are available in the day-time and from the outdoor shade, which corresponds to about 1000 photons/cm^2 per second. This is, of course, a very small quantity of energy, but very high if compared to the number of molecules adsorbed on the surface. This suggested the potential applications in the self-cleaning material technology. One of the first commercially applications was the self-cleaning glass-cover for tunnel light. The largely used high pressure sodium lamps show an UV emission of about 3 mW/cm^2, enough to keep efficient the photo-oxidative self-cleaning process.

Nowadays, various self-cleaning materials are used in commercial applications and intense researches are aimed to improve these materials (Parkin & Palgrave, 2005).

During the studies on the self-cleaning surfaces, it was found a marked change in the water wettability of the TiO_2 surface before and after UV light irradiation (Wuang et al., 1997). During the UV light exposition the contact angle of TiO_2 coated surface showed a strong decrease from typically initial several tens of degrees (depending on the surface roughness) to nearly 0° (Sakai et al., 2003; White et al., 2003; Zubkov et al., 2005). This discover widened the commercial applications of TiO_2 coated materials. The limitations of the self-cleaning efficiency, due to the low quantity of UV-light present in sunlight spectrum and outdoor shade, were resolved. The stains adsorbed onto TiO_2 surface can be easily washed only by using water, because water is adsorbed between stain and the highly hydrophilic TiO_2 surface. Moreover, even if the quantity of light is too low to decompose the stains, the surface is maintained clean by supplying water. Thus, outdoor coated surfaces can be kept clean by rainwater. Such called "Photocatalytic Building Materials" have found use in outdoor application, for example, an exterior glass of 20000 m² was installed in the terminal of Chubu International Airport in 2005.

The same UV-induced high hydrophilic behaviour of TiO_2 coated surface has been applied in the development of anti-fogging treatment (Gan et al., 2007; Tricoli et al., 2009). Drops are formed when there is low affinity between water and the surface; on a highly hydrophilic surface, no water drops are formed but a uniform liquid thin film.

3.3 Applications of other semiconductors

Despite the widespread use TiO_2 in many applications, the relatively high cost of the photoactive anatase has proved to be uneconomical for large-scale water treatment operations. Thus, several other ways for the photocatalytic degradation of pollutants have been explored (Chatterjee & Dasgupta, 2005; Mills & Le Hunte, 1997).

Because of their narrower band-gaps, metal calcogenides, such as CdS, ZnO and CdSe, show good sensitivities toward incident light in the visible spectrum. However, these narrower band-gap, make the semiconductor suitable of the photo-corrosion process, which can be partly suppressed by the addition of sulfide and sulfite to the contacting solution.

Because of its similar band-gap (3.2 eV), ZnO photocatalytic capacity should be comparable to that of TiO_2. Due to its high light absorption in the region between 300 and 400 nm, ZnO is found to be as reactive as TiO_2 toward the degradation of phenol (Dindar & Içli, 2001). However, as previously reported, photo-corrosion phenomenon can occurs under UV light, and this is one of the main reasons for the decrease of ZnO photocatalytic activity in aqueous solutions. Recently, ZnO nanoparticles have been reported as better photocatalysts in degrading common organic contaminants as compared to bulk ZnO and commercial TiO_2 Degussa P25 (Hariharan, 2006).

Similarly to ZnO, nanostructured ZnS particles show good catalytic activity and are used in the removal of organic pollutants and toxic water pollutants (Hu, 2005).

Enhanced photocatalytic activities may result from doping semiconductors by transition metals. These techniques generally influence the optical and electronic properties of the semiconductors, and can induce a shift of the optical absorption toward the visible region (Pouretedal et al., 2009; Ullah & Dutta, 2008).

Water splitting is another important field of application of semiconductor-sensitized systems. In this area, as already reported for TiO_2, many efforts are aimed to the development of efficient photoelectrochemical devices, in which illuminated semiconductors are the process promoter (Mills & Le Hunte, 1997).

Often, platinum group metal, deposited on the surface of the semiconductor, are used to facilitate the reduction of water by the photogenerated electrons and platinum group metal oxides are often used to mediate the oxidation of water by the photogenerated holes. Of course, in the presence of a large band-gap semiconductors (TiO_2 as example), with band-gap potential much higher than the oxygen one, there is a sufficiently large over-potential for the reaction to proceed readily without an oxygen redox catalyst.

4. Photo-induced organic synthesis

Photocatalysis in organic synthesis concerns the use of light to induce chemical transformations onto organic substrates which are transparent in the wavelength range employed. Radiation is absorbed by a photocatalyst whose electronically excited states induce electron- or energy-transfer reactions able to trigger the chemical reactions of interest. Significant examples of photocatalytic processes employed for synthetic purposes are oxidation and reduction processes, isomerization reactions, C-H bond activations, and C-C and C-N bond-forming reactions. The use of solar light as a reagent in oxidative catalysis is particularly relevant to realize innovative and economically advantageous processes for the conversion of hydrocarbons into oxygenates compounds and, at the same time, to move toward a "sustainable chemistry" that has a minimal environmental impact. The main reasons are because the sunlight represents a totally renewable source of energy; the photochemical excitation requires milder conditions than thermal activation and allows one to conceive shorter reaction sequences and to minimize undesirable side reactions. Generally, an important role in the photocatalyzed processes is played by O_2. It is important to underline that the search for new catalysts capable of inducing the oxofunctionalization of hydrocarbons with this environmentally friendly and cheap reagent represents a major target from the synthetic and industrial points of view. On the basis of pure thermodynamic considerations, most organic compounds are not stable with respect to oxidation by O_2. There are, however, kinetic limitations in this process mainly imposed by the triplet ground state of the O_2 molecule, which is not consistent with the singlet states of many organic substrates. Activation of both O_2 and the organic substrate may be achieved by photochemical excitation with light of the visible or of the near-ultraviolet regions (> 300 nm). The use of heterogeneous and organized systems is a suitable way to control efficiency and selectivity of catalytic processes through the control of the microscopic environment surrounding the catalytic centre. In particular, the nature of the reaction environment may affect numerous physical and chemical functionalities of the photocatalytic system, such as the absorption of light, the generation of elementary redox intermediates, the rate of competitive chemical steps, and the adsorption-desorption equilibria of substrates, intermediates, and final products. Moreover, another fundamental role of a solid support is to make the photocatalyst more easily handled and recycled.

4.1 Role of O_2

In dispersed semiconductor photocatalytic processes oxygen acts as electron acceptor. On illuminated TiO_2 surfaces, in the presence of air (O_2), hydrogen peroxide and reactive oxygen-centred radical species are formed via oxygen reduction (Karraway, 1994 & Kormann, 1988). The first step of O_2 action consists in the capture of the electron promoted in the CB by light irradiation of semiconductor surface (Eq. 1).

$$TiO_2 \xrightarrow{\hphantom{aa} h\nu \hphantom{aa}} (e^-/h^+)$$

$$O_2 + e^- \longrightarrow O_2^{\bullet-}$$

$$2\,O_2^{\bullet-} + 2H_2O \longrightarrow H_2O_2 + 2\,OH^- + O_2$$

$$H_2O_2 + e^- \longrightarrow {}^\bullet OH + OH^-$$

$$(1)$$

The role of O_2 is not just that of scavenging the photogenerated electrons: it produces the so-called active oxygen species while the simultaneous oxidation of an organic substrate yields radical intermediates (Eq. 2).

$$RH_2 + h^+ \longrightarrow {}^\bullet RH + H^+$$

$$(OH^-)_s + h^+ \longrightarrow ({}^\bullet OH)_s$$

$$RH_2 + ({}^\bullet OH)_s \longrightarrow {}^\bullet RH + H_2O$$

$$(2)$$

$${}^\bullet RH + ({}^\bullet OH)_s \longrightarrow (HROH)_{ads}$$

$$(HROH)_{ads} + 2({}^\bullet OH)_s \longrightarrow RO^\bullet + 2H_2O$$

(s = surface, ads = adsorbed)

Therefore, through a Kisch type B photocatalysis mechanism, coupling reactions of oxidation and reduction intermediates lead to products of partial or total oxidation, depending on experimental conditions (Eq. 3) (Macyk & Kisch, 2001).

$${}^\bullet RH + O_2 \longrightarrow HROO^\bullet$$

$$HROO^\bullet + RH_2 \longrightarrow {}^\bullet RH + HROOH$$

$$HROO^\bullet + e^- \longrightarrow RO^\bullet + OH^-$$

$$(3)$$

$${}^\bullet RH + O_2^{\bullet-} \longrightarrow RO^\bullet + OH^-$$

$${}^\bullet RH + HOO^\bullet \longrightarrow RO^\bullet + H_2O$$

$$HROO^\bullet + HOO^\bullet \longrightarrow HROOH + O_2$$

As shown above, oxygen represents the source of highly reactive radical intermediates, which are the "engine" of the photocatalyzed reaction mechanisms.

4.2 Redox potentials of semiconductors

The relevant redox potential for some photocatalysts and H_2O are compared in Fig. 7 (Kisch, 2001; Ravelli et al., 2011). An excited photocatalyst (C*) oxidizes a substrate (S) when E(C*) is more positive than the E(S), whereas there is reduction when E(C*) is lower than E(S). For example as shown in Fig. 6, excited TiO_2, as previously reported, is capable of causing water splitting, since the valence/conduction band-gap is sufficiently large for encompassing the H_2O redox potentials (at least at low pH values) (Montalti, 2006). Single crystal catalysts in aqueous systems usually shift cathodically with higher pH value by approximately 0.06 V per pH unit as reported for TiO_2 (Ward, 1983), CdS (White & Bard, 1985), and ZnS (Fan et al., 1983). In addition to this pH dependence, surface impurities, adsorbed compounds and the change to organic solvents may induce strong shift in the redox potential. In the case of CdS, the removal of traces of elemental sulphur and cadmium from the surface induces a cathodic shift of even almost 1 V! (Meissner & Memming, 1988).

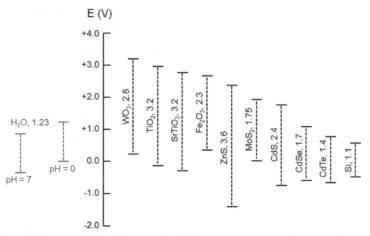

Fig. 7. Band-gap and redox potentials (*vs*. NHE) at pH = 1 vs H_2O at pH = 0 and 7

4.3 Influence of catalyst shape on photoactivity

The rate of products formation is influenced by the specific surface area of the catalyst. Typically, two opposite effects play a key role: the rate of electron-hole recombination and the concentration of adsorbed substrates (Heller et al., 1987). The recombination rate increases linearly with surface area, and accordingly the reaction rate should decrease. On the contrary, there is a linear increase on the redox process rate due to increasing concentration of adsorbed substrates, which should also increase the product formation rate. Moreover, it is expected that, depending on the nature of semiconductor and substrates, the reaction rate may increase, decrease or remain constant with increasing surface area. For example, Reber founds that in the photoreduction of H_2O, only CdS with very large (>> 100 m^2/g) or very low (< 6.7 m^2/g) specific surface areas produce hydrogen at a significant rate (Reber & Rusek, 1986).

The chemoselectivity may also depend on the surface area, usually related to the particle size. Maira and co-workers have studied the effect of particle size in the gas-phase photooxidation of toluene over TiO_2. Particles from 6 to 20 nm have show to influence the activity and selectivity of the photocatalyst; in particular, selectivity toward benzaldehyde production increases as the particle size increases (Maira et al., 2001). These results can be explained considering that smaller particles are characterized by a larger surface area and also by a larger proportion of edges and corners which increases as size decreases. Edges and corners are expected to exhibit different catalytic and adsorption properties compared with the more planar surface sites. The involvement of Ti^{4+}-O^- radicals could be the reason of the increase of selectivity to benzaldehyde as the particle size of the TiO_2 samples increases from 6 to 20 nm.

Matra has found that size can affect the acid-base properties of surface hydroxyl groups (Martra, 2000). In the Merk TiO_2 these groups are electron acceptor centres, while in Degussa P25 they present a nucleophilic character. These differences can explain the different selectivity toward benzaldehyde formation in the gas-phase photocatalytic oxidation of toluene, in particular, the lower selectivity observed with P25 is ascribed to a stronger adsorption of the aldehyde.

5. Organic synthesis in the presence of illuminated semiconductors

Although semiconductor materials like as TiO_2 have been extensively applied in the environmental field in order to mineralize organic pollutants, their application as tool for the organic synthesis has been also investigated. Recently, some reviews on this field have been published (Gambarotti, 2010; Pastori, 2009; Shiraishi, 2008; Anpo & Kamat, 2010). Herein we report some examples taken from the literature which can be used to take a view on the recent state of the art.

5.1 TiO$_2$ in organic photosynthesis

The selective oxidation of hydrocarbons under mild conditions is an important industrial and scientific challenge because relatively cheap feedstocks are converted into highly value-added products.

5.1.1 Synthesis of phenol derivatives

Phenol is an highly important chemical intermediate because it is used in a large number of different sectors including the production of phenolic resins, caprolactam, aniline, alkylphenols, diphenols, and salicylic acid (Weber et al., 2005). It is produced from benzene mainly through the cumene process (Hock process). Currently, in the USA and Western Europe, about 20 % of the total benzene production is consumed for the synthesis of phenol (Weissermel & Arpe, 1997). In the last years interests have aroused for the direct production of phenol from benzene, despite these proposed processes are still not effective and competitive if compared with the classic procedures (Bal et al., 2006; Guo et al., 2009; Tani et al., 2005; Tlili et al., 2009). Among them, photocatalytic processes, in particular based on the use of semiconductors, gained attention both in industriy and academy as an economically and environmentally favourable approach (Chen et al., 2009). Recently, Molinari and co-workers reported the direct benzene conversion to phenol in a hybrid photocatalytic membrane reactor (Molinari et al., 2009a). The authors used TiO_2 Degussa P25 as catalyst to promote the oxidation of benzene to phenol in water under UV irradiation (λ = 366 nm, power = 6.0 mW cm^{-2}). The particularity was the adoption of a membrane reactor having two compartments, one containing the aqueous phase, in which TiO_2 and benzene were dispersed, and the other one containing the organic phase, consisting on benzene, whose role was to remove phenol, produced during the photocatalysis, from the aqueous phase. The membrane was a hydrophobic polypropylene porous flat sheet. The authors claimed that the system allowed the production of the phenol and its separation, although the formation of intermediate oxidation by-products, like benzoquinone, hydroquinone and other oxidized molecules was observed.

Recently Zhang et al. (Zhang et al., 2011) reported the selective photocatalytic conversion of benzene to phenol by using titanium oxide nanoparticles incorporated in hydrophobically modified mesocellular siliceous foams (MCF) as shown in Fig. 8.

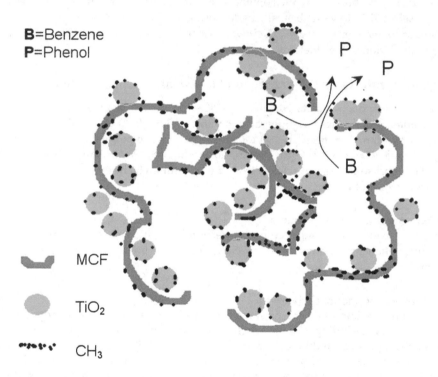

Fig. 8. Titanium oxide entrapped in the cagelike mesopores of hydrophobically modified MCF for the hydroxylation of benzene

Generally, the hydroxyl radicals generated during the photo-illumination of the TiO_2 crystals are highly reactive and unselective causing a large formation of secondary by-products. The main characteristic of this method is based on the hydrophobicity of the MCF, for that benzene is preferentially attracted into the hydrophobic mesopores while the more hydrophilic phenol is rapidly release from the cavity before it undergoes further oxidation. In this way an increase in the selectivity is obtained. In particular, the authors compared the catalytic activity of three materials ($TiO_2@MCF$, $TiO_2@MCF/CH_3$, $TiO_2@MCF/CH_3/UV$) with the corresponding activity of TiO$_2$ in the anatase form performing the oxidation of benzene in water under 300 W Xe-arc lamp irradiation for 24 h. The catalyst indicated with TiO$_2$@MCF consists of TiO$_2$ crystals incorporated in the MCF structure. Such crystals had a size of 9 nm, lower than the corresponding ones of the simply TiO$_2$ (16 nm). The surface area (S$_{BET}$) of TiO$_2$@MCF (29.5 wt % of TiO$_2$) was 637 m^{2}g^{-1}. In order to increase the hydrophobicity of the catalyst, TiO$_2$@MCF was treated with triethoxymethylsilane (TMS) getting the material TiO$_2$@MCF/CH$_3$ (TiO$_2$ wt %: 21.7, S$_{BET}$: 450 m^{2}g^{-1}). Finally, the methylsilyl groups grafted on the TiO$_2$ surface were selectively removed by post UV-irradiation treatment obtaining the catalyst TiO$_2$@MCF/CH$_3$/UV (TiO$_2$ wt %: 23.3, S$_{BET}$: 491 m^{2}g^{-1}). The authors found that TiO$_2$@MCF had better performances than TiO$_2$ with regard to both the phenol yield (Y$_P$, mmol phenol / g TiO$_2$) and the phenol selectivity (S$_P$, %). In particular Y$_P$ increased from about 10 mmol/g TiO$_2$ for TiO$_2$ to about 35 mmol/g TiO$_2$ for TiO$_2$@MCF. Similarly an increase of S$_P$ was observed (from 15.8 to 22.2 %). Upon silylating TiO$_2$@MCF, a reduction of the catalytic performance is experienced (Y$_P$ is lower than the correspond value of TiO$_2$, while S$_P$ remains practically similar to the value of TiO$_2$@MCF, essentially because the active TiO$_2$ surface sites were blocked by the methylsilyl groups in TiO$_2$@MCF/CH$_3$. The removal of such groups by UV irradiation allows to recover the catalytic properties. In fact TiO$_2$@MCF/CH$_3$/UV gave a Y$_P$ above 50 mmol/g TiO$_2$ while S$_P$ is about 35 %.

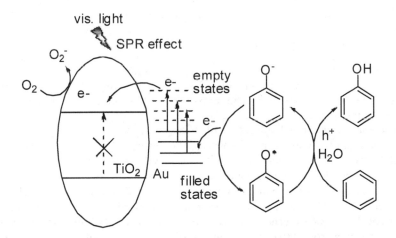

Fig. 9. Proposed mechanism for the photooxidation of benzene

The photocatalytic activity of TiO$_2$ can be modified by depositing noble metal nanoparticles on the particles surface. This is the case of the work of Zheng et al (Zheng et al., 2011). The authors developed a noble-metal plasmonic photocatalyst effective for the oxidation of benzene to give phenol under visible light irradiation. Three different noble metal were used: Pt, Au and Ag. Between them Au@TiO$_2$ microsphere having 2 wt % of Au showed the best catalytic performance (Au > Pt >> Ag) giving high yield (above 60 %) and selectivity (about 90 %). Such results were obtained irradiating the reacting mixture (phenol (0.06 g) dissolved in 50 mL of deionized water (50 mL) with benzene (0.07 mL) and the catalyst (50 mg)) with a 300 W Xe arc lamp for a variable time (up to 5 h). No reaction was observed when metal-free TiO_2-microspheres were used because the visible light ($\lambda \geq 400$ nm) cannot excite TiO_2 (band-gap 3.2 eV). In Fig. 9 it is shown the mechanism proposed by the authors to explain the photooxidation of benzene into phenol in the presence of an initial amount of phenol under visible light irradiation. An initial concentration of phenol is of paramount importance and the higher is the concentration, the higher are the yield and the selectivity. The O_2 involved in the mechanism is the oxygen already dissolved in the reacting mixture because this was not purged with nitrogen. The visible light excites the electrons in the Au nanoparticles to move fast toward the conduction band minimum of the TiO_2 where they are withdrawn by the O_2 so reducing it. The phenoxy ions are oxidized to phenoxy free radicals through the release of electrons to the electron-depleted Au nanoparticles. The phenoxy radicals are finally involved in the oxidation of the benzene to phenol becoming phenoxy ions again. The importance of this work relies on the possibility of a modulation of the catalytic properties of the TiO_2 by depositing different noble metals on the base of the different spectral characteristic of the light and with regard to a wider class of organic substrates.

Toluene is another aromatic hydrocarbon having relevancy in the industry. It is mainly converted to benzene by catalytic dealkylation and sometimes it is used for production of xylenes by transalkylation and disproportionation (Weber et al., 2005). The oxidation of toluene with O_2 to produce benzaldehyde throughout TiO_2 photocatalytic reactions has been investigated by different authors in the last years. Recently Ouidri and Khalaf (Ouidri & Khalaf, 2009) reported the oxidation of toluene by using TiO_2-pillared montmorillonite in acetonitrile or water or mixture of them and irradiating the reaction mixture by a 125 W Hg lamp for 1 h. The authors found that reactions performed in water gave the higher yield of benzaldehyde. Furthermore, the catalyst prepared according to this protocol showed a superior catalytic activity than the Degussa P25. The authors explained this experimental evidence with the higher capacity of the TiO_2-pillared montmorillonite to adsorb toluene compared with the Degussa P25. The presence of the catalyst is mandatory for the effectiveness of the process.

A further method is described by Cao et al. (Cao et al., 2011) who prepared TiO_2 hollow spheres by the hydrothermal reaction between TiF_4 and H_2O. Such material was used in the photocatalytic oxidation of toluene to give benzaldehyde in water under irradiation with a light of 310 nm. The TiO_2 spheres had a diameter of about 1 μm as shown by the FESEM and TEM images. The S_{BET} was in the range of 6.6 – 19 m^2g^{-1} while the anatase crystals size was in the range of 31 – 56 nm depending on the reaction time adopted for the synthesis (hydrothermal time). Such crystals covered the surface of the hollow spheres exposing the {001} face. The authors found that the conversion of toluene gradually increased from 9.0 % to 21 %, when the hydrothermal time was increased from 20 min to 6 h while the selectivity was about 90 %. In particular the catalytic activity was in the following order: TiO_2-20 min < TiO_2-40 min << TiO_2-6 h < TiO_2-12 h < TiO_2-72 h. The TiO_2-72 h microspheres showed activity two times higher than the commercial Degussa P25. As this last one has a S_{BET} of 45 m^2g^{-1} (higher than 6.6 m^2g^{-1} associated to the TiO_2-72h) and the {101} facets are more developed than that of TiO_2-72 h, the difference in the reactivity could be ascribed to the wider {001} facets in TiO_2-72h. The superior catalytic activity of the {001} facets has been already documented for different reactions in particular with regard to the degradation processes of water pollutants

(Liu et al., 2010). Indeed, it has been reported that water molecules undergo spontaneous dissociative adsorption on clean anatase {001} forming hydrogen peroxide and peroxyl radicals which are then involved in the degradative process (Selloni, 2008).

5.1.2 Other hydrocarbon functionalizations

More complex functionalizations of aromatic hydrocarbons have also been promoted by TiO$_2$. For example the perfluoroalkylation of arenes and α-methylstyrene derivatives in acetonitrile using titanium oxide as photocatalyst and in presence of alcohols and NaBF$_4$ has been reported (Eq. 4, Iizuka & Yoshida, 2009). The electrons in the conduction band of TiO$_2$ can reduce the perfluoroalkyl iodide forming a perfluoroalkyl radical and iodide ion. The resulting radical reacts with the arene forming an arenium radical, which is successively oxidized to the corresponding arenium cation by the action of the holes providing at the end the product. Different alcohols were considered. The best performances were obtained with methanol and ethanol while 2-propanol gave lower conversions and selectivities in the desidered products. Both alcohol and NaBF$_4$ may act as modulators of the red-ox network in particular reducing the hole-electron recombination process.

$$CF_3(CF_2)_nI \; + \; \underset{X}{\overset{X}{\bigcirc}} \quad \xrightarrow[\text{MeCN - MeOH}]{h\nu, TiO_2, NaBF_4} \quad \underset{X}{\overset{X}{\bigcirc}}-(CF_2)_nCF_3 \tag{4}$$

$$CF_3(CF_2)_nI \; + \; Ph\overset{CH_3}{\underset{H}{\diagup}}Y \quad \xrightarrow[\text{MeCN - MeOH}]{h\nu, TiO_2, NaBF_4} \quad Ph\overset{CH_2}{\underset{H}{\diagup}}\overset{Y}{\underset{(CF_2)_nCF_3}{}}$$

Heterocyclic compounds have also been synthesized through the use of TiO$_2$. Shiraishi and co-workers investigated a one-pot synthesis of benzimidazoles by simultaneous photocatalytic and catalytic reactions on Pt@TiO$_2$ nanoparticles (Shiraishi et al., 2010). A number of *ortho*-arylenediamines were converted to their corresponding benzimidazoles by Pt@TiO$_2$ catalyzed reaction with different alcohols at 303 K for 30 min under UV irradiation (xenon lamp, 2 kW; light intensity, 18.2 Wm^{-2} at 300 – 400 nm) using, in some cases, MeCN as co-solvent. A comparison between Pt@TiO$_2$ and a commercial TiO$_2$ similar to Degussa P25 (JRC-TIO-4 TiO$_2$ from the Catalyst Society of Japan with a anatase/rutile = 8:2) was reported, showing a significant improvement in both conversions (> 99%) and yields (generally above 80 % with different cases where the yield was above 95 %) after using Pt@TiO$_2$ (Eq. 5). The UV-light excite TiO$_2$ creating e$^-$/h$^+$ pairs. The hole pairs oxidize the alcohol to the corresponding aldehyde on the surface of the TiO$_2$. The aldehyde spontaneously condense with the arylenediamine forming an imine intermediate, which undergoes cyclization furnishing a benzimidazoline. This last one is finally fast oxidized to the product by the catalytic action of Pt. In this way Pt@TiO$_2$ has a double role: it works as *photocatalyst* in the oxidation of the alcohol and works as *catalyst* (due the presence of Pt) in the conversion of the benzimidazoline to the benzimidazole. Indeed, if Pt were not present the fast step from **5** to **2** (Eq. 6) would not be possible, allowing the shifting of the equilibrium **4-5** toward **4**. So, a further aldehyde molecule would condense giving the diimine **3** which finally furnishes the by-product **6**.

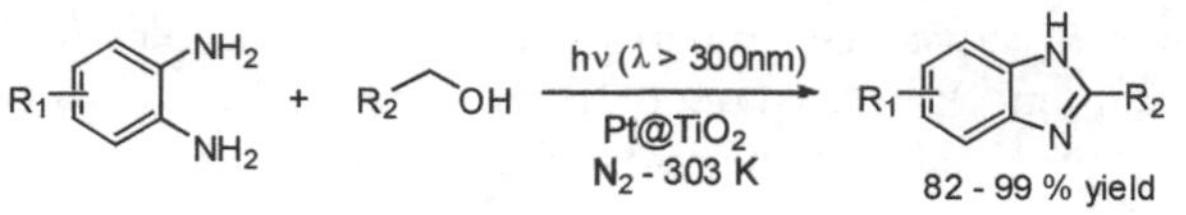

(5)

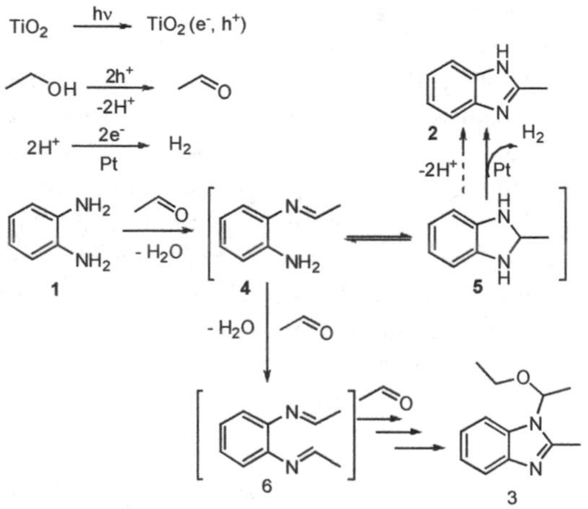

(6)

A similar approach for the synthesis of benzimidazoles and indazole has been reported (Selvam & Swaminathan, 2010). The authors found that 2-nitrophenyl azide, in the presence of different alcohols, undergoes a combined redox reaction and condensation after irradiation in the presence of TiO_2, giving 2-alkylbenzimidazoles as the main product, whereas, using 2-nitrobenzyl azide, the reduced amine does not react with aldehyde but undergoes cyclization to form indazole. The photoactivity was significantly enhanced by doping TiO_2 with Ag or Pt especially under solar light irradiation. In this case, for example, the yield of 2-methylbenzimidazole increased from 85 % by using Degussa P25 to 98 % with both Ag-TiO_2 and Pt-TiO_2 (Eq. 7)

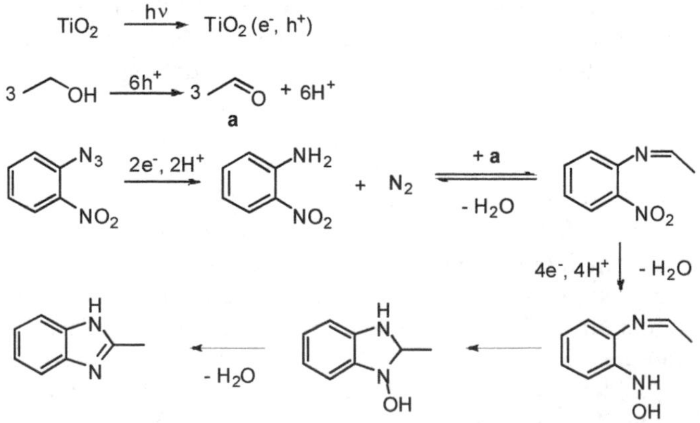

(7)

The substitution of protonated heteroaromatic bases by nucleophilic carbon-centered radicals, commonly reported as "Minisci Reaction" (Minisci et al., 1986), is one of the main general reactions of this class of aromatic compounds as a result of the large variety of successful radical sources, the high regio- and chemoselectivity and the simple experimental conditions. The well known Friedel–Crafts aromatic substitutions are unfavored when carried out on aromatic substrates bearing electron withdrawing groups. 6-Membered mono and poli N-heteroaromatic bases behave as electron poor substrates toward "classic" ionic electrophylic substitutions and only carbon atoms in b-position to nitrogen are suitable of substitution. However the reaction rate is much lower than the one carried out on benzene-systems. Otherwise C-centered radicals show nuchleophilic character and easily react with deactivated bases reproducing most of the Friedel–Crafts aromatic substitutions, but with opposite reactivity and selectivity. One of the main goal of the free-radical nucleophilic substitution is represented by the almost total absence of by-product formation during the reaction. In fact only catalytic amount of catalysts or initiators, depending on the reaction types, are needed to promote the reactions, whereas the classic Friedel–Crafts protocol needs stoichiometric quantity of Lewis acids. TiO_2 photocatalysis offers an alternative to the free-radical functionalization of heteroaromatic bases in the presence of classic metal-peroxide system (Gambarotti et al., 2010; Augugliaro et al., 2010). Caronna and co-workers reported the carboxyamidation of bases in the presence of H_2O_2 or air under sunlight (Eq. 8 a) (Caronna et al., 2003; Caronna et al., 2007a). A similar protocol has been successfully applied to the etherification of bases, affording, in the case of 1,3,5-trioxane and 1,3-dioxolane, a green-route to heteroaromatic aldehydes (Eq. 8 b) (Caronna et al., 2005). Moreover, in the presence of aliphatic aldehydes the bases are both acylated and alkylated (Caronna et al., 2007b). Particularly, it is interesting the fact that under the reaction conditions, the decarbonylation rate of the acyl radical intermediate is extraordinarily high, and good conversion to the corresponding alkyl derivatives are obtained also with primary aldehydes (Eq. 8 c).

Recently Selvam reported the synthesis of quinaldines from photocatalytic conversion of aniline and its derivatives in ethanol under mild conditions in the presence TiO_2 or Au-TiO_2 (Eq. 9) (Selvam & Swaminathan, 2010). The authors found that Au-TiO_2 has superior catalytic properties compared to TiO_2. Using Au-TiO_2, after only 4 hours of photo-irradiation at 365 nm the yield of quinaldine was about 75 % with a 10 % yield of 2,3-dimethylindole and 5% yield of other by-products, whereas, with TiO_2 the yield of quinaldine was about 50 % after 6h with 15 % of 2,3-dimethylindole. The presence of water reduces drastically the yield of quinaldine (5 % yield with a 96/4 ethanol-water ratio).

$$\text{(structure with NH}_2\text{)} + C_2H_5OH \xrightarrow[\substack{\text{Au-TiO}_2 \\ N_2}]{h\nu\ (\lambda > 300\text{nm})} \text{(quinoline structure)} \qquad (9)$$

The partial oxidation of alcohols for the production of fine and specialty chemicals represents a further field of investigation having high scientific and industrial impact. The use of TiO_2 to promote the oxidation of alcohols to the corresponding aldehydes has been investigated by different authors in the last years. Pillai and Sahle–Demessie described the oxidation of various primary and secondary alcohols to aldehydes in a gas-phase photochemical reactor with immobilized TiO_2 (Pillai & Sahle–Demessie, 2002). Enache reported the solvent-free oxidation of primary alcohols to aldehydes using Au-Pd/ TiO_2 catalysts without light irradiation (Enache et al., 2006).

Molinari investigated the photooxidation of geraniol to citral (Eq. 10), which is used in perfumes, flavourings and in the manufacture of other chemicals (Molinari et al, 2009b).

$$\text{(geraniol structure with OH)} \xrightarrow[\substack{P25\ TiO_2 \quad MeCN}]{O_2 \quad h\nu\ (\lambda > 320\ nm)} \text{(citral structure with O)} \qquad (10)$$

In this work, dispersions of Degussa P25 TiO_2 in CH_3CN containing geraniol (0.01 M) were irradiated by UV-light ($\lambda > 320$ nm), at room temperature and under 760 Torr of O_2 for different times (up to 140 min). The formation of citral occurs as consequence of the adsorption of the geraniol on the surface of the TiO_2 forming an alkoxide ion, RCH_2O^-. This latter is then oxidized to its corresponding alkoxide radical $RCH_2O\cdot$ and finally to the aldehyde RCHO by the direct electron transfer to the positive holes. Hence, the mechanism proposed by the authors is not based on the intervention of the hydroxyl radicals and it is supported by some ESR experiments. Furthermore, the experiments showed that water has negative effect on the oxidation of the -OH group of geraniol. In fact, water reduces the adsorption of the alcohol on the TiO_2 photoactive surface both due to the increase of the polarity which keep the geraniol preferentially in the solution and also due the competitive adsorption of water on the TiO_2. Finally, this inhibits the direct electron transfer from geraniol to the electron holes of TiO_2 reducing its catalytic effect. In order to investigate the effect of the nature of alcohol, the authors compared the reactivity of geraniol with that of trans-2-penten-1-ol, an allylic alcohol similar to geraniol but with a shorter chain of carbon atoms. As expected, trans-2-penten-1-ol is adsorbed on TiO_2 much better than geraniol and hence higher amount of aldehyde is obtained. Furthermore, if primary aliphatic alcohol with chains of the same lengths are considered like as citronellol (as corresponding of geraniol) and 1-pentanol (as corresponding of trans-2-penten-1-ol), it is observed that each primary alcohol is always better adsorbed than the corresponding allylic alcohol getting an increase of the yield of aldehyde.

The conversion of p-anisyl alcohol to p-anisaldehyde, an important intermediate for pharmaceutical industry, through the use of N-doped mesoporous titania (meso-$TiO_{2-x}N_x$) under sunlight as well as UV-lamp irradiation has been reported (Sivaranjani & Gopinath, 2011). The catalyst was prepared by a SCM technique introducing an aqueous solution of titanyl nitrate and urea into a muffle furnace at 400 °C. The fast evaporation of water was followed by a smouldered combustion of the titanyl nitrate (acting as precursor of Ti) in ammonia atmosphere (acting as fuel) obtained by the decomposition of the urea to NH_3 and CO_2. Different N-doped solid catalysts were prepared changing the precursor/fuel ratio. The oxidations were performed in aqueous solutions and without bubbling oxygen, obtaining yields of p-anisaldehyde of 25 % under sunlight and 30 % under laboratory light source in 7 hours. Despite the absence of oxygen, these results are comparable to that reported by Palmisano, who used a home-prepared TiO_2 as well as TiO_2 Merck (100 % anatase) and Degussa P25 (Palmisano et al., 2007b). In this work, 10 % yields of p-anisaldehyde was obtained in the presence of TiO_2 Merck or Degussa P25 and 30 - 40 % yields in the presence of home-made catalyst (depending on its amount, phase constitution and morphology).

More recently Chen described the photo-catalytic acetylation of 2-phenylethanol by acetic anhydride in the presence of TiO_2 nanoparticles (Chen, 2011). The synthesis was performed by suspending TiO_2 nanoparticles (32 nm grade) in H_2O_2 (30 % wt) for 30 min. under UV light (254 nm, 15 W) and then concentrating it in order to photoactivate the TiO_2 surface (Eq. 11). The solid was suspended in a solution of 2-phenylethanol in CH_2Cl_2 and irradiated for 10 h. The authors investigated the effects of the oxidant (O_2, t-butyl hydroperoxide, H_2O_2) finding that hydrogen peroxide was the most effective (99 % conversion) despite, even in absence of oxidant, a 88 % conversion was obtained. Moreover, different solvents were used finding that CH_2Cl_2 was the best one (> 99% conversion after 10 h) while reactions performed in toluene, hexane, ether, THF, acetone, ethyl acetate and CH_3CN gave lower or null conversions also during longer times. A screening of different TiO_2 catalyst (anatase, rutile) having different morphological parameters as well as other oxides like as Y_2O_3, WO_3, ZrO_2, was also carried out, finding that 32 nm TiO_2 (anatase/rutile = 4/1) was the optimal photo-catalyst both with regard the yield (> 99%) and the reaction time (10 h).

$$\text{(11)}$$

5.2 Other semiconductors in organic photosynthesis

As described before, TiO_2 is undoubtedly the most commonly used catalyst in organic photosynthesis. It is often metal-doped, in order to increase the wavelength radiation adsorption, and supported over inert materials (silica or zeolites), with the unique scope of increasing the surface area and consequently enhancing the reaction rate.

In the last decades, several other inorganic semiconductors have been investigated for the development of innovative organic photosynthetic strategies, including metal sulfides (ZnS and CdS) (Kisch, 2001), metal oxides (ZrO_2, ZnO, V_2O_5, SnO_2, Sb_2O_4, CeO_2, WO_3 and Sn/Sb mixed oxides) (Maldotti et al., 2002a), and polyoxometalates.

One of the reasons, which often make TiO_2 the photocatalyst of choice, is the higher protocol efficiency usually observed, in terms of both conversion and selectivity, when it is compared with other semiconductors. This is, for example, the case for the selective photoactivated radical addition of tertiary amines to electron deficient alkenes such as α,β-unsaturated lactones (Marinković & Hoffmann, 2001, 2003), conducted in the presence of TiO_2, ZnS or SiC. When reacting the N-methylpyrrolidine, the best results were achieved with TiO_2 (Eq.

12 a), while the reaction of N-methylpiperidine led to a different product distribution (Eq. 12 b). In particular, TiO_2 favoured the selective formation of product **A**, deriving from a classical Michael addition, while ZnS promoted the formation of product **B** in poor yields, probably because of its less oxidative properties.

$$(12)$$

Another significative example is furnished by the selective synthesis of 2-methylpiperazine from N-(β-hydroxtpropyl)-ethylenediamine by means of semiconductor-zeolite composite photocatalysts (Subba Rao & Subrahmanyam, 2002). Also in this case, zeolites modified with semiconductors ZnO and CdS were not very effective, while TiO_2-zeolites composites considerably facilitated the intramolecular cyclization (Eq. 13).

$$(13)$$

Nevertheless, in many cases inorganic semiconductors different from TiO_2 have recently allowed to afford new and intriguing photosynthetic approaches. In this section we aim to show a few very recent examples of the application of these photocatalysts for selective oxidation, reduction and C-C bond forming reactions.

5.2.1 Oxidation reactions

One of the main drawbacks in semiconductor catalyzed photoxidations in the presence of O_2 is the low selectivity in the desired product. In fact, in many cases the photomineralization of the substrate occurs, leading to the formation of CO_2.

A successful route to overcome this limitation is to employ the semiconductor, usually in the form of metal oxide, fixed or dispersed on a suitable inorganic support. Additional advantages in the use of semiconductors in their heterogenized form are the availability of wide range of supports and the easy recovery and recycle of the catalyst. Once again, TiO_2 is often the first choice, but in many cases other metal oxides provide better results.

Highly dispersed $(Me^{n+}O)$-Si binary oxides, prepared by both conventional impregnation and sol-gel procedures, have been widely employed under irradiation for the aerobic oxidation of alkanes and alkenes (Maldotti, 2002a). Propylene was converted to different oxidized species, varying the final product on the basis of the photocatalyst of choice: the corresponding epoxide in the presence of ZnO-SiO_2 (Yoshida et al., 1999), acetaldehyde with V_2O_5-SiO_2 (Tanaka et al., 1986), a mixture of both by means of CrO_x-SiO_2 (Murata et al., 2001).

Alkanes too were shown to undergo partial oxidation, when illuminated in the presence of silica-supported vanadium oxide, affording the corresponding carbonylic products in good yields with only traces of CO_x (Tanaka et al., 2000).

Supported polyoxometalates based on tungsten afforded high photocatalytic activity in the oxidation of different substrates. $(n\text{-}Bu_4N)_4W_{10}O_{32}$ (TBADT), supported on silica by impregnation procedure, was successfully employed at room temperature and atmospheric pressure to promote by irradiation ($\lambda > 300$ nm) the aerobic oxidation of cyclohexane to a 1:1 mixture of cyclohexanol and cyclohexanone (Molinari et al., 1999), while under analogous conditions cyclohexene was converted to the corresponding cyclohexenyl hydroperoxide with 90 % of selectivity (Molinari et al., 2000). This decatungstate was also immobilized on a mesoporous MCM-41-type material, which allows to obtain a better dispersion of the semiconductor, and the new material was tested in the oxidation of the cyclohexane for comparison with the silica supported material (Maldotti et al., 2002b). These studies showed the key role of the solid matrix in controlling the ketone/alcohol ratio of the products. In fact, the higher surface area of the mesoporous material favored the conversion of cyclohexanol to the corresponding cyclohexanone, affording a higher final ratio ketone/alcohol (2.4).

Another tungsten-based heteropolyoxometalate supported on amorphous silica ($H_3PW_{12}O_{40}/SiO_2$) was successfully used for the selective aerobic photocatalytic oxidation of benzylic alcohols to the corresponding aldehydes and ketones, with yields up to 97 % (Farhadi et al., 2005). Even more interesting, in spite of the higher reactivity of carbonylic derivatives, no overoxidation was observed, and no traces of carboxylic acids were found in the final products (Eq. 14).

$$\underset{Ar}{\overset{OH}{\underset{R}{|}}} \quad \xrightarrow[{H_3PW_{12}O_{40}\text{-}SiO_2\,/\,h\nu}]{O_2\,/\,CH_3CN\,/\,r.t.} \quad \underset{Ar}{\overset{O}{\underset{R}{||}}} \qquad (14)$$

R = H, aryl, alkyl

5.2.2 Reduction reactions

If we exclude the conversion of CO_2 into more useful hydrogenated chemicals, which is however characterized by low energy conversion efficiencies, the main application of the photoinduced reduction reaction is the transformation of nitro-aromatic derivatives.

Being the most powerful reducing semiconductor, CdS usually provides highest yield and selectivity by converting nitro-benzenes into corresponding anilines, if compared with the common TiO$_2$ (Maldotti et al., 2000). Nevertheless, it requires UV light to be activated, this limiting the range of applications.

Very recently, Pfitzner and coworkers (Füldner et al., 2011) have reported that the blue light irradiation of PbBiO$_2$X (X = Cl or Br), in the presence of triethanolamine (TEOA) as an electron donor, provides the selective and complete reduction of nitrobenzene derivatives to their corresponding anilines (Eq. 15). By replacing bismuth with antimony in the oxide halide no conversion was observed. The same unreactivity resulted by operating in the presence of PbBiO$_2$I. The authors suggest that the different catalytic activity should be ascribed to multiple factors, including the crystal structures and their optical and redox properties. Photocatalyst recycling experiments showed that sonication of PbBiO$_2$X before reuse was essential to retain its activity up to five catalytic cycles.

$$NO_2 \xrightarrow[\lambda = 440 \text{ nm } / 20 \text{ h}]{\text{PbBiO}_2X \text{ / TEOA}} NH_2 \tag{15}$$

> 99 %

Anilines can be also obtained in 50 % yields and complete selectivity by photoreduction of aryl azides catalyzed by CdS or CdSe nanoparticles (Warrier et al., 2004). The reaction, which occurs under very mild conditions (room temperature, atmospheric pressure, neutral pH and aqueous medium), requires the presence of sodium formate as sacrificial electron donor (Eq. 16).

$$N_3 \xrightarrow[\text{hv}]{\text{CdS}} NH_2 \tag{16}$$

$$HCO_2Na \qquad\qquad CO_2$$

The high efficiency of this photocatalyzed reduction is attributed to the large driving force for electron transfer to the azide, which in turn arises from the much more negative potential of excited CdS nanoparticles electrons relative to the azide reduction potential.

More recently, the same photocatalytic approach was successfully applied to the reduction of aryl azide-terminated, self-assembled monolayers on gold to the corresponding arylamine species (Radhakrisham et al., 2006).

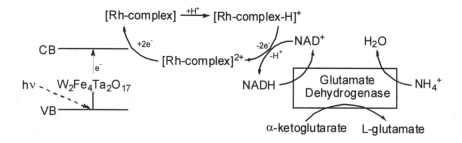

Fig. 10. Schematic diagram of photocatalyst $W_2Fe_4Ta_2O_{17}$ mediated bioreactor for the enzymatic synthesis of L-glutamate.

Photoreduction has been also applied to the design of a novel photobioreactor capable to couple a redox enzyme biocatalysis (Glutamate Dehydrogenase) with the new visible-light active heterogeneous photocatalyst $W_2Fe_4Ta_2O_{17}$ for the production of L-glutamate (Park et al., 2008). The idea arises from the necessity to develop an efficient and industrially feasible method for the in situ regeneration of nicotinamide (NAD) co-factor, on which depend the activity of many enzymes. In fact, the high cost of NAD often limits the industrialization potentials of many promising enzymatic processes. The overall schematic mechanism, which regulates this photobioreactor, is reported in Fig. 10. In the photocatalytic cycle, the $W_2Fe_4Ta_2O_{17}$ upon band-gap excitation by visible light ($\lambda > 420$ nm) promotes electrons to the conduction band, which are in turn easily transferred to an organometallic Rh complex. The latter, after undergoing hydrogen abstraction from the aqueous medium, transfers electrons and hydride to NAD^+, forming NADH.

5.2.3 C-C bond forming reactions

Among semiconductor photocatalysts, Ru(II)polypyridine complexes are attracting increasing interest because of their stability at room temperature and enhanced photoredox properties (Narayanam & Stephenson, 2011). In particular, $Ru(bpy)_3Cl_2$ has been widely employed, as its irradiation with visible light leads to the excited species $Ru(bpy)_3^{2+*}$, which in turn can be employed as strong single electron oxidant and reductant, depending upon reaction conditions. Besides classical oxidative reactions introducing oxygen atoms and reductive hydrogenations, $Ru(bpy)_3Cl_2$ has found large use in the photocatalytic promotion of C-C bond formation.

Irradiation of aryl enones, in the presence of a mixture of $Ru(bpy)_3Cl_2$, i-Pr$_2$NEt and LiBF$_4$ in acetonitrile, allowed to develop a highly diastereoselective intramolecular [2 + 2] cycloaddition reaction, which, in all the examples reported, led to the formation of the corresponding cis-cyclobutanes (Eq. 17) (Ischay & al., 2008). In the reaction mechanism, i-Pr$_2$NEt has the role to reduces the excited species $Ru(bpy)_3^{2+*}$ to $Ru(bpy)_3^+$, which seems to be the real initiator of the process, while LiBF$_4$, being a Lewis acid, favours the solubility of Ru complex in acetonitrile.

The same system was applied to promote the intermolecular homo-dimerization of aryl enones (Eq. 18) and the crossed intermolecular cyloaddition (Eq. 19) (Du & Yoon, 2009).

More recently, a photoredox strategy based on $Ru(bpy)_3Cl_2$ photocatalyst was successfully employed both for the selective α-trifluoromethylation upon in situ or pre-generated enolsilanes and silylketenes (Eq. 20) (Pham et al., 2011) and to promote an oxidation/[3 + 2] cycloaddition/aromatization cascade reaction (Zou et al., 2011), the latter leading to the formation of pyrrolo[2,1-α]isoquinolines from ethyl 2-(3,4-dihydroisoquinolin-2(1H-yl) acetate (Eq. 21)

We have already reported the photocatalytic activity of tetrabutylammonium decatungstate salt ((n-Bu$_4$N)$_4$W$_{10}$O$_{32}$) in selective photoxidations. TBADT has been also used by Albini and coworkers to promote the photocatalytic radical conjugate addition of electron-poor olefins by cycloakanes (Eq. 22) (Dondi et al., 2006) and the acylation of α,β-unsaturated nitriles, ketones and esters (Eq. 23) (Esposti et al., 2007), affording the desired products in good yields.

The same group has recently shown that irradiated TBADT can also effectively catalyze the alkylation at position 2 of 1,3-benzodioxoles, making this moiety more biological active and enzyme-specific (Eq. 24) (Ravelli et al., 2011).

Another significant example of potentials of semiconductor photocatalysis is represented by the artificial photosynthesis design, that is the fixation of CO$_2$ molecules to afford higher organic compounds (Hoffmann et al., 2011).

For example, many studies have concentrated on the fixation of CO$_2$ in carboxylic acids to produce intermediates in key cellular processes. Recently Guzman and Martin have reported that a glyoxylate can be methylated to produce the corresponding lactate, directly involved in the reductive tricarboxylic acid cycle, by photocatalytic fixation of CO$_2$ mediated by ZnS (Guzman & Martin, 2010).

Nevertheless, it has been recently outlined (Yang et al., 2010) that many results reported in the literature and related to these studies could be influenced by the presence of carbon residues left over from the synthesis of metal oxide semiconductors. In other words, there could be experimental artefacts affecting reports and final conclusions, so that more investigations in the field of artificial photosynthesis is still mandatory.

6. Conclusions

In the last decades a growing interest has been devoted to the development of photocatalytic processes both in the homogeneous and in the heterogeneous phase. Particularly, concerning the heterogeneous systems, great interest has aroused the use of photosensitive semiconductors as catalysts for organic processes, due to their ease to use, recycle and low environmental impact. Although most of the actual applications are restricted to the decomposition of organic pollutants, semiconductors are becoming more and more important for the development of new photocatalyzed organic protocols, as an alternative to the conventional metal-catalyzed thermal processes. Generally, TiO_2 has a dominant role in all the semiconductor-phtocatalyzed applications, including the organic synthesis, however, in the last decades many others transition metals photocatalysts, have been developed. Actually, in the scientific landscape, big challenges are represented by the reduction of energy consumption and environmental impact, and photocatalysis could be one of the winning answers in the chemistry field.

7. References

Albini, A., & Fagnoni, M. (2008). 1908: Giacomo Ciamician and the Concept of Green Chemistry, ChemSusChem, vol. 1, pp. 63–66, ISSN 1864-564X

Anpo, M., & Kamat, P.V. (Eds.). (2010) Environmentally Benign Photocatalysts, Springer, ISBN 978-0-387-48441-9, Dordrecht Heidelberg London New York

Augugliaro, V., Caronna, Di Paola, A., Marcì, G., Pagliaro, M., Palmisano, G., & Palmisano, L. (2010). TiO_2-Based Photocatalysis for Organic Synthesis. In: Environmentally Benign Photocatalysts, Anpo, M., & Kamat, P.V. (Eds.), pp. 623-645, Springer, ISBN 978-0-387-48441-9, Dordrecht Heidelberg London New York

Bal, R., Tada, M., Sasaki, T., & Iwasawa, Y. (2006). Direct Phenol Synthesis by Selective Oxidation of Benzene with Molecular Oxygen on an Interstitial-N/Re Cluster/Zeolite Catalyst, Angewandte Chemie International Edition, Vol. 45, pp. 448–452, ISSN: 1521-3773

Balzani,V., Credi, A., & Venturi, M. (2008). Photochemical Conversion of Solar Energy. ChemSusChem, Vol. 1, pp. 26-58, ISSN 1864-564X

Bard, A.J. (1979). Photoelectrochemistry and heterogeneous photocatalysis at semiconductors. Journal of Photochemistry, Vol. 10, pp. 59-75, ISSN 1010-6030

Becquerel, C R. (1839). Mémoire sur les effets électriques produits sous l'influence des rayons solaires. Comptes Rendus de l'Académie des Sciences, Vol. 9, pp. 561-567

Brattain, W.H., Garret, C.G.B. (1955). Bell System Technical Journal, Vol. 34, p. 129

Cao, F.L., Wang, J-G., Lv, F-J., Zhang, D-Q., Huo, Y-N., Li, G-S., Li, H-X., & Zhu, J. (2011). Photocatalytic oxidation of toluene to benzaldehyde over anatase TiO_2 hollow spheres with exposed {001} facets, Catalysis Communications, Vol. 12, pp. 946–950, ISSN 1566-7367

Caronna, T., Gambarotti, C., Palmisano, L., Punta, C., & Recupero, F. (2003). Sunlight induced functionalisation of some heterocyclic bases in the presence of polycrystalline TiO_2, Chemical Communications, Vol. 18, pp. 2350-2351, ISSN 1364-548X

Caronna, T., Gambarotti, C., Palmisano, L., Punta, C., & Recupero, F. (2005) Sunlight-induced reactions of some heterocyclic bases with ethers in the presence of TiO$_2$ - A green route for the synthesis of heterocyclic aldehydes, *Journal of Photochemistry And Photobiology A: Chemistry*, Vol. 171, pp. 237-242, ISSN 1010-6030

Caronna, T., Gambarotti, C., Mele, A., Pierini, M., Punta, C., & Recupero, F. (2007a). A green approach to the amidation of heterocyclic bases: the use of sunlight and air, *Research on Chemical Intermediates*, Vol. 33, pp. 311-317, ISSN 1568-5675

Caronna, T., Gambarotti, C., Palmisano, L., Punta, C., Pierini, M., & Recupero, F. (2007b). Sunlight-induced functionalisation reactions of heteroaromatic bases with aldehydes in the presence of TiO$_2$: A hypothesis on the mechanism, *Journal of Photochemistry And Photobiology A: Chemistry*, Vol. 189, pp. 322-328, ISSN 1010-6030

Carp, O., Huisman, C.L., & Reller, A. (2004). Photoinduced reactivity of titanium dioxide. *Progress in Solid State Chemistry*, Vol. 32, pp. 33-177, ISSN 0079-6786

Chatterjee, D., & Dasgupta, S. (2005). Visible light induced photocatalytic degradation of organic pollutants. *Journal of Photochemistry and Photobiology C: Photochemistry Reviews*, Vol. 6, pp. 186-205, ISSN 1389-5567

Chattopadhyay, D., & Rakshit, P.C. (Eds.) (2010). *Electronics: Fundamentals and Applications*, New Age International (P) Ltd., Publishers ISBN : 978-81-224-3147-6, New Delhi-110 002, India

Chen, C.T., Kao, J.Q., Liu, C.Y., & Jiang, L.Y. (2011). Oxidative, photo-activated TiO$_2$ nanoparticles in the catalytic acetylation of primary alcohols, *Catalysis Science & Technology*, Vol. 1, pp. 54–57

Chen, X., Zhang, J., Fu, X., Antonietti, M., & Wang, X. (2009). Fe-g-C$_3$N$_4$-Catalyzed Oxidation of Benzene to Phenol Using Hydrogen Peroxide and Visible Light, *Journal of the American Chemical Society*, Vol. 131, pp. 11658–11659, ISSN 1520-5126

Ciamician, G. & Silber, P. (1900). *Chemische Berichte*, Vol. 33, p. 2911-2913

Deanna, H.C., Agrios, A.G & Gray, A.k. (2003). Explaining the Enhanced Photocatalytic Activity of Degussa P25 Mixed-Phase TiO$_2$ Using EPR. *The Journal of Physical Chemistry B*, Vol. 107, pp. 4545-4549, ISSN 1520-5207

Dindar, B., & Içli, S. (2001). Unusual photoreactivity of zinc oxide irradiated by concentrated sunlight. *Journal of Photochemistry and Photobiology A: Chemistry*, Vol. 140, pp. 263-268, ISSN 1010-6030

Dondi, D., Cardarelli, A.M., Fagnoni, M., & Albini, A. (2006). Photomediated synthesis of β-alkylketones from cycloalkanes. *Tetrahedron*, Vol.62, pp. 5527-5535, ISSN 0040-4020

Du, J., & Yoon, T.P. (2009). Crossed Intermolecular [2+2] Cycloadditions of Acyclic Enones via Visible Light Photocatalysis. *Journal of the American Chemical Society*, Vol.131, 14604-14605, ISSN 1520-5126

Enache, D.I., Edwards, J.K., Landon, P., Solsona-Espriu, B., Carley, A.F., Herzing, A.A., Watanabe, M., Kiely, C.J., Knight, D.W., & Hutchings, G.J. (2006). Solvent-Free Oxidation of Primary Alcohols to Aldehydes Using Au-Pd/TiO$_2$ Catalysts, Science, Vol. 311, pp. 362-365, ISSN 1095-9203

Esposti, S., Dondi, D., Fagnoni, M., & Albini, A. (2007). Acylation of Electrophilic Olefins through Decatungstate-Photocatalyzed Activation of Aldehydes. *Angewandte Chemie. International Edition*, Vol.46, pp. 2531-2534, ISSN 1521-3773

Esswein, A.J., & Nocera, D.G. (2007). Hydrogen Production by Molecular Photocatalysis. *Chemical Review*, Vol. 107, pp. 4022-4047, ISSN 1520-6890

Evans, P., & Sheel, D.W. (2007). Photoactive and antibacterial TiO_2 thin films on stainless steel. *Surface and Coatings Technology*, Vol. 201, pp. 9319–9324, ISSN 0257-8972

Fan, F.-R.F., Leempoel, P., & Bard, a.J. (1983). Semiconductor electrodes. *Journal of The Electrochemical Society*, Vol. 130, pp. 1866-75, ISSN 0013-4651

Farhadi, S., Afshari, M., Maleki, M., & Babazadeh, Z. (2005). Photocatalytic Oxidation of Primary and Secondary Benzylic Alcohols to Carbonyl Compounds Catalyzed by $H_3PW_{12}O_{40}/SiO_2$ under an O_2 Atmosphere. *Tetrahedron Letters*, Vol.46, pp. 8483-8486, ISSN 0040-4039

Fox, M.A., & Dulay, M. (1993). Heterogeneous Photocatalysis. *Chemical Review*, Vol. 83, pp. 341-357, ISSN 1520-6890

Fox, M.A. (2001). Synthetic Applications of Photocatalytic Oxidation and Reduction Reactions of Organic Reactants on Irradiated Semiconductor Surfaces, In: *Electron Transfer in Chemistry*, Balzani V. (Ed.), pp. 272-311, Wiley-VCH, ISBN 3-527-29912-2, Weinheim (Germany)

Frank, S.N., & Bard, A.J. (1997). Heterogeneous photocatalytic oxidation of cyanide ion in aqueous solutions at titanium dioxide powder. *Journal of the American Chemical Society*, Vol. 99, pp. 303-304, ISSN 1520-5126

Friedrich, W. (1828). Ueber künstliche Bildung des Harnstoffs. *Annalen der Physik und Chemie*, Vol. 88, No. 2, pp. 253–256

Fu, G., Vary, P.S., & Lin, C.T. (2005). Anatase TiO_2 Nanocomposites for Antimicrobial Coatings. *The Journal of Physical Chemistry B*, Vol. 109, pp. 8889-8898, ISSN 1520-5207

Fujishima, A., & Honda, K. (1972). Electrochemical Photolysis of Water at a Semiconductor Electrode. *Nature*, Vol. 238, p. 37, ISSN 0028-0836

Füldner, S., Pohla, P., Bartling, H., Dankesreiter, S., Stadler, R., Gruber, M., Pfitzner, A., & König, B. (2011). Selective Photocatalytic Reductions of Nitrobenzene Derivatives Using $PbBiO_2X$ and Blue Light. *Green Chemistry*, Vol.13, pp.640-643, ISSN 1463-9270

Gambarotti, C., Punta, C., Recupero, F., Caronna, T., & Palmisano L. (2010). TiO_2 in Organic Photosynthesis: Sunlight Induced Functionalization of Heterocyclic Bases. *Current Organic Chemistry*, Vol. 14(11), pp. 1153-1169, ISSN 1385-2728

Gan, W.Y., Lam, S.W., Chiang, K., Amal, R., Zhao, H., & Brungs, M.P. (2007). Novel TiO_2 thin film with non-UV activated superwetting and antifogging behaviours. *Journal of Materials Chemistry*, Vol. 17, pp. 952–954, ISSN 1364-5501

Gaya, U.I., & Abdullaha, A.H. (2008). Heterogeneous photocatalytic degradation of organic contaminants over titanium dioxide: A review of fundamentals, progress and problems. *Journal of Photochemistry and Photobiology C: Photochemistry Reviews*, Vol. 9, pp. 1–12, ISSN 1389-5567

Guo, Y., Zhang, X., Zou, H., Liu, H., Wang, J., & Yeung, K.L. (2009). Pd–silicalite-1 composite membrane for direct hydroxylation of benzene. *Chemical Communications*, pp. 5898–5900, ISSN 1364-548X

Guzman, M. I., & Martin, S. T. (2010). Photo-production of lactate from glyoxylate: how minerals can facilitate energy storage in a prebiotic world. *Chemical Communications*, pp. 2265-2267, ISSN 1359-7345

Hariharan, C. (2006). Photocatalytic degradation of organic contaminants in water by ZnO nanoparticles: Revisited. *Applied Catalysis A: General*, Vol. 304, pp. 55-61, ISSN 0926-860X

Hashimoto, K., Irie, H., & Fujishima, A . (2005). TiO_2 Photocatalysis: A Historical Overview and Future Prospects. *Japanese Journal of Applied Physics*, Vol. 44, pp. 8269-8285, ISSN 1347-4065

Heller, A., Degani, Y., Johnson, D.W., & Gallagher, P.K. (1987). Controlled Suppression and Enhancement of the Photoactivlty of Titanium Dioxide (Rutile) Pigment. *The Journal of Physical Chemistry*, Vol. 91, pp. 5987-5991, ISSN 0022-3654

Heller, A. (1995). Chemistry and Applications of Photocatalytic Oxidation of Thin Organic Films. *Accounts of Chemical Research*, Vol. 28, pp. 503-508, ISSN 1520-4898

Hoffmann, M.R, Martin, S.T, Choi, W., & Bahnemann, D.W. (1995). Environmental Applications of Semiconductor Photocatalysis. *Chemical Review*, Vol. 95, pp. 69-96, ISSN 1520-6890

Hoffmann, M.R., Moss, J.A., & Baum, M. M. (2011). Artificial Photosynthesis: Semiconductor Photocatalytic Fixation of CO_2 to Afford Higher Organic Compounds. *Dalton Transactions*, Vol.40, pp. 5151-5158, ISSN 1364-5447

Hu, J-S., Ren, L-L., Guo, Y-G., Liang, H-P., Cao, A-M., Wan, L-J., & Bai, C-L. (2005). Mass Production and High Photocatalytic Activity of ZnS Nanoporous Nanoparticles. *Angewandte Chemie International Edition*, Vol. 44, pp. 1269 -1273, ISSN 1521-3773

Iizuka, M., & Yoshida, M. (2009). Redox system for perfluoroalkylation of arenes and a-methylstyrene derivatives using titanium oxide as photocatalyst, *Journal of Fluorine Chemistry*, Vol. 130, pp. 926–932, ISSN 0022-1139

Ischay, M.A., Anzovino, M.E., Du, J., & Yoon, T. P. (2008). Efficient Visible Light Photocatalysis of [2+2] Enone Cycloadditions. *Journal of the American Chemical Society*, Vol.130, 12886-12887, ISSN 1520-5126

Karraway, E.R., Hoffman, A.J., & Hoffmann, M.R. (1994). Photocatalytic Oxidation of Organic Acids on Quantum-Sized Semiconductor Colloids. *Environmental Science & Technology*, Vol. 28, pp. 786-793, ISSN 1520-5851

Kisch, H. (2001). Semiconductor photocatalysis for organic synthesis, In: *Advances in Photochemistry*, Vol. 26, Neckers, D.C., Bünau, G.V., Jenks, W.S. (Eds.), 93-143, John Wiley & Sons, Inc., ISBN 0-471-39467-X, USA

Kormann, C., Bahnemann, D.W., & Hoffmann, M.R. (1988). Photocatalytic Production of H_2O_2 and Organic Peroxides in Aqueous Suspensions of TiO_2, ZnO, and Desert Sand. *Environmental Science & Technology*, Vol. 22, pp. 798-806, ISSN 1520-5851

Li, C., Yuan, J., Han, B., Jiang, L., & Shangguan, W. (2010). TiO_2 nanotubes incorporated with CdS for photocatalytic hydrogen production from splitting water under visible light irradiation. *International Journal of Hydrogen Energy*, Vol. 35, pp. 7073-7079, ISSN 0360-3199

Linsebigler, A.L., Lu, G., & Yates, J.T. (1995). Photocatalysis on TiO_2 Surfaces: Principles, Mechanisms, and Selected Results. *Chemical Review* , Vol. 95, pp. 735-758, ISSN 1520-6890

Liu, M., Piao, L., Zhao, L., Ju, S., Yan, Z., He, T., Zhou, C., & Wang, W. (2010). Anatase TiO_2 single crystals with exposed {001} and {110} facets: facile synthesis and enhanced photocatalysis, *Chemical Communications*, Vol. 46, pp. 1664-1666, ISSN 1364-548X

Liu, Y., Wanga, X., Yanga, F., & Yang, X. (2008). Excellent antimicrobial properties of mesoporous anatase TiO_2 and Ag/TiO_2 composite films. *Microporous and Mesoporous Materials*, Vol. 114, pp. 431–439, ISSN 1387-1811

Łukasiak, L., & Jakubowski, A. (2010). History of Semiconductors. *JTIT*, vol. 1, pp. 3-9, ISSN 1899-8852

Macyk, W., & Kisch, H. (2001). Photosensitization of Crystalline and Amorphous Titanium Dioxide by Platinum(iv) Chloride Surface Complexes. *Chemistry - A European Journal*, Vol. 7, pp. 1862-1867, ISSN 1521-3765

Maeda, K., & Domen, K. (2007). New Non-Oxide Photocatalysts Designed for Overall Water Splitting under Visible Light. *The Journal of Physical Chemistry C*, Vol. 111, pp. 7851-7861, ISSN 1932-7455

Maira, A. J., Yeung, K.L., Soria, J., Coronado, J.M., Belver, C., Lee, C.Y., & Augugliaro, V. (2001). Gas-phase photo-oxidation of toluene using nanometer-size TiO_2 catalysts. *Applied Catalysis B: Environmental*, Vol. 29, pp. 327-336, ISSN 0926-3373

Malato, S., Blanco, J., Alarcón, D.C., Maldonado, M.I., Fernández-Ibáñez, P., & Gernjak, W. (2007). Photocatalytic decontamination and disinfection of water with solar collectors. *Catalysis Today*, Vol. 122, pp. 137-149, ISSN 0920-5861

Maldotti, A., Andreotti, L., Molinari, A., Tollari, S., Penoni, A., & Cenini, S. (2000). Photochemical and Photocatalytic Reduction of Nitrobenzene in the Presence of Cyclohexene. *Journal of Photochemistry and Photobiology A: Chemistry*, Vol.133, pp. 129-133, ISSN 1010-6030

Maldotti, A., Molinari, A., & Amadelli, R. (2002a). Photocatalysis with Organized Systems for the Oxofunctionalization of Hydrocarbons by O_2. *Chemical Reviews*, Vol. 102, pp. 3811-3836, ISSN 1520-6890

Maldotti, A., Molinari, A., Varani, G., Lenarda, M., Storaro, L., Bigi, F., Maggi, R., Mazzacani, A., & Sartori, G. (2002b). Immobilization of $(n-Bu_4N)_4W_{10}O_{32}$ on Mesoporous MCM-41 and Amorphous Silicas for Photocatalytic Oxidation of Cycloalkanes with Molecular Oxygen. *Journal of Catalysis*, Vol. 209, pp. 210-216, ISSN 0021-9517

Marinković, S., & Hoffmann, N. (2001). Efficient Radical Addition of Tertiary Amines to Electron-Deficient Alkenes Using Semiconductors as Photochemical sensitisers. *Chemical Communications*, pp. 1576-1578, ISSN 1364-548X

Marinković, S., & Hoffmann, N. (2003). Semiconductors as Sensitizers for the Radical Addition of Tertiary Amines to Electron Deficient Alkenes. *Internationa Journal of Photoenergy*, Vol.5, pp. 175-182, ISSN 1687-529X

Martra, G. (2000). Lewis acid and base sites at the surface of microcrystalline TiO2 anatase: relationships between surface morphology and chemical behavior. *Applied Catalysis A: General*, Vol. 200, pp. 275-285, ISSN 0926-860X

Mccullagh, C., Robertson, J.M.C., Bahnemann, D.W., & Robertson, P.K.J. (2007). The application of TiO$_2$ photocatalysis for disinfection of water contaminated with pathogenic micro-organisms: a review. *Research on Chemical Intermediates*, Vol. 33, pp. 359-375, ISSN 1568-5675

Meissner, D., & Memming, R. (1988). Photoelectrochemistry of Cadmium Sulfide. 1. Reanalysis of Photocorrosion and Flat-Band Potential. *The Journal of Physical Chemistry*, Vol. 92, pp. 3476-3483, ISSN 0022-3654

Mills, A., & Le Hunte, S. (1997). An Overview of semiconductor photocatalysis. *Journal of Photochemistry and Photobiology A: Chemistry*, Vol. 108, pp. 1-35, ISSN 1010-6030

Mills, A., & Lee, S-K. (2002). A web-based overview of semiconductor photochemistry-based current commercial applications. *Journal of Photochemistry and Photobiology A: Chemistry*, Vol. 152, pp. 233-247, ISSN 1010-6030

Minisci, F., Vismara, E., Fontana, F., Morini, G., Serravalle, M., & Giordano, C. (1987). Polar effects in free-radical reactions. Solvent and isotope effects and effects of base catalysis on the regio- and chemoselectivity of the substitution of protonated heteroaromatic bases by nucleophilic carbon-centered radicals. *The Journal of Organic Chemistry*, Vol. 52, pp. 730-736, ISSN 1520-6904

Molinari, A., Amadelli, R., Mazzacani, A., Sartori, G., & Maldotti, A. (1999). Heterogeneous Photocatalysis for Synthetic Purposes: Oxygenation of Cyclohexane with H$_3$PW$_{12}$O$_{40}$ and (nBu$_4$N)$_4$W$_{10}$O$_{32}$ Supported on silica. *Journal of Chemical Society, Dalton Transactions*, pp. 1203-1204, ISSN 1472-7773

Molinari, A., Amadelli, R., Carassiti, V., & Maldotti, A. (2000). Photocatalyzed Oxidation of Cyclohexene and Cyclooctene with (nBu$_4$N)$_4$W$_{10}$O$_{32}$ and (nBu$_4$N)$_4$W$_{10}$O$_{32}$/FeIII [meso-Tetrakis(2,6-dichlorophenyl)-porphyrin] in Homogeneous and Heterogeneous Systems. *European Journal of Inorganic Chemistry*, pp. 91-96, ISSN 1099-0682

Molinari, R., Caruso, A., & Poerio, T. (2009a). Direct benzene conversion to phenol in a hybrid photocatalytic membrane reactor, Catalysis Today, Vol. 144, pp. 81–86, ISSN 0920-5861

Molinari, A., Montoncello, M., Rezala, H., & Maldotti A. (2009b). Partial oxidation of allylic and primary alcohols with O$_2$ by photoexcited TiO$_2$, *Photochemical & Photobiological Sciences*, 2009, 8, 613–619, ISSN 1474-905X

Montalti, M., Credi, A., Prodi, L., Gandolfi, M-T. (2006) *Handbook of Photochemistry* (3rd edition), Taylor & Francis Group, LLC, ISBN 0-8247-2377-5, Boca Raton, FL (USA)

Murata, C., Yoshida, H., & Hattori, T. (2001). Visible light-induced photoepoxidation of propene by molecular oxygen over chromia–silica catalysts. *Chemical Communications*, pp. 2412-2413, ISSN 1364-548X

Narayanam, J. M. R., & Stephenson, C. R. J. (2011). Visible Light Photoredox Catalysis: Applications in Organic Synthesis. *Chemical Society Reviews*, Vol.40, pp. 102-113, ISSN 0306-0012

Ohama, Y., & Van Gemert, D. (Eds.). (2011). *Application of Titanium Dioxide Photocatalysis to Construction Materials*, Springer, ISBN 978-94-007-1296-6, Dordrecht Heidelberg London New York

Ouidri, S., & Khalaf, H. (2009). Synthesis of benzaldehyde from toluene by a photocatalytic oxidation using TiO_2-pillared clays, *Journal of Photochemistry and Photobiology A: Chemistry*, Vol. 207, pp. 268-273, ISSN 1010-6030

Palmisano, G., Augugliaro, V., Pagliaro, M., & Palmisano, L. (2007a). Photocatalysis: a promising route for 21st century organic chemistry. *Chemical Communications*, pp. 3425-3437, ISSN 1364-548X

Palmisano, G., Yurdakal, S., Augugliaro, V., Loddo, V., & Palmisano, L. (2007b). Photocatalytic Selective Oxidation of 4-Methoxybenzyl Alcohol to Aldehyde in Aqueous Suspension of Home-Prepared Titanium Dioxide Catalyst, *Advanced Synthesis & Catalysis*, Vol. 349, pp. 964 – 970, ISSN 1615-4169

Park, C.B., Lee, S.H., Subramanian, E., Kale, B.B., Lee, S.M., & Baeg, J.-O. (2008). Solar Energy in Production of L-Glutamate through Visible Light Active Photocatalyst-Redox Enzyme Coupled Bioreactor. *Chemical Communications*, pp. 5423-5424, ISSN 1364-548X

Parkin, P.I., & Palgrave, R.G. (2005). Self-cleaning coatings. *Journal of Materials Chemistry*, Vol. 15, pp. 1689-1695, ISSN 1364-5501

Pastori, N., Gambarotti, & C., Punta, C. (2009). Recent Developments in Nucleophilic Radical Addition to Imines: the Key Role of Transition Metals and the New Porta Radical-Type Version of the Mannich and Strecker Reactions. *Mini-Reviews In Organic Chemistry*, Vol. 6(3), pp. 184-195, ISSN 1570-193X

Pham, P.V., Nagib, D.A., & MacMillan, W. C. (2011). Photoredox Catalysis: A Mild, Operationally Simple Approach to the Synthesis of α-Trifluoromethyl Carbonyl Compounds. *Angewandte Chemie. International Edition*, Vol.50, pp. 6119-6122, ISSN 1521-3773

Pillai, U.R., & Sahle-Demessie, E. (2002). Selective Oxidation of Alcohols in Gas Phase Using Light-Activated Titanium Dioxide, *Journal of Catalysis*, Vol. 211, pp. 434-444, ISSN 0021-9517

Pouretedal, H.R., Norozi, A., Keshavarz, M.H., Semnani, A. (2009). Nanoparticles of zinc sulfide doped with manganese, nickel and copper as nanophotocatalyst in the degradation of organic dyes. *Journal of Hazardous Materials* Vol. 162, pp. 674-681, ISSN 0304-3894

Radhakrishnan,C., Lo, M.K.F., Warrier, M.V, Garcia-Garibay, M., A., & Monbouquette, H.G. (2006). Photocatalytic Reduction of an Azide-Terminated Self-Assembled Monolayer Using CdS Quantum Dots. *Langmuir*, Vol.22, pp. 5018-5024, ISSN 1520-5827

Ravelli, D., Dondi, D., Fagnoni, M., & Albini, A. (2009). Photocatalysis. A multi-faceted concept for green chemistry. *Chemical Society Reviews*, Vol. 38, pp. 1999-2011, ISSN 0306-0012

Ravelli, D., Albini, A., & Fagnoni, M. (2011). Smooth Photocatalytic Preparation of 2-Substituted 1,3-Benzodioxoles. *Chemistry - A European Journal*, Vol.17, pp. 572-579, ISSN 1521-3765

Reber, J.F., & Rusek, M. (1986). Photochemical Hydrogen Production with Platinized Suspensions of Cadmium Sulfide and Cadmium Zinc Sulfide Modified by Silver Sulfide. *The Journal of Physical Chemistry*, Vol. 90, pp. 824-834, ISSN 0022-3654

Sakai, N., Fujishima, A., Watanabe, T., & Hashimoto, K.(2003). Quantitative Evaluation of the Photoinduced Hydrophilic Conversion Properties of TiO₂ Thin Film Surfaces by the Reciprocal of Contact Angle. *The Journal of Physical Chemistry B*, Vol. 107, pp. 1028-1035, ISSN 1520-5207

Selloni, A. (2008). Crystal growth: Anatase shows its reactive side, *Nature Materials*, Vol. 7, pp. 613-615, ISSN 1476-1122

Selvam, K., & Swaminathan, M. (2010). Au-doped TiO₂ nanoparticles for selective photocatalytic synthesis of quinaldines from anilines in ethanol, *Tetrahedron Letters*, Vol. 51, pp. 4911–4914, ISSN 0040-4039

Selvam, K., Krishnakumar, B., Velmurugan, R., & Swaminathan, M. (2009). A simple one pot nano titania mediated green synthesis of 2-alkylbenzimidazoles and indazole from aromatic azides under UV and solar light, *Catalysis Communications*, Vol. 11, pp. 280–284, ISSN 1566-7367

Shiraishi, Y., & Hirai, T. (2008). Selective organic transformations on titanium oxide-based photocatalysts, *Journal of Photochemistry and Photobiology C: Photochemistry Reviews*, Vol. 9, pp. 157–170, ISSN 1389-5567

Shiraishi, Y., Sugano, Y., Tanaka, S., & Hirai, T. (2010). One-Pot Synthesis of Benzimidazoles by Simultaneous Photocatalytic and Catalytic Reactions on Pt@TiO₂ Nanoparticles, *Angewandte Chemie International Edition*, Vol. 49, pp. 1656 –1660, ISSN 1521-3773

Sivaranjani, K., & Gopinath, C.S. (2011). Porosity driven photocatalytic activity of wormhole mesoporous TiO₂₋ₓNₓ in direct sunlight, *Journal of Materials Chemistry*, Vol. 21, pp. 2639-2647, ISSN 1364-5501

Skorb, E.V., Antonouskaya, L.I., Belyasova, N.A., Shchukin, D.G., Möhwald, H., & Sviridov D.V. (2008). Antibacterial activity of thin-film photocatalysts based on metal-modified TiO₂ and TiO₂:In₂O₃ nanocomposite. *Applied Catalysis B: Environmental*, Vol. 84, pp. 94-99, ISSN 0926-3373

Subba Rao, K. V., & Subrahmanyam, M. (2002). Synthesis of 2-Methylpiperazine by Photocatalytic Reaction in a Non-Aqueous Suspension of Semiconductor-Zeolite Composite Catalysts. *Photochemical & Photobiological Sciences*, Vol.1, pp. 597-599, ISSN 1474-905X

Sunada, K., Kikuchi, Y., Hashimoto, K., & Fujishima, A. (1998). Bactericidal and Detoxification Effects of TiO₂ Thin Film Photocatalysts. *Environmental Science & Technology*, Vol. 32, pp. 726–728, ISSN 1520-5851

Sunada, K., Watanabe,†T., & Hashimoto, K. (2003). Bactericidal Activity of Copper-Deposited TiO₂ Thin Film under Weak UV Light Illumination. *Environmental Science & Technology*, Vol. 37, pp. 4785-4789, ISSN 1520-5851

Tanaka, T., Ooe, M., Funabiki, T., & Yoshida, S. (1986). Formation of an Epoxide Intermediate in the Photo-Oxidation of Alkenes over Silica-Supported Vanadium Oxide. *Journal of the Chemical Society, Faraday Transactions 1: Physical Chemistry in Condensed Phases*, Vol.82, pp. 35-43, ISSN 0956-5000

Tanaka, T., Takenaka, S., Funabiki, T., & Yoshida, S. (2000). Photocatalytic Oxidation of Alkane at a Steady Rate Over Alkali-Ion-modified Vanadium Oxide Supported on Silica. *Catalysis Today*, Vol.61, pp.109-115, ISSN 0920-5861

Tani, M., Sakamoto, T., Mita, S., Sakaguchi, S., & Ishii, Y. (2005). Hydroxylation of Benzene to Phenol under Air and Carbon Monoxide Catalyzed by Molybdovanadophosphoric Acid. *Angewandte Chemie International Edition*, Vol. 44, pp. 2586–2588, ISSN: 1521-3773

Tlili, A., Xia, N., Monnier, F., & Taillefer, M. (2009). A Very Simple Copper-Catalyzed Synthesis of Phenols Employing Hydroxide Salts, *Angewandte Chemie International Edition*, Vol. 48, pp. 8725–8728, ISSN: 1521-3773

Tricoli, A., Righettoni, M., & Pratsini, S.E. (2009). Anti-Fogging Nanofibrous SiO_2 and Nanostructured SiO_2–TiO_2 Films Made by Rapid Flame Deposition and In Situ Annealing. *Langmuir*, Vol. 25, pp. 12578–12584, ISSN 1520-5827

Ullah, R., & Dutta, J. (2008). Photocatalytic degradation of organic dyes with manganese-doped ZnO nanoparticles. *Journal of Hazardous Materials* Vol. 156, pp. 194-200, ISSN 0304-3894

Wang, R., Hashimoto, K., Fujishima, A., Chikuni, M., Kojima, E., Kitamura, A., Shimohigoshi, M., & Watanabe, T. (1997). Light-induced amphiphilic surfaces. *Nature*, Vol. 388, pp. 431-432, ISSN 0028-0836

Ward, D., White, J.R., & Bard, A.J. (1983). Electrochemical Investigation of the Energetics of Particulate Titanium Dioxide Photocatalysts. The Methyl Viologen- Acetate System. *Journal of the American Chemical Society*, Vol. 105, pp. 27-31, ISSN 1520-5126

Warrier, M., Lo, M. K. F., Monbouquette, H., & Garcia-Garibay, M. A. (2004). Photocatalytic Reduction of Aromatic Azides to Amines Using CdS and CdSe Nanoparticles. *Photochemical &Photobiological Sciences*, Vol.3, pp. 859-863, ISSN 1474-905X

Watanabe, T. et al. (1992): Proc. 1st Int. Conf. TiO_2 Photocatalyst, ed. H. Al-Ekabi

Weber, M., Weber, M., & Kleine-Boymann, M. (2005). Phenol, In: *Ullmann's Encyclopedia of Industrial Chemistry*, Wiley-VCH, ISBN: 9783527306732, Weinheim (Germany)

Weissermel, K., & Arpe, H.J. (1997). *Industrial organic chemistry* (3rd edition), Wiley-VCH, ISBN 3-527-28838-4 Gb, Weinheim (Germany)

White, J.M., Szanyi, J., & Henderson, M.A. (2003). The Photon-Driven Hydrophilicity of Titania: A Model Study Using TiO_2(110) and Adsorbed Trimethyl Acetate. *The Journal of Physical Chemistry B* , Vol. 107, pp. 9029-9033, ISSN 1520-5207

White, J.R., & Bard, A.J. (1985). Electrochemical Investigation of Photocatalysis at CdS Suspensions in the Presence of Methylviologen. *The Journal of Physical Chemistry*, Vol. 89, pp. 1947-1954, ISSN 0022-3654

Yang, C-C., Yu, Y-H., von der Linden, B., Wu, J.C.S., & Mul, G. (2010). Artificial Photosynthesis over Crystalline TiO_2-Based Catalysts: Fact or Fiction?. *Journal of the American Chemical Society*, Vol.132, pp. 8398-8406, ISSN 1520-5126

Yoshida, H., Murata, C., & Hattori, T. (1999). Photooxidation of Propene to Propene Oxide by Molecular Oxygen over Zinc Oxide Dispersed on Silica. *Chemistry Letters*, Vol.28, pp. 901, ISSN 0366-7022

Zhang, G., Yi, J., Shim, J., Leea, J., & Choi, W. (2011). Photocatalytic hydroxylation of benzene to phenol over titanium oxide entrapped into hydrophobically modified siliceous foam, *Applied Catalysis B: Environmental*, Vol. 102, pp. 132–139, ISSN 0926-3373

Zhang, Z., Hossain, Md F., & Takahashi, T. (2010). Photoelectrochemical water splitting on highly smooth and ordered TiO$_2$ nanotube arrays for hydrogen generation. *International Journal of Hydrogen Energy*, Vol. 35, pp. 8528-8535, ISSN 0360-3199

Zheng, Z., Huang, B., Qin, X., Zhang, X., Dai, Y., & Whangbo, M-H. (2011). Facile in situ synthesis of visible-light plasmonic photocatalysts M@TiO$_2$ (M=Au, Pt, Ag) and evaluation of their photocatalytic oxidation of benzene to phenol, *Journal of Materials Chemistry*, Vol. 21, pp. 9079-9087, ISSN 1364-5501

Zou, Y-Q., Lu L-Q., Fu, L., Chang, N.J., Rong, J., Chen, J-R., & Xiao, W-J. (2011) Visible-Light-Induced Oxidation/[3+2] Cycloaddition/Oxidative Aromatization Sequence: A Photocatalytic Strategy to Construct Pyrrolo[2,1-α]isoquinolines. *Angewandte Chemie International Edition*, Vol.50, pp. 1-6, ISSN 1521-3773

Zubkov, T., Stahl, D., Thompson, T.L., Panayotov, D., Diwald, O., & Yates, J.T. Jr. (2005). Ultraviolet Light-Induced Hydrophilicity Effect on TiO$_2$(110)(1×1). Dominant Role of the Photooxidation of Adsorbed Hydrocarbons Causing Wetting by Water Droplets. *The Journal of Physical Chemistry B*, Vol. 109, pp. 15454-15462, ISSN 1520-5207

5

Manganese Compounds as Water Oxidizing Catalysts in Artificial Photosynthesis

Mohammad Mahdi Najafpour

Chemistry Department, Institute for Advanced Studies in Basic Sciences (IASBS), Zanjan,
Iran

1. Introduction

Artificial photosynthesis is an umbrella term but it could be introduced as a research field that attempts to mimic the natural process of photosynthesis and uses sunlight to oxidizing and reducing different compounds. In this process, we could assume water as one of the compounds that could be reduced and (or) oxidized to hydrogen and (or) oxygen, respectively. Water splitting is the general term for a chemical reaction in which water is decomposed to oxygen and hydrogen (Pace, 2005).

Production of hydrogen fuel from electrolysis of water would become a practical strategy if we could find a "super catalyst" for water oxidation reaction (Bockris, 1977). Super catalyst means a stable, low cost, efficient and environmentally friendly catalyst. The water oxidation half reaction in water splitting is overwhelmingly rate limiting and needs high over-voltage (~1V) that results the low conversion efficiencies when working at current densities required, also at this high voltage, other chemicals will be oxidized and this would be environmentally unacceptable for large-scale hydrogen production (Bockris, 1977). Thus, a significant challenge in the sustainable hydrogen economy is to design a water oxidizing catalyst.

2. Water Oxidizing Catalysts in artificial photosynthesis

In past few years, there has been a tremendous surge in research on the synthesis of various metal compounds aimed at simulating water oxidizing complex (WOC) of photosystem II (PSII) (Liu et al., 2008; Kanan, M.W. & Nocera, 2008; Cady et al., 2008; Yagi & Kaneko, 2001; Ruttinger & Dismukes, 1997). Of these materials, the Co, Ru and Ir compounds have been shown to be an effective catalyst for water oxidation. However, most of the compounds are expensive and often relate to potentially carcinogenic salts.

Particular attention has been given to the manganese compounds aimed at simulating the WOC of PSII (Umena et al., 2011) not only because it has been used by *Nature* to oxidize water but also because manganese is cheap and environmentally friendly. In this chapter we consider manganese compounds as structural or (and) functional models for the WOC of PSII

2.1 Structural models for biological Water Oxidizing Complex

There are many mono, di, tri and tetra nuclear manganese complexes as structural models for the WOC in PSII (Mullins & Pecoraro, 2008).

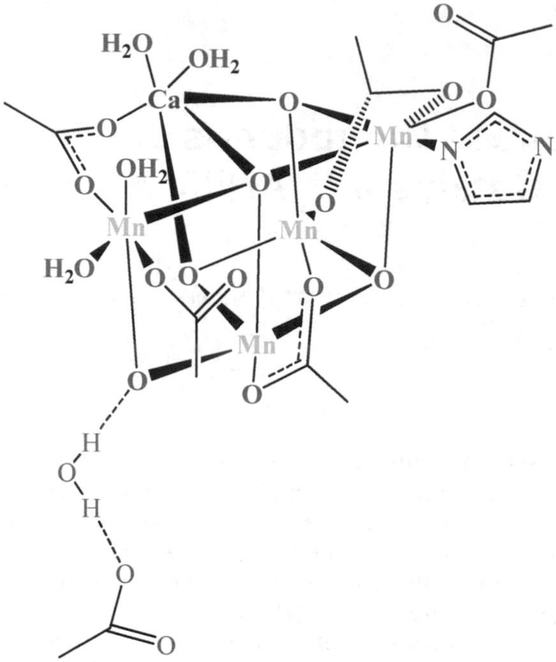

Fig. 1. The WOC and the localization of the substrate water binding sites on the WOC (Umena et al., 2011).

The WOC in PSII is a tetranuclear manganese complex (Fig.1) (Umena et al., 2011). However, mononuclear models are useful and simple complexes for the isolation of high-valent complexes. Regarding these mononuclear manganese complexes, we could obtained many information about spectroscopic properties of Mn(V) compounds as there are no Mn(V) synthetic examples of dinuclear or higher nuclearity structures that have been crystallographically characterized (Fig. 2.) (Mullins & Pecoraro, 2008).

Water and terminal hydroxo ligands in Mn(IV) complexes are very important as there are suggested as one of the substrates for oxygen production in the WOC of PS II. Busch and coworkers reported the first structurally characterized example of a mononuclear Mn (IV) complex with two terminal hydroxo ligands (Fig. 2) (Yin et al., 2006). Using the pH titration of aqueous solutions of the complex, it revealed two acid–base equilibria with pK_1 = 6.86 and pK_2 = 10, the latter apparently being associated with dimer formation. The complex has shown as a catalyst in olefin epoxidation and hydrogen atom abstraction reactions (Yin et al., 2006).

The Mn(V) = O are very important and it has been proposed as an intermediate in Natural water oxidation. Miller et al have reported the first structurally characterized Mn(V) = O complex (Fig. 2a) (Miller et al., 1998). The complex could not oxidize water but give to us important spectroscopic information of Mn(V) = O group that proposed as an intermediate in biological water oxidation (Miller et al. 1998). However, as considered by Pecoraro, these synthetic Mn(V) = O complexes are stabilized by special ligand(s) and the activity of Mn(V) = O could be completely different from Mn(V) = O of the WOC in PSII (Mullins & Pecoraro, 2008).

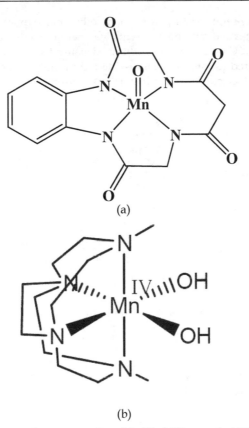

(a)

(b)

Fig. 2. Schematic structures of a mononuclear Mn(V) (Miller et al., 1998) (a) and a Mn(IV) complex terminal hydroxo ligands (b) (Yin et al., 2006).

Many dinuclear manganese complexes have been studied with different oxo bridge (μ-O) as possible models for the WOC. The relationship between Mn-Mn distance and Mn–O–Mn angle has been considered for such complexes (Mullins & Pecoraro, 2008).

In 1988, there was suggestion that WOC had a mononuclear manganes center in close proximity to a trinuclear center and this suggestion was emerged basedon magnetic, spectroscopic, and crystallographic studies in 2000 (Mullins & Pecoraro, 2008). Regarding this issue, study of magnetic, spectroscopic, and crystallographic properties of trinuclear clusters has re-emerged and many trinuclear manganese complexes with linear or bent structure and different oxidation states for manganese ions have been synthesized and characterized (Mullins & Pecoraro, 2008).

Tetra nuclear manganese complexes are very interesting group of complexes as they could be studied as structural models for the WOC. These complexes show different structures regarding manganese ions from linear to cubane (Mullins & Pecoraro, 2008) (Fig. 3).

In 2005, Christou and co-workers have reported the first high oxidation state manganese-calcium cluster. The structure contains [Mn$_4$CaO$_4$] sub-units similar to that found in the WOC in PSII (Fig. 4) (Mishra et al., 2005). Recently, Kotzabasaki et al. (2011) synthesized a heterometallic polymeric complex {[Mn$^{III}_6$Ca$_2$O$_2$(Me-saO)$_6$(prop)$_6$(H$_2$O)$_2$].2MeCN.0.95H$_2$O}$_n$

(prop = propionate; Me-saOH$_2$ = 2-hydroxyphenylethanone oxime) (Fig. 4). Nayak et al. (2011) have also synthesized two new polynuclear heterometallic cluster complexes with [Mn^{III}3M^{II}Na] (M= Mn, Ca) core were synthesized using two in situ formed Schiff bases. This compound appeared to catalyse water oxidation in the presence of NaOCl which was followed by using Clark electrode and online mass spectrometry (Nayak et al., 2011).

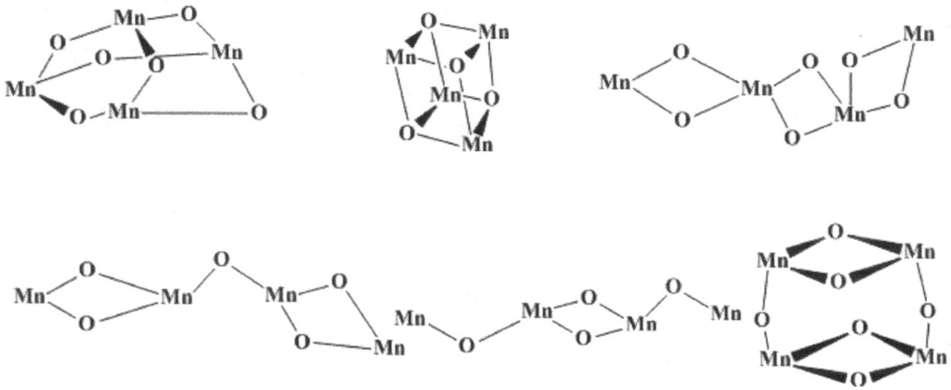

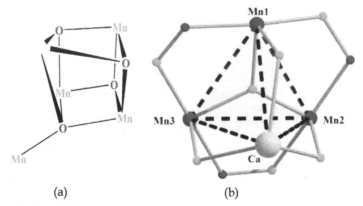

Fig. 3. Different core types observed in Mn-oxo tetramers.

Fig. 4. The first high oxidation state manganese–calcium cluster reported by Christou and co-workers. The structure contains [Mn$_4$CaO$_4$] sub-units similar to that found in the WOC in PSII (a). The heterometallic polymeric complex {[Mn^{III}$_6$Ca$_2$O$_2$(MesaO)$_6$(prop)$_6$(H$_2$O)$_2$]·2MeCN·0.95H$_2$O}$_n$ reported by Kotzabasaki et al. (2011) (b) (figure was reproduced from Kotzabasaki et al. (2011)).

In 1997, Styring's group reported a molecule containing a sensitizer covalently linked to a manganese complex (Fig. 5). In this compound, the ruthenium chromophore could donate an electron to an external acceptor and consequently oxidize a coordinated manganese ion with rate constants ~ 50 ns to 10 μs. Then this group showed that how a Ru–Tyr molecular dyad could be used to power the light driven oxidation of a dinuclear Mn$_2$^{III,III} complex (Lomoth et al., 2006). Similar to PSII, it was shown that upon light absorption, a chain reaction of first electron transfer could be achieved whereby the Ru(III) species, obtained in presence of an electron acceptor, was quenched through an intermolecular electron transfer leading to the formation of the tyrosyl radical.

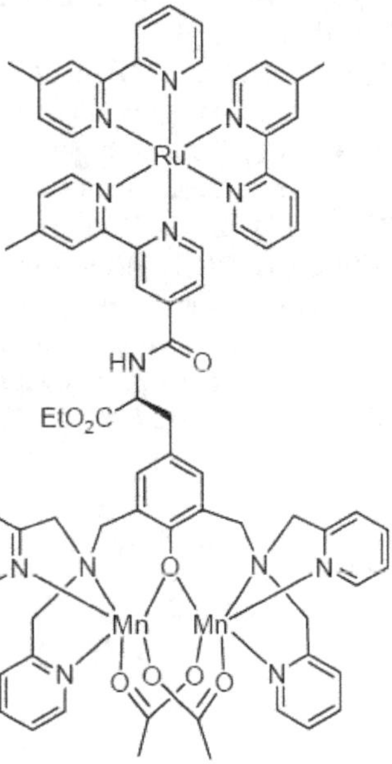

Fig. 5. Molecule containing a sensitizer covalently linked to a manganese complex.

2.2 Functional models for biological Water Oxidizing Complex

There are a few manganese complexes that produce oxygen in the presence of different oxidants (Cady et al., 2008; Yagi & Kaneko, 2001; Ruttinger & Dismukes 1997). It is important to know that even if the oxidants may do donate O-atoms that end up in the product oxygen, can nevertheless play an important role in identifying potentially useful catalytic components for water photolysis and electrolysis.

2.2.1 Terpy and bpy manganese complexes

It is shown that $[(bpy)_2Mn^{III}(\mu\text{-}O)_2Mn^{IV}(bpy)_2]^{3+}$ and $[(OH_2)(terpy)Mn^{III}(\mu\text{-}O)_2Mn^{IV}(terpy)(OH_2)]^{3+}$ (bpy: 2,2'-Bipyridine; terpy: 2,2';6',2''-terpyridine) (Tagore et al., 2008; Yagi & Narita, 2004) have water oxidation activity in the presence of $(NH_4)_2[(Ce(NO_3)_6]$, NaClO and KHSO$_5$. Yagi and Narita (2004) observed when comparable amount of $[(bpy)_2Mn^{III}(\mu\text{-}O)_2Mn^{IV}(bpy)_2]^{3+}$ or $[(OH_2)(terpy)Mn^{III}(\mu\text{-}O)_2Mn^{IV}(terpy)(OH_2)]^{3+}$ were adsorbed onto Kaolin clay, the addition of a large excess of $(NH_4)_2[(Ce(NO_3)_6]$ to its aqueous suspension produced a significant amount of oxygen. The rate of oxygen evolution increased linearly with the amount of $[(bpy)_2Mn^{III}(\mu\text{-}O)_2Mn^{IV}(bpy)_2]^{3+}$ indicating unimolecular oxygen evolution in contrast with bimolecular catalysis of $[(OH_2)(terpy)Mn^{III}(\mu\text{-}O)_2 Mn^{IV}(terpy)(OH_2)]^{3+}$.

The unimolecular oxygen evolution might be explained by either O-O coupling of di-μ-O bridges or attack of outer-sphere water onto a μ-O bridge in high oxidation species, probably including μ-O- radical bridges (Yagi & Narita, 2004). When [(bpy)$_2$Mn^{III}(μ-O)$_2$Mn^{IV}(bpy)$_2$]$^{3+}$ was dissolved in water containing excess (NH$_4$)$_2$[(Ce(NO$_3$)$_6$], no evolution of gas was observed, and analysis of the gas phase confirmed that no oxygen was formed; however oxygen evolution   was observed when these complex as a solid was added to a (NH$_4$)$_2$[(Ce(NO$_3$)$_6$] solution (Yagi & Narita, 2004). Isotopic labeling of the solvent water showed that indeed water was oxidized (Yagi & Narita, 2004). It was also reported that [(OH$_2$)(terpy)Mn^{III}(μ-O)$_2$Mn^{IV}(terpy)(OH$_2$)]$^{3+}$ has oxygen evolution activity in the presene of KHSO$_5$ or NaOCl, as primary oxidants. A proposal mechanism was reported by Brudvig group for the reaction (Fig. 6) (Cady et al., 2008). Recently, Brudvig's group have reported the attachment of [(OH$_2$)(terpy)Mn^{III}(μ-O)$_2$Mn^{IV}(terpy)(OH$_2$)]$^{3+}$ onto TiO$_2$ nanoparticles via direct adsorption. The resulting surface complexes were characterized by EPR and UV-visible spectroscopy, electrochemical measurements and computational modeling. Their results showed that the complex attaches to near-amorphous TiO$_2$ by substituting one of its water ligands by the TiO$_2$, as suggested by EPR data (Li et al. 2009).

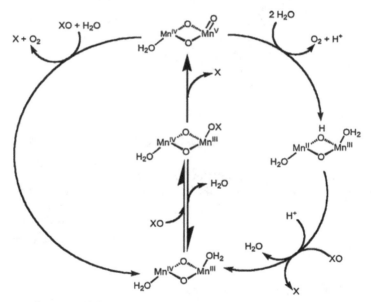

Fig. 6. Schematic diagram of the proposed mechanism of water oxidation by [(OH$_2$)(terpy)Mn^{III}(μ-O)$_2$Mn^{IV}(terpy)(OH$_2$)]$^{3+}$ (figure was reproduced from Cady et al. (2008)).

2.2.2 Schiff base complex

A number of manganese (III) complexes of the type [{MnL(H$_2$O)}]$^{2+}$ (L = dianion of O,N,tetradentate Schiff base), in aqueous solution, have been shown to liberate dioxygen and reduce p-benzoquinone to hydroquinone when irradiated with visible light (Ashmawy et al., 1985). The photoactivity is critically dependent on the structure of the ligand, the complex [{Mn(salpd) (H$_2$O)}][ClO$_4$], (salpd = propane-l,3-diylbis(salicylideneiminate)) being the most active (Ashmawy et al., 1985). All the active complexes exhibit a band at 590 nm in the electronic spectrum, which is absent for the inactive complexes (Ashmawy et al., 1985). The rate of dioxygen evolution is dependent on the manganese (III) complex (first order) and quinone concentrations (order, 0.5) and the pH of the reaction medium, but is independent of solvent. The activation energy for dioxygen evolution in [{Mn(salpd) (H₂O)}][ClO₄], is 80 kJ mol⁻¹, and the evidence points to homolytic, rather than heterolytic, fission of water (Ashmawy et al., 1985). The wavelength dependence of the reaction rate shows a maximum for photolysis in the 450-600 nm region where the quinone does not absorbed (proposal mechanism is shown in Fig. 7) (Ashmawy et al., 1985).

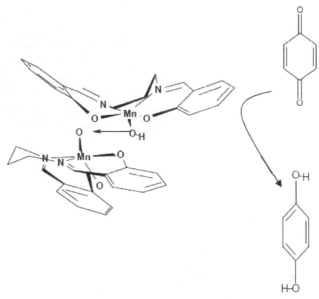

Fig. 7. Proposal mechanism for water oxidation by reduce p-benzoquinone to Hydroquinone (Ashmawy et al., 1985).

Oxygen evolution was also observed upon mixing solid manganese(III) bidentate Schiff base complexes with aqueous solutions of (NH₄)₂[(Ce(NO₃)₆] (Najafpour & Boghaei, 2009). However, oxygen evolution was not observed upon mixing solutions of the complexes (in acetonitrile) with Ce(IV). Electron-withdrawing substituents on the Schiff base ligands (NO₂, Br) enhanced the reactivity of the manganese complexes toward oxygen evolution. Oxygen evolution was also affected by R groups on the ligands, in the order Me>Et > Bz (Najafpour & Boghaei, 2009).

It was reported that hydrogen peroxide was produced when HClO₄ was added in stoichiometric amounts to solutions of the Mn(IV) Schiff base dimer [Mn₂(IV)(BuSalen)₂(O)₂], H₂O in acetone at 0 °C (Fig. 8) (Boucher & Coe 1975). The complex was converted in 62% yield to the Mn(III) monomer from which the dimer was initially synthesized by air oxidation in chloroform solution. However, Pecoraro suggested that as [Mn₂ᴵⱽ(BuSalen)₂(O)₂], H₂O has catalase activity, if hydrogen peroxide was produced, dioxygen would evolve from solution (Pecoraro et al, 1994).

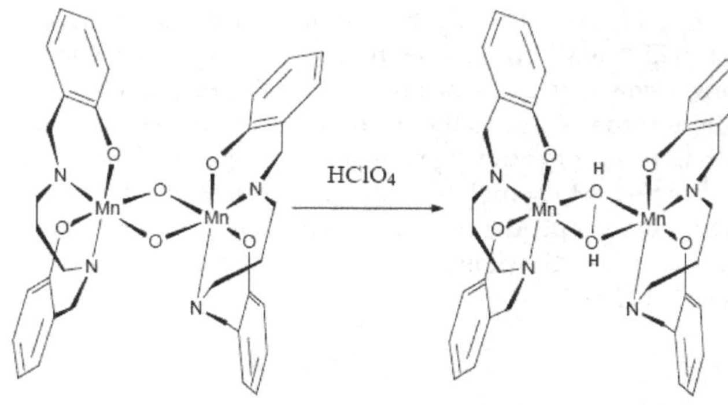

Fig. 8. Proposal mechanism for hydrogen peroxide producing when HClO$_4$ was added in stoichiometric amounts to solutions of the Mn(IV) Schiff base dimmer (Boucher & Coe 1975).

Fujiwara et al. have reported the preparation and characterization of a series of dichloromanganese (IV) Schiff base complexes (Fujiwara et al., 1985). They have shown that the manganese(IV) complex dichlorobis(N-R-3-nitrosalicylideneaminato) manganese(IV) reacts with water to liberate molecular oxygen (Fig. 9) (Fujiwara et al., 1985).

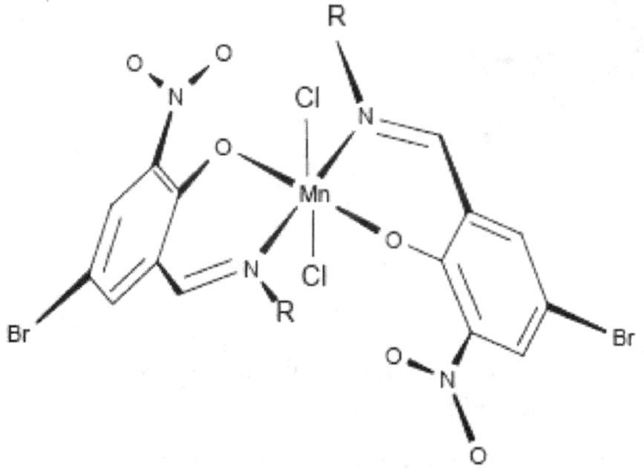

Fig. 9. Structure of the complex *trans*-Mn(IV)L$_2$Cl$_2$ (L = *N*-alkyl-3-nitrosalicylimide) (Fujiwara et al., 1985).

Absorbed spectrometry using an alkaline pyrogallol solution and measurement of dissolved oxygen by an oxygen electrode were employed to detect and determine dioxygen liberated during the reaction of manganese(IV) complexes with water (Fujiwara et al., 1985). It could be seen that the reactivity is affected by the alkyl groups of the complexes: the reaction with water is retarded in order of (mol of O$_2$ per mol of complex) n-C$_3$H$_7$ (0.27)< n-C$_8$H$_{17}$ (0.2)< n-C$_{12}$H$_{25}$ (0.12). These results indicate that the long-chain alkyl groups such as n-C$_8$H$_{17}$ and n-C$_{12}$H$_{25}$ can protect the central manganese(IV) ion from attack by water molecules (Fujiwara et al., 1985). This may arise from hydrophobicity of these groups. In other words, the reactivity of the manganese(IV) complexes with water can be controlled by the choice of alkyl groups. Also they have found that the pH values of reaction decrease in the course of the reaction of the manganese(IV) complexes in the presence of water (without any buffer solution) (Fujiwara et al., 1985). Using ^{18}O-labeled water showed that indeed water was oxidized in this reaction (Fujiwara et al., 1985).

2.2.3 Porphyrin complexes

Shimazaki et al. (2004) have reported dimanganese complexes of dimeric tetraarylporphyrins linked by 1,2-phenylene bridge (Fig. 10). The catalyst can oxidize olefins such as cyclooctene to form epoxide with stiochiometric amount of m-chloroperbenzoic acid. It is proposed that the oxidation of a dimanganese (III) tetraarylporphyrin dimer could give the corresponding high valent Mn(V)=O complex, which is the active species in these oxidation. They reported on the oxidation of the dimanganese porphyrin dimer by employing meta-Chloroperoxybenzoic acid as an oxidant, and the characterization of the resulting Mn(V)=O species by spectroscopic methods.

Furthermore, oxygen evolution was observed from the Mn(V)=O species when a small excess of trifluoromethanesulfonic acid was added (Shimazaki et al., 2004). Mn(V)=O was detected by EPR, UV/VIS, and Raman spectrum (Shimazaki et al., 2004).

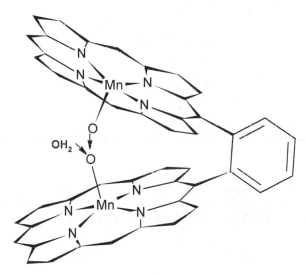

Fig. 10. Dimeric tetraarylporphyrins linked by 1,2-phenylene bridge as a model for the WOC in PSII (Shimazaki et al., 2004).

2.2.4 Cubane like model

Several types of experimental evidence have demonstrated that the synthetic complexes $Mn_4O_4(O_2PR_2)_6$, R = Ph and 4–MePh, containing the $[Mn_4O_4]^{6+}$ core surrounded by six facially bridging bidentate phosphinate anions, produce dioxygen following removal of one phosphinate ligand to form the reactive butterfly complex $[Mn_4O_4(O_2PR_2)_5]$ (Fig. 11) (Maniero et al., 2003).

Dissociation of a phosphinate ligand is achieved using light absorbed by a charge transfer O-Mn transition, producing dioxygen in high quantum high yield (46–100%) (Maniero et al., 2003).

The redox potential of $[Mn_4O_4(O_2PR_2)_6]/[Mn_4O_4(O_2PR_2)_6]^+$ is 1.38 V vs NHE, which is considerably greater than those found for the dimanganese(III,IV)/(III,III) couple and the majority of known (IV,IV)/(III,IV) couples. The $[Mn_4O_4(O_2PR_2)_6]$ cubane complex reacted with the hydrogen-atom donor, phenothiazine in a CH_2Cl_2 solution, forming $[Mn_4O_4(O_2PR_2)_6]$ and $[Mn_4O_4(O_2PR_2)_6]^+$ as well as releasing two water molecules from the core. This result shows that two of the corner oxos of the cubane can be converted into two labile water molecules. The evolution of oxygen molecule from Mn_4O_4 cubane core was corroborated by the detection of $^{18}O_2$ from $[Mn_4O_4(O_2PR_2)_6]$ (Maniero et al., 2003).

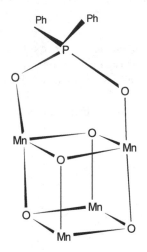

Fig. 11. Synthetic complexes Mn_4O_4 $(O_2PR_2)_6$, R = Ph and 4–MePh (only one O_2PR_2 is shown).

2.2.5 Mn(II) complexes of monoanionic pentadentate ligands

McKenzie's group (Seidler-Egdal et. al., 2011) reported that tert-Butyl hydroperoxide oxidation of Mn(II) complexes of **1** (Fig. 12), in large excesses of the tert-Butyl hydroperoxide,

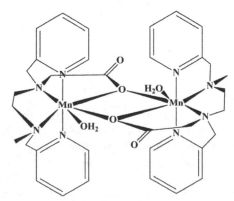

Fig. 12. The Mn(II) complex (**1**) of monoanionic pentadentate ligands reported by McKenzie's group.

is concurrent with an oxygen evolution with turnovers of up to 10^4 mol of oxygen per mol of [Mn] and calculated rate constants from two series of experiments of 0.039 and 0.026 mol [O_2] s^{-1} M^{-2}. A 1:1 reaction of tert-Butyl hydroperoxide with [Mn] is rate determining and the resultant species is proposed to be the mononuclear, catalytically competent, Mn(IV)=O (Seidler-Egdal et. al., 2011).

2.2.6 Manganese Oxides and Hydroxides

Oxides and Hydroxides of transition metals cations like Fe(III), Co(III), Mn(III), Ru(IV), and Ir(IV) appear to be efficient catalysts for water oxidation in the presence of Ce(IV), $S_2O_8^{-2}$ and Fe(bpy)$_3^{3+}$ as oxidants.

Shilov and Shafirovich in 1965 have shown that colloidal MnO_2 catalyzes the oxidation of water to dioxygen in the presence of strong oxidants like Ce(IV) and Ru(bpy)$_3^{3+}$) (Shilov & Shafirovich, 1979 (translation)). Suggested mechanism is shown in Scheme 1.

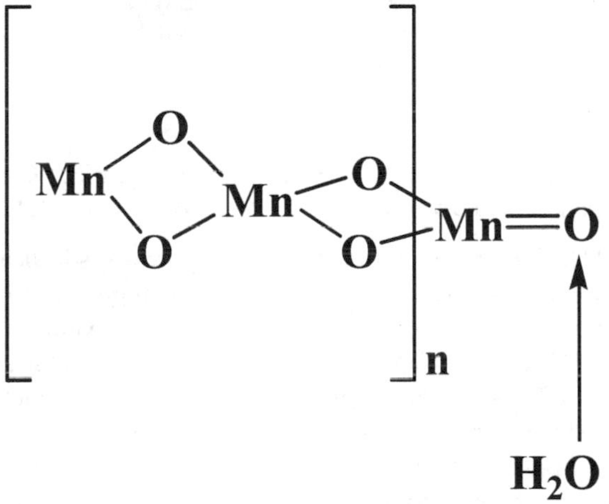

Scheme 1. Suggested mechanism for water oxidation by Oxides and Hydroxides of transition metals.

Recently, we introduced amorphous calcium - manganese oxide as efficient and biomimetic catalysts for water oxidation (Najafpour et al., 2010). These oxides are very closely related to the WOC in PSII not only because of similarity in the elemental composition and oxidation number of manganese ions but also because of similarity of structure and function (Najafpour 2011a,b; Zaharieva et al., 2011) (Fig. 13).

The structure of these amorphous powders have been evaluated, using extended-range X-ray absorption spectroscopy (XAS), X-ray absorption near-edge structure (XANES) and Extended X-Ray Absorption Fine Structure (EXAFS) (Zaharieva et al., 2011). These results reveal similarities between the amorphous powders and the water oxidizing complex of PSII. Two different Ca-containing motifs were identified in these amorphous manganese – calcium oxides (Zaharieva et al., 2011). One of them results in the formation of Mn$_3$Ca cubes, as also proposed for the WOC of PSII. Other calcium ions likely interconnect oxide-layer fragments. It was concluded that these readily synthesized manganese-calcium oxides are the closest structural and functional analogs to the native the WOC of PSII found so far

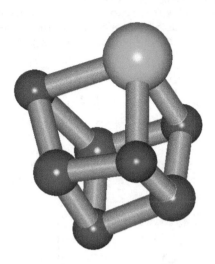

Fig. 13. Motif of edge-sharing MnO_6 octahedra (di-μ-oxido bridging) in the manganese – calcium oxides (manganese: blue, oxygen: red and calcium: green) (Zaharieva et al., 2011).

(Zaharieva et al., 2011). Oxygen evolution formation pathways indicated by ^{18}O-labelling studies showed that water oxidation by these layer manganese oxides in presence of cerium (IV) ammonium or $[Ru(bpy)_3]^{+3}$ is a "real" water oxidation reaction and both oxygen atoms of formed dioxygen molecules originated from water (Shevela et al., 2011). In continuation of our efforts to synthesize an efficient and biomimetic catalyst for water oxidation, we synthesized nano - size amorphous calcium - manganese oxides that are very closely related to the WOC in PSII not only because of similarity in the elemental composition, oxidation number of manganese ions and similarity of structure and function to Mn_4O_5Ca cluster in PSII but also because of nearer on the of catalyst particle size as compared to previously reported micro - size amorphous calcium manganese oxides (Fig. 14.).

2.2.7 Genetic engineering models
Wydrzynski and co-workers, in a different approach, introduced a reverse engineering approach to build a simple, light-driven photo-catalyst based on the organization and function of the donor side of the PSII reaction centre and a bacterioferretin molecule is being 'coaxed' using genetic engineering to include the chromophores for light absorption and Mn complexes for water oxidation; they could observed oxidation of manganese in this engineered system (Conlan et al., 2007). Recently, Nam and co-workers imtroduced a biologically templated nanostructure for visible light driven water oxidation that uses a genetically engineered virus scaffold to mediate the co-assembly of zinc porphyrins (photosensitizer) and iridium oxide clusters (water oxidizing catalyst) (Nam et al., 2010). Their results suggested that the biotemplated nanoscale assembly of functional components is a promising route to improved photocatalytic water-splitting systems.

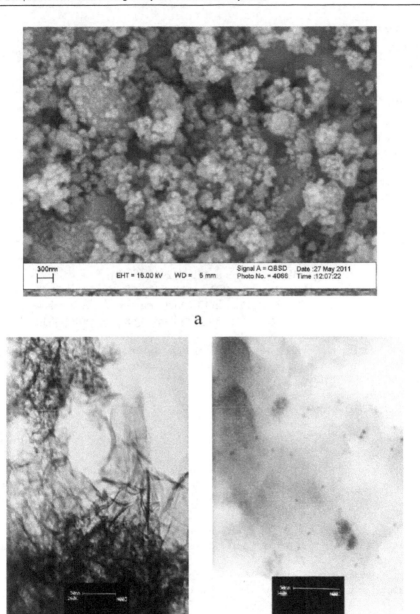

Fig. 14. Scanning electron microscope (SEM) (a) and transmission electron microscopy (TEM) (b,c) images of amorphous calcium - manganese oxides. These readily synthesized manganese-calcium oxides are the closest structural and functional analogs to the native the WOC of PSII found so far.

3. Acknowledgment

The author is grateful to Institute for Advanced Studies in Basic Sciences for financial support.

4. References

Ashmawy, F.T.; McAuliffe, C.A.; Parish, R. V. & Tames, J. (1985) Water Photolysis. Part 1. The Photolysis of Co-ordinated Water in [{Mn(salpd) (H₂O)}][ClO₄], (L = Dianion of Tetradentate O,N,-Donor Schiff Bases). A Model for the Manganese Site in Photosystem II of Green Plant Photosynthesis, J. Chem. Soc. Dalton Trans., p.1391-1397

Bockris, J.O.M. (1977) Energy-the solar hydrogen alternative, Wiley&Sons, New York.

Boucher, L.J. & Coe, C.G. (1975) Manganese-Schiff base complexes. VI. Synthesis and spectroscopy of aquo [N,N'-ethylenebis(4-sec-butylsalicylaldiminato)]manganese (III) perchlorate and μ-dioxo-bis[N,N'-ethylenebis(4-sec butylsalicylaldiminato)] dimanganese(IV) and related N,N'-trimethylenebis(4-sec-butylsalicylaldimine) complexes, Inorg. Chem., v.14 (6), p. 1289–1294

Cady, C.W.; Crabtree, R.H. & Brudvig, G.W. (2008) Functional models for the oxygen-evolving complex of photosystem II, Coord. Chem. Rev., v.252, p.444–455

Conlan, B.; Hillier, W. & Wydrzynski, T. (2007) Engineering model proteins for Photosystem II function, Photosynth. Res., v.94, p. 225-233

Fujiwara, M.; Matsushita, T. & Shono T. (1985) Reaction of dichloromanganese(IV) Schiff-base complexes with water as a model for water oxidation in photosystem II, v4(11), p.1895-1900

Kanan, M.W. & Nocera, D.G. (2008) In situ formation of an oxygen-evolving catalyst in neutral water containing phosphate and Co²⁺, Science, v.321, p.1072-1075

Kotzabasakia, V.; Siczek,M.; Lis T.& Milios, C.J. (2011)The first heterometallic Mn–Ca cluster containing exclusively Mn(III) centers, Inorg.Chem.Commun.,v.14(11), p.213-216

Li, G.; Sproviero &E. M.; Snoeberger III, R.C.; Iguchi, N.; Blakemore, J.D. Crabtree, R.H.; Brudvig, G.W. & Batista, V.S. (2009) Deposition of an oxomanganese water oxidation catalyst on TiO₂ nanoparticles: computational modeling, assembly and characterization, Energy Environ. Sci., v.2, p. 230–238

Liu, F.; Concepcion, J.J.; Jurss, J.W.; Cardolaccia, T.; Templeton, J.L. & Meyer, T.J. (2008) Mechanisms of water oxidation from the blue dimer to photosystem II, Inorg. Chem., v.47, p.1727- 1752

Lomoth, R.; Magnuson, A.; Sjödin, M.; Huang, P.; Styring, S.& Hammarström, L. (2006). Mimicking the electron donor side of Photosystem II in artificial photosynthesis, Photosynth. Res., v.87, p.25–40

Maniero, M.; Ruettinger, W.F.; Bourles, E.; McLendon, G. & Dismukes, G. C. (2003) "Kinetics of proton-coupled electron-transfer reactions to the manganese-oxo "cubane" complexes containing the Mn₄O⁶⁺₄ and Mn₄O⁷⁺₄ core types, PNAS, v.100(7), p.3703-3712

Miller, C.G.; Gordon-Wylie, S.W.; Horwitz, C.P.; Strazisar, S.A.; Peraino, D.K.; Clark, G.R.; Weintraub, S.T. & Collins, T.J. (1998) A Method for Driving O-Atom Transfer:

Secondary Ion Binding to a Tetraamide Macrocyclic Ligand, J. Am. Chem. Soc., v.120 p.11540-11541

Mishra, A.; Wernsdorfer, W.; Abboud, K.A. & Christou, G. (2005) The first high oxidation state manganese–calcium cluster: relevance to the water oxidizing complex of photosynthesis, Chem. Commun., p.54-56

Mullins, C.S. & Pecoraro, V.L. (2008) Reflections on small molecule manganese models that seek to mimicphotosynthetic water oxidation chemistry, Coord. Chem. Rev., v.252, p.416-443

Najafpour, M.M. & Boghaei, D.M. (2009) Heterogeneous water oxidation by bidentate Schiff base manganese complexes in the presence of cerium (IV) ammonium nitrate, Transition Met. Chem., v.34, p.367-372

Najafpour, M.M.; Ehrenberg, T.; Wiechen, M. & Kurz, P. (2010) Calcium -manganese (III) oxides (CaMn$_2$O$_4$•xH$_2$O) as biomimetic oxygen- evolving catalysts, *Angew Chem. Int. Ed.*, v.49, p.2233-2237

Najafpour, M. M. (2011a) Mixed-Valence Manganese Calcium Oxides as Efficient Catalysts for Water Oxidation, *Dalton Trans.*, v.40, p.3793-3795

Najafpour MM (2011b) Calcium manganese oxides as structural and functional models for the active site in the oxygen evolving complex in photosystem II: Lessons from simple models. *J. Photochem. and Photobiol B*, v.104, p.111-117

Nam, Y.S.; Magyar, A.P.; Lee, D.; Kim, J.W.; Yun, D.S.; Park, H.; Pollom Jr, T.S.; Weitz, D.A. & Belcher, A.M. (2010) Biologically templated photocatalytic nanostructures for sustained light-driven water oxidation. Nat. Nanotechnol., v.5, p. 340-344

Nayak, S.; Nayek,H.P.; Dehnen,S.; Powellc, A.K. & Reedijk, J. (2011) Trigonal propeller-shaped [Mn^{III}$_3$M^{II}Na] complexes (M = Mn, Ca): structuraland functional models for the dioxygen evolving centre of PSII, Dalton Trans., v.40, p.2699-2702

Pace, R. (2005) An Integrated Artificial Photosynthesis Model. In: Collings AF, Critchley C (eds) Artificial photosynthesis: From basic biology to industrial application, 1edn. Wiley-VCH, Weinheim, p. 13-34

Pecoraro, V.L.; Baldwin, M.J. & Gelasco, A. (1904) Interaction of Manganese with Dioxygen and Its Reduced Derivatives, *Chem. Rev.*, v.94 (3), pp 807–826

Ruttinger, W. & Dismukes, G.C., (1997) Synthetic water oxidation catalysts for artificial photosynthetic water oxidation. *Chem. Rev.*, v.97, p. 1–24

Shafirovich, V. Y. & Shilov, A. E. (1979) Catalytic oxidation of water with the participation of manganese compounds in neutral and slightly acid media, *Kinetika i Kataliz*, **v.20**, p.1156-1162

Seidler-Egdal, R.K.; Nielsen, A.; Bond, A.D.; Bjerrum, M.J. & McKenzie C.J. (2011) High turnover catalysis of water oxidation by Mn(II) complexes of monoanionic pentadentate ligands, Dalton Trans., v.40, p.3849-3858

Shevela, D.; Koroidov, S.; Najafpour, M. M.; Messinger, J. & Kurz, Ph. (2011) Manganese Oxides as Oxygen Evolution Catalysts: O2 Formation Pathways Indicated by 18O-Labelling Studies, *Chem. Eur. J.*, v.17, p. 5415-5423

Shimazaki, Y.; Nagano T.; Takesue, H. ;Ye, B.H.; Tani F. & Naruta,Y.(2004) Characterization of a dinuclear Mn(V)=O complex and its efficient evolution of O$_2$ in the presence of water, Angew. Chem. Int. Ed., v.43, p.98-100

Tagore, R.; Crabtree, R.H. & Brudvig, G.W. (2008) Oxygen evolution catalysis by a dimanganese complex and its relation to photosynthetic water oxidation, Inorg. Chem., v.47, p.1815-1823

Umena, Y.; Kawakami, K.; Shen, J.R. & Kamiya, N. (2011) Crystal structure of oxygen-evolving photosystem II at a resolution of 1.9Å. Nature, v.473, p.55-60

Yagi, M. & Narita, K. (2004) Catalytic O_2 evolution from water induced by adsorption of [(OH$_2$)(Terpy)Mn(μ-O) $_2$Mn(Terpy)(OH$_2$)]$^{3+}$ complex onto clay compounds, J. Am. Chem. Soc., v.126, p.8084-8085

Yagi, M. & Kaneko, M. (2001) Molecular catalysts for water oxidation, Chem. Rev., v.101, p.21-36

Yin, G.; McCormick, J.M.; Buchalova, M.; Danby, A.M.; Rodgers, K.; Day, V.W.; Smith, K.; Perkins, C.M.; Kitko, D.; Carter, J.D.; Scheper, W.M.& Busch, D.H. (2006) Synthesis, Characterization, and Solution Properties of a Novel Cross-Bridged Cyclam Manganese(IV) Complex Having Two Terminal Hydroxo Ligands, Inorg. Chem., v.45, p. 8052- 8061

Zaharieva, I.; Najafpour, M. M.; Wiechen, M.; Haumann, M. ; Kurz, P. & Dau, H. (2011) Synthetic Manganese-Calcium Oxides Mimic the Water-Oxidizing Complex of Photosynthesis Functionally and Structurally Energy Environ Sci., DOI: 10.1039/c0ee00815j

6

The Enhancement of Photosynthesis by Fluctuating Light

David Iluz[1,2], Irit Alexandrovich[1] and Zvy Dubinsky[1]
[1]The Mina and Everard Goodman Faculty of Life Sciences,
Bar-Ilan University, Ramat-Gan
[2]The Department of Geography and Environment, Bar-Ilan University, Ramat-Gan,
Israel

1. Introduction

As early as 1953, reports documenting the enhancement of photosynthesis by plants when exposed to flashing light, as compared to the same photon dose under continuous light, were published (Kok, 1953; Myers, 1953). Using the unicellular green alga *Chlorella*, the effects of varying frequencies and dark/light duration ratios on photosynthetic rates were described. Such results have kindled research on their role in nature and their application in various photobioreactors and algal mass culture facilities, aimed at the production of valuable carotenoids, lipids, and additional products of commercial interest, such as biodiesel. The present review discusses the characteristics of the fluctuating underwater light field in natural waters, bioreactors, and ponds, and summarizes their effects on photosynthesis and growth.

Based on the advances in the understanding of the various mechanisms and processes affecting the efficiency and yields of photosynthesis, we discuss their interaction with continuous and intermittent light. The following are examined:

a. Post-illumination enhanced respiration
b. Photodynamic damage to the 32kd protein of Photosystem II (PSII)
c. The xanthophyll cycle
d. Thermal-energy dissipation
e. In addition, the enhancement phenomenon is examined in relation to the intensity of the ambient light

In the context of the present review, it is appropriate to list and compare the different terms and definitions used in the description and study of fluctuating light fields (Table 1). In the following review, the term 'fluctuating light' is used for clarity and consistency.

1.2 Fluctuating light in terrestrial ecosystems
1.2.1 Canopy

In many forests with closed canopies, only a small fraction (0.5-5%) of the solar radiation incident above the canopy reaches the understory. Understory plants in these forests experience a highly dynamic light environment, with brief, often unpredictable periods of direct solar irradiance (sunflecks), punctuating the dim and diffuse background irradiance.

These sunflecks are combined with diurnal fluctuations in irradiance, ranging from sunflecks lasting only a few seconds or less in heavily shaded sites to cloud-induced fluctuations ("shadeflecks") lasting up to an hour or longer in open sites (Knapp & Smith, 1987). The nature of sunflecks, their size, shape, duration, and peak photon flux density depends on the height and precise arrangement of vegetation within the forest canopy as well as the position of the sun in the sky. The occurrence of a sunfleck at a particular location and time in the forest understory depends on different, often interacting, factors: the coincidence of the solar path with a canopy opening; the movement of clouds that obscure or reveal the sun; and the wind-induced movement of foliage and branches (in the canopy or in the understory plants themselves). These factors interact to yield a highly dynamic light environment in which the photon flux reaching leaves can increase or decrease over orders of magnitude in a matter of seconds. The effects of sunflecks on understory photosynthesis were studied in several works and revealed conflicting results (Leakey et al., 2002, 2005). In some cases, photosynthesis as CO_2 assimilation by understory seedlings was enhanced, whereas at elevated temperatures (38°C), it decreased.

Subject	Condition	Frequency	Organism	Reference
Sunflecks	Caused by leaf movement in forest understory	Seconds to minutes	Dipterocarp seedlings	Leakey et al. (2002, 2005)
Flickering light	Underwater high-frequency light fluctuation resulting from lens effect on the water surface	High frequency Less than one second	endosymbiont photosynthesis of reef-building corals	Yamasaki and Nakamura (2008); Veal et al. (2010)
Fluctuating light	Sunlight focusing by short sea-surface waves; appears in peculiar form of irradiance pulses, termed 'flashes'	At depth of 1 m, up to 200 min^{-1}	Reef-building coral *Acropora digitifera*	Stramski (1986)
Intermittent light	Short flashes	In values of a few milliseconds. The critical flash time is a function of the incident intensity. The dark time must be 10 times as long as the flash time	*Chlorella pyrenoidosa* cultures	Kok (1953)
Flashing light		In photobioreactors, known as 'flashing light effect'		Myers (1953)

Table 1. Some historical observations of fluctuating light on the water surface and underwater

1.2.2 In the aquatic environment

In natural aquatic systems, there are several factors that cause variations in irradiance. Irregular variations are caused by surface wave movement (Walsh & Legendre, 1983), cloud cover (Marra & Heinemann, 1982), and the vertical movement of phytoplankton (Falkowski & Wirick, 1981).

The high-frequency (less than 1 Hz) light fluctuations, known as 'flicker light', are produced by a lens effect of moving water surface, or waves, that simultaneously focuses and diffuses sunlight in the few upper meters (Hieronymi & Macke, 2010) (Fig. 1a). Because flicker light potentially produces excessively strong light as well as dimmer light, such fluctuations may have profound effects on the photosynthesis of benthic algae, seagrasses, and zooxanthellate corals. We will limit our discussion to shallow-water sessile organisms that are particularly prone to be influenced by a recurrent light provision anomaly such as these flickers. The effect diminishes with depth due to the shape- and hence the focal length of waves and scattering by particles (Fig. 1).

Waves act as lenses because of the differences in refractive indices between air and water, focusing light below the wave for a brief period. In shallow water, this effect can be seen clearly by eye (Fig. 1), appearing against a dark background as flickering bands of focused light. The location of the focusing events depends upon the shape of the waves. Large, rounded waves focus into deeper regions than small, sharply curved ripple waves (Kirk, 1994; Schubert et al., 2001) (Fig. 1).

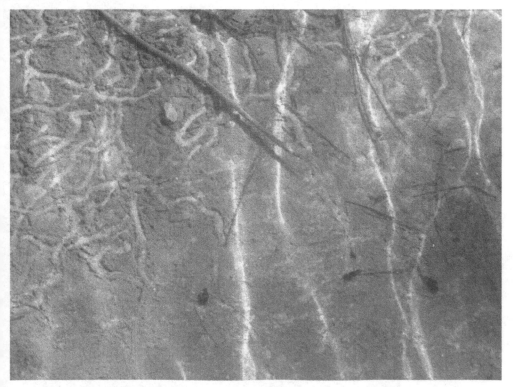

Fig. 1. Underwater light patterns on a shallow sandy bottom due to surface waves

Underwater irradiance fluctuations result from temporal, non-linear variations of sea-surface topography. When solar radiation is broken, the intensity and light penetration depth depend on the wavelength of light and general and local water features. Phytoplankton, which live at different depths, must adjust and be acclimated to the features of the underwater field to which they are exposed for doing photosynthesis (Kirk, 1994).

Ripples on the water surface cause considerable heterogeneity of subsurface light (Figs. 2, 3) through the lens effect, which simultaneously focuses and diffuses sunlight in the upper few meters, producing a constantly moving pattern of interspersed light and shadows on the substrate. Due to that effect, light intensity in shallow water environments sometimes reaches more than 9,000 μmol quanta $m^{-2}s^{-1}$, corresponding to 300-500% of the surface light intensity (Fig. 1) (Schubert et al., 2001). The lens effect produced by waves generates narrow belts of supersaturating light that pass over the bottom surface for less than a second (Figs. 2, 3). In addition to the focusing effect of light, the same curvature of the water surface also produces lower light intensity intervals interspersed between the intense peaks, namely, producing a light-diffusing effect (Schubert et al., 2001). This light-diffusing effect may serve as a relaxation period for algal photosynthesis.

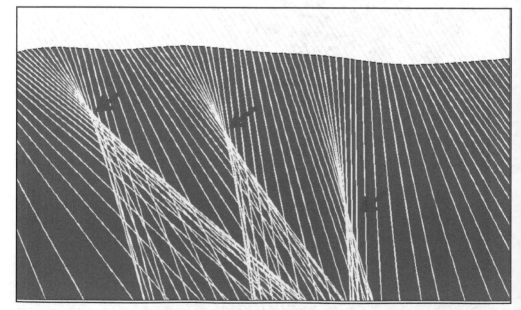

Fig. 2. Focusing beams beneath wave crests and scattering under troughs. The arrows indicate the dependence of maximal-effect depth on wave radius (after Grosser et al., 2008).

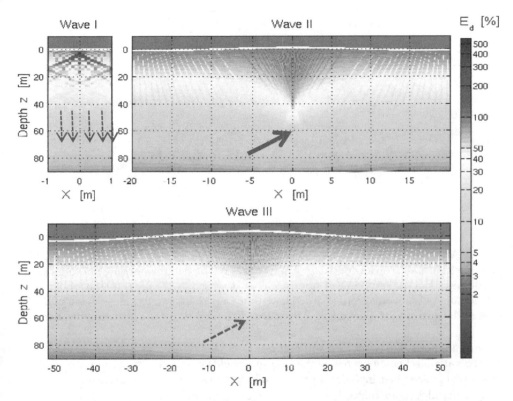

Fig. 3. Downward irradiance patterns below three waves with respect to 100% at the surface (note the logarithmic color scale and the different scales of the x axes) Wave I, flat surface, no lensing; Wave II, strong lensing; wave III, large radius, weak lensing (after Hieronymi and Macke (2010)).

Due to its dependence on wave geometry and on their horizontal movement, there is a tight coupling between wind velocity, wave height, surface smoothness and the underwater light field (Fig. 4). The effect of small-scale roughness on preventing wave lensing was applied in the study by Veal et al. (2010), who used water sprinklers to obtain non-lensing controls.

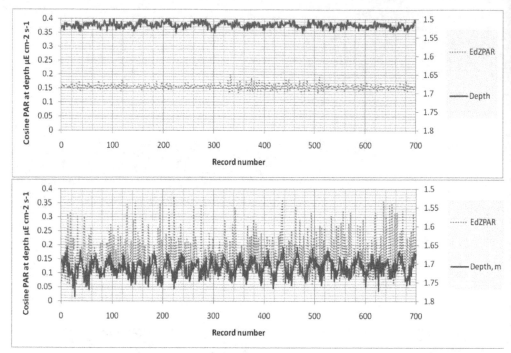

Fig. 4. PAR fluctuations as measured by PRR800 at 0.8 m depth: A) on a very calm day (0.4 m/sec wind); and B) on a windy day (7 m/sec wind). The depth values are not calibrated but do represent the amplitude of surface waves as sensed by the pressure sensor of the PRR800 unit. Data were recorded at 15 Hz without any shading effect, downwind of the IUI pier, in the gulf of Eilat. Coefficient of variance=40% (Dishon, personal communication)

1.3 Applications in bioreactors and cultures

Microalgal biotechnology is one of the emerging fields in our era. In recent years, there has been great interest in using microalgae as sources of a wide range of fine chemicals, oils, and polysaccharides (Borowitzka, 1992; Jensen, 1993; Munro et al., 1999). Many laboratory scale photobioreactors have been reported but most of them are extremely difficult to scale up due to the phenomenon of mutual shading at high cell densities. Light, which is an essential factor in the phototrophic growth of microalgae, cannot be stored, so it must be supplied continuously. Due to the high light-harvesting efficiency of chlorophyll in microalgae, algae absorb all the light that reaches them even though they cannot use all the photons. This phenomenon causes a dramatic decrease in light utilization efficiency since the photons cannot penetrate into the depth of the culture, even when enough photons are supplied at the surface. Many photobioreactors have been developed to overcome this problem (Park & Lee, 2000; Qiang & Richmond, 1996; Richmond et al., 2003).

Much recent research effort has been devoted to finding a means for shifting the group of cells exposed to the light in such a manner that each cell will receive just its quanta of light and then will immediately be replaced by another one, so that none of the light will be wasted (Richmond, 2004). As the algal cells are propelled by the mixing of the culture between the surface of the pond, the transparent wall of the reactor, or the dark depth of the culture, the cells are exposed to a fluctuating light regime whose properties depend on the light source and its intensity, spectral distribution and beam geometry, culture density and depth, reactor architecture, and the hydrodynamics of mixing. These parameters define the range of light intensities to which the cells are exposed, as well as their frequency.

There are two major types of photobioreactors that are considered the most common production systems: outdoor open ponds (Fig. 5) and enclosed photobioreactors (Fig. 6). Open ponds have been the most widely used system for large-scale outdoor microalgae cultivation for food and medicine supplements during the last few decades (Borowitzka, 1993). They were built as a single unit or multiple joint units, with agitation by means of paddlewheels, propellers, or airlift pumps.

Open ponds

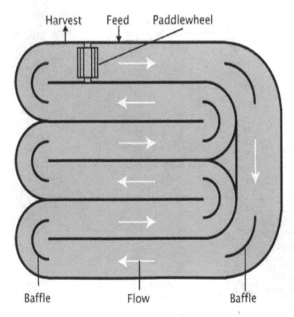

Fig. 5. Open ponds

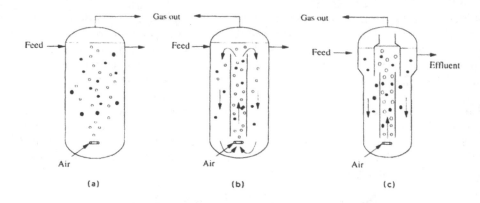

Fig. 6. Tubular photobioreactors (*left: Cal Poly CEA Energy Working Group,* right:
http://userwww.sfsu.edu/~art511_h/emerging10/natasha/domef/project1algae.html)

Due to the susceptibility of open pond systems to contamination by opportunistic algal species, enclosed photobioreactors have evolved. Two major types of enclosed photobioreactors are tubular and plate types. Due to the enclosed structure and relatively controllable environment, enclosed photobioreactors can reach high cell densities and easy-to-maintain monoculture (Lee, 2001; Ugwu et al., 2008). Tubular photobioreactors (PBRs), constructed of transparent glass or plastic, is one of the popular outdoor systems for mass algae cultivation. It can be horizontal, vertical, conical, and inclined in shape. By mixing, it can be an airlift or pump system (Ugwu et al., 2008). The advantages of tubular and plate types of PBR are a narrow light path (1.2-12.3 cm) that allows much higher cell concentration than in the open pond system, a larger illuminating area, and fewer contamination issues.

Both the spectral quality and the intensity of light are important for algal growth and metabolism. In high-density algal cultures, the light delivery becomes restricted as the cell concentration increases. This mutual shading or self-shading will shield the cells that are apart from the illumination surface from receiving light. As a result, the light penetration depth should be calculated in order to achieve a successful photobioreactor (Lee, 1999).

One of the solutions was to create a turbulent flow by installing static mixers. Turbulent-flow conduction in the reactor exposes the cells at short intervals to high light intensities at the reactor surface and they can, therefore, process the light energy collected in the subsequent dark phases.

A further way of generating turbulence in the photobioreactor is to provide a gassing device that achieves the desired effect at an appropriate gassing rate. The provision of flow-conducting intervals can also improve the flashing-light effect if a defined frequency is established for the illumination time (6509188, 2001)

The flashing light effect should be considered because turbulent flow in the reactor gives microalgae a chance to come close to the irradiated surface in the opaque medium at an irregular frequency, and this intermittent illumination enhances photosynthesis of the algae (Sato et al., 2010).

In high cell-density culture, a light gradient inside the photobioreactor will always occur due to light absorption and mutual shading by the cells. Depending on the mixing characteristics of the system, the cells will circulate between the light and dark zones of the reactor. The main limiting factor for the development of microalgal biotechnology is that light energy cannot be stored and homogenized inside the reactor. In photobioreactors, the light regime is determined by the light gradient and liquid-circulation time. Strong light can be efficiently used by working at the optimal cell density in combination with a proper liquid-circulation time (Richmond, 2000).

The increase in turbulence distributes the light optimally in the reactor chamber of the photobioreactor. Operation as an airlift loop ensures high turbulence with low sharing forces acting on the algal cells. Given high turbulence and, at the same time, high radiation intensity, the flashing-light effect can be utilized, and the cells would not have to be continuously illuminated (Fig. 7) (Patent 6509188, 2001).

The principle of the airlift loop reactor is that it achieves optimal light supply due to a low layer thickness and the directed conduction of flow in the reactor via static mixers. The solution lies in directed transportation of the cells, which is forced by the installation of the static mixers: the rising gas bubbles are diverted to the static mixes and set the culture medium flowing in a stationary, circular current. The algal cells are, thereby, transported to the light and then back into the shaded zone at a rhythm of approximately 1 Hz. The average light intensity, which is evenly distributed onto all algal cells in the reactor, can be varied by changing the spacing between the reactors outdoors (Fig. 8) (Patent 6509188, 2001).

Fig. 7. Flat-panel airlift (FPA) reactor with static mixers for defined transport of algae to the light. Up) Fraunhofer Institute for Interfacial Engineering and Biotechnology; and Down) Subitec - Sustainable Biotechnology

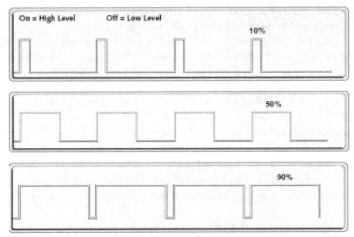

Fig. 8. The duty cycle indicates the 'on' or 'off' time in percent, ranging from 0% to 100% of period for repetitive pulse train. The duty cycle is the proportion of time in which the pulse is 'on', and is expressed as a percentage: duty cycle = 100% * (pulse time high)/(total pulse period).(http://www.dprg.org/tutorials/2005-11a/index.html)

1.4 The enhancement of photosynthesis by fluctuating light

Reports documenting the enhancement of algal photosynthesis when exposed to flashing light, described the effects of varying frequencies and dark/light duration ratios (Kok, 1953; Myers, 1953). These authors had shown that short flashes of high intensity can be used by single celled algae with high efficiency - if separated by sufficiently long dark periods. Kok (1953) has shown that for *Chlorella pyrenoidosa*, the dark time must be at least ten times as long as the flash time for fully efficient utilization of the incident light in photosynthesis. This means that if an algal cell is exposed to high-intensity light for a short time, it can absorb all that light in the 'light' stage of photosynthesis and then utilize it in succeeding stages in the dark. Contrary to the above, Grobbelaar et al. (1996) reported that a longer dark period relative to the preceding light period does not necessary lead to higher photosynthetic rates/efficiencies. He showed that the effect of altering the L/D (Light/Dark) ratio was clearly evident, where at any given L/D frequency the highest photosynthetic rates were measured at a ratio of 1L/1D.

Light/dark cycles have been proven to determine the light efficiency and productivity of photobioreactors. Very fast alternations between high light intensities and darkness (from less than 40 μs to 1 s) can greatly enhance photosynthetic efficiency (Kok, 1953; Matthijs et al., 1996; Nedbal et al., 1996; Phillips & Myers, 1954; Terry, 1986).

Such results have kindled research on their application in various photobioreactors and algal mass culture facilities aimed at the production of valuable carotenoids, lipids, and additional products of commercial interest, such as biodiesel and hydrogen (Grobbelaar et al., 1996; Park et al., 2000; Terry, 1986; Yoshimoto et al., 2005).

Several mechanisms were postulated to explain the enhancement effects of flashing light on photosynthesis. It has been suggested that at high frequencies, there are non-linear effects related to the limited availability of quinone pool receptors under continuous high light (Kok et al., 1970), whereas other works hypothesized Rubisco particle circulation in turbulent flow (Yoshimoto et al., 2005), and mutual shading of algal cells thereby increasing the time they are exposed to optimal light (Park et al., 2000).

The enhancement of photosynthesis by flashing light (both the rate and the efficiency) has been known for many years. Richmond and Vonshak (1978) and Laws et al. (1983) reported increased growth rates when the stirring speeds of their mass cultures were doubled. They ascribed the increase to a more favorable dark/light cycle with increase turbulence. According to Terry (1986), relatively high photon flux densities are necessary for the enhancement of photosynthesis in a modulating light field.

Furthermore, according to the findings of Grobbelaar et al. (1996), the enhancement of photosynthesis in L/D environments depends on a number of conditions, the most important being that the L/D cycle be less than 1 Hz. Under such conditions, photosynthetic rates and light utilization efficiencies are increased. The prime conclusion of this study was that the photosynthetic rates enhanced exponentially with increasing light/dark frequencies.

Kübler and Raven (1996) studied the enhancement of photosynthesis under fluctuating light simulating marine shallow water conditions in the rhodophytes *Palmaria palmata* and *Lomentaria articulate*. They determined the dependence of the enhancement on flash frequency.

Pons and Pearcy (1992) studied the enhancement of photosynthesis on leaves of soybean plants exposed to constant and flashing light regimes with light flecks of different frequencies, durations, and photon flux density (PFD). The net CO_2 fixation from 1 s duration of light flecks was 1-3 times higher than predicted from steady-state measurements in constant light at the light-fleck and background PFD. This light-fleck utilization efficiency (LUE) was somewhat higher at a high than at a low frequency of 1 s light flecks. LUE in flashing light with very short light flecks (0-2 s) and single 1 s light flecks was as high as 2, but decreased sharply with increasing duration of light flecks.

1.5 Zooxanthellate corals

In a coral endosymbiont exposed to supersaturating light intensities at flicker light conditions, reduced photoinhibition was recorded compared with exposure to constant light of the same supersaturating intensity. Reduction in photoinhibition by flicker light was reported to be more pronounced at high water temperatures in a study on the zooxanthellate coral *Acropora digitifera* (Yamasaki & Nakamura, 2008). Veal et al. (2010) studied the effects of wave lensing on two Red Sea zooxanthellate corals at ambient (27°C) and elevated (31°C) temperatures. They used water sprinklers that broke the water surface, thereby eliminating the lensing effect while having minimal effect on downwelling PAR and UV. In their experiments, they did not find any biological effect of the intense light flashes caused by lensing, and concluded that these flashes do not contribute to the triggering of bleaching episodes.

Regular variations are due to the circulatory motion within internal waves (Savidge, 1986). Regularly fluctuating irradiance has been shown to alter the rate of photosynthesis of marine algae (Savidge, 1986). Enhancement of photosynthesis to values as high as 180% of control have occasionally been recorded (Jewson & Wood, 1975; Marra, 1978). At supersaturating light intensities, photosynthesis was less inhibited by flicker light than by constant light in the reef-building coral *Acropora digitifera* (Yamasaki & Nakamura, 2008).

The effect of light acclimation on the P/I curve is shown in Figure 9: the open arrows show the direction of the response to high light and the solid arrows show the response to low light acclimation. At high light P_{max} increases, while at low light it decreases. Alpha (α) increases with low light acclimation and decreases under high light exposure. The onset of photoinhibition occurs at lower light intensities for low light-acclimated algae compared to high light-acclimated algae that can tolerate much higher light intensities. I_k is lower for low light-acclimated algae than for high light-acclimated algae. Even dark respiration (R_d) varies considerably depending on the light history (Grobbelaar & Soeder, 1985), where high light-acclimated algae will have higher dark respiration rates than low light-acclimated ones. However, the question of the effect of flashing light on the photoacclimation process has not been studied. It remains to be seen whether algae exposed to flashing or fluctuating light acclimate to the peak intensity or to some intermediate value. In Figure 10, the effect of flashing light frequency and intensity on the photosynthesis versus energy relationship is shown. In all cases, a 1/1 light/dark duty cycle was used.

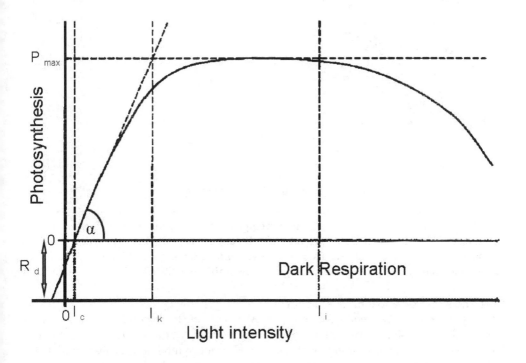

Fig. 9. Light-response curve of photosynthesis versus light intensity (Grobbelaar, 2006).

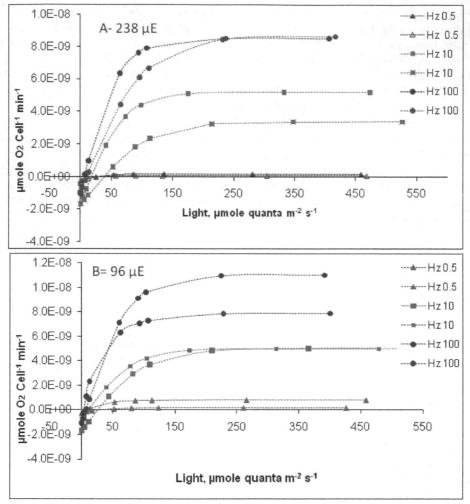

Fig. 10. Photosynthetic rates of the green microalga *Chlorella* sp. subjected to increasing frequencies of light in relation to constant light. Following two weeks of growth, measurements were made at each frequency, two replicates = algae at high light intensity (238 µmole quanta m^{-2} s^{-1}) (A), and B = algae at low light intensity (96 µmole quanta m^{-2} s^{-1}).

1.5.1 The effect on zooxanthellate corals

Effects of flicker light on endosymbiont photosynthesis of the reef-building coral *Acropora digitifera* (Dana, 1846) were evaluated with pulse amplitude modulation chlorophyll fluorometry. At supersaturating light intensities, photosynthesis was less inhibited by flicker light than by constant light. Reduction in photoinhibition by flicker light was pronounced at high water temperatures. Flicker light may strongly influence endosymbiont photosynthesis of corals inhabiting shallow reef habitats, especially during periods of strong solar irradiance and high water temperature.

1.6 Mechanisms

Since fluctuating light enhancement has been observed only under high light intensities, we offer the following putative explanations to that phenomenon. Under high light conditions, photosynthetic organisms recruit several processes to deal with excessive absorbed light and reduce oxidative damage. In terms of energy efficiency, these processes are responsible for losses in photosynthetic production.

These processes are:

a. Photoacclimation - Plants must maintain balance between the energy derived from the light reactions in the chloroplast and the amount of energy utilized during carbon fixation and other metabolic processes. Changes in environmental conditions upset this balance, requiring algae to physiologically adjust, or acclimate (Hüner et al., 1998). In the natural environment, there are daily and seasonal changes in light intensity, creating stress that induces a physiological response (photoacclimation) to optimize growth. Such responses minimize damages stemming from super-optimal intensities, while maximizing light harvesting and quantum yields under dim light. For instance, an increase in light intensity can generate damaging reactive oxygen intermediates that may induce photoinhibition and limit growth (Osmond, 1994). Conversely, when light intensity decreases, cells are unable to generate enough energy via photosynthesis to fulfill their metabolic requirements, again limiting primary production (Falkowski & LaRoche, 1991). Short-term (seconds to minutes) photoacclimation includes the dissipation of excess light energy via the xanthophyll-cycle carotenoids (Demmig-Adams & Adams, 1996) and/or state transitions that can change the excitation energy distribution between photosystems I and II (PSI, PSII) (Bennett, 1983; Bonaventura & Myers, 1969). Long-term photoacclimation (hours) occurs when the short-term changes are insufficient for coping with the changes in light intensity, thus resorting to extensive changes in enzyme activity and gene expression that lead to alterations in the concentration of photosynthetic complexes, leading to changes in antenna composition and photosystem stoichiometry (Anderson et al., 1995; Falkowski & LaRoche, 1991). High light-acclimated cells typically have less photosynthetic units or smaller antennae if photosystem-unit (PSU) numbers remain unchanged, lower cellular content of chlorophylls and other light-harvesting pigments such as fucoxanthin and peridinin, and high concentrations of photoprotective ones, like □□carotene and astaxanthin. The opposite is true for cells exposed to low light intensities. Overall, photoacclimation attempts to maintain constant photosynthetic efficiency under a variety of light intensities by adjusting the capacity of the plant to harvest and utilize light (Chow et al., 1990). The question is: to what irradiance level do cells acclimate under fluctuating light.

b. Photoinhibition (see Powles, 1984; Lidholm et al., 1987; Long et al., 1994). This process results in a significant loss of photosynthetic production. It involves degradation of the 32-kD D1 polypeptide (e.g., Adir et al. 2003), as well as PSI-A, PSI-B, and acceptor-side-located small photosystem I polypeptides (Tjus et al., 1999). Its extent and kinetics depend on the previous light-history of the organism, and the intensity and duration of the photoinhibitory illumination. We assume that under suitable frequencies, photoinhibition is reduced either since exposure to the high light is too short to bring about damage or that the subsequent dark interval allows for damage repair.

c. Enhanced post-illumination respiration (EPIR) was first reported by Falkowski et al. (1985) and reexamined by Beardall et al. (1994). However, since that phenomenon that by steeply increasing respiration losses reduces net production and yields is dose dependent, the brevity of the high light exposure minimizes its effects.

d. Thermal dissipation, also termed NPQ (nonphotochemical quenching [of fluorescence]) in variable fluorescence-based studies, may account for the loss of from one half to nearly all of the absorbed light energy. Thermal dissipation, measured directly by photoacoustics, depends on the light history of the organism, with losses increasing under high light, and any stress may account for losses of 30-80% of absorbed light energy (Dubinsky et al., 1998; Pinchasov et al., 2005).

e. The xanthophyll cycle allows the fine tuning of the photosynthetic apparatus to ambient light by switching between two states of the pigment couples constituting the xanthophyll cycle. When exposed to low light, most of its energy is used in the photochemical photolysis of water and the subsequent reduction of CO_2 to high energy photosynthate. Under high light, the xanthophyll cycle pigments undergo epoxidation and now divert light energy to harmless heat rather than damaging excess light (Adams et al., 1999; Demmig-Adams, 1998). Again, the relatively slow activation of the cycle, or its 'switching' to the high light state, prevents, or at least minimizes, its beneficial, albeit costly, activity under fluctuating light.

f. RuBisCO - A dynamic model for photosynthesis was developed to elucidate the effect of flashing light that enhances the efficiency of photosynthesis. One particular feature of the model is that discrete RuBP particles circulate in the Calvin cycle and their speeds in the cycle are determined by the amount of ATP generated in the photon reception process. This can realize the light saturation under continuous light and the flashing light effect under fluctuating illumination. Laboratory experiments were conducted on *Chaetoceros calcitrans* (Yoshimoto et al., 2005).

All of the above lead us to the following conclusions:

a. Short duration periods of intense light are too brief to result in damage to the photosynthetic apparatus leading to photoinhibition of photosynthetic rates.

b. Short light periods do not allow sufficient time for the full activation of the xanthophyll-driven thermal energy dissipation of absorbed light, which would reduce the energy allocated to photochemistry.

c. Short light periods reduce the biomass losses incurred due to enhanced post-illumination respiration.

d. Dark intervals allow the regeneration and reoxidation of intermediate electron and CO_2 acceptors in the quinone pool and the Calvin cycle.

In summary, we can conclude that present developments allow better understanding of the mechanisms involved in the long-known enhancement of photosynthesis under fluctuating and flashing light. Future research will focus on defining the boundaries of the process in terms of irradiance levels, frequencies, and duty cycle. Regarding the mechanisms, these are likely to be explored in detail in order to understand their relative contribution under different conditions, and with more careful examination of the taxonomic differences in both characteristics and mechanisms.

2. Acknowledgments

This study was supported by NATO SfP 981883. We thank Sharon Victor for the English editing.

3. References

Adams, W.W., Demmig-Adams, B., Logan, B.A., Barker, D.H. & Osmond, C.B. (1999). Rapid changes in xanthophyll cycle-dependent energy dissipation and photosystem II efficiency in two vines, *Stephania japonica* and *Smilax australis*, growing in the understory of an open Eucalyptus forest. *Plant Cell and Environment*, Vol.22, No.2, pp. 125-136, ISSN 0140-7791

Adir, N., Zer, H., Shochat, S. & Ohad, I. (2003). Photoinhibition - a historical perspective. *Photosynthesis Research*, Vol.76, No.1-3, pp. 343-370, ISSN 0166-8595

Anderson, J.M., Chow, W.S. & Park, Y.I. (1995). The grand design of photosynthesis: Acclimation of the photosynthetic apparatus to environmental cues. *Photosynthesis Research*, Vol.46, No.1-2, pp. 129-139, ISSN 0166-8595

Beardall, J., Burgerwiersma, T., Rijkeboer, M., Sukenik, A., Lemoalle, J., Dubinsky, Z. & Fontvielle, D. (1994). Studies on enhanced post-illumination respiration in microalgae. *Journal of Plankton Research*, Vol.16, No.10, pp. 1401-1410, ISSN 0142-7873

Bennett, J. (1983). Regulation of photosynthesis by reversible phosphorylation of the light harvesting chlorophyll a/b protein. *Biochemical Journal*, Vol.212, pp. 1-13

Bonaventura, C. and Myers, J. (1969). Fluorescence and oxygen evolution from *Chlorella pyrenoidosa*. *Biochimica et Biophysica Acta*, Vol.189, No.3, pp. 366-383, ISSN 0006-3002

Borowitzka, M.A. (1992). Algal biotechnology products and processes - Matching science and economics. *Journal of Applied Phycology*, Vol.4, No.3, pp. 267-279, ISSN 0921-8971

Borowitzka, M.A. (1993). Large-scale algal culture systems: The next generation. In: J. Sargeant, S. Washer, M. Jones & M.A. Borowitzka (Eds.), *11th Australian Biotechnology Conference, 1993, Perth, West Australia*, p. 61

Chow, W.S., Melis, A. & Anderson, J.M. (1990). Adjustments of photosystem stoichiometry in chloroplasts improve the quantum efficiency of photosynthesis *Proceedings of the National Academy of Sciences of the United States of America*, Vol.87, No.19, pp. 7502-7506, ISSN 0027-8424

Demmig-Adams, B. (1998). Survey of thermal energy dissipation and pigment composition in sun and shade leaves. *Plant and Cell Physiology*, Vol.39, No.5, pp. 474-482, ISSN 0032-0781

Demmig-Adams, B. & Adams, W.W. (1996). The role of xanthophyll cycle carotenoids in the protection of photosynthesis. *Trends in Plant Science*, Vol.1, No.1, pp. 21-26, ISSN 1360-1385

Dubinsky, Z., Feitelson, J. & Mauzerall, D.C. (1998). Listening to phytoplankton: Measuring biomass and photosynthesis by photoacoustics. *Journal of Phycology*, Vol.34, No.5, pp. 888-892

Falkowski, P.G. & Wirick, C.D. (1981). A simulation model of the effects of vertical mixing on primary production. *Marine Biology*, Vol.65, pp. 69-75

Falkowski, P.G. & LaRoche, J. (1991). Acclimation to spectral irradiance in algae. *Journal of Phycology*, Vol.27, pp. 8-14

Falkowski, P.G., Dubinsky, Z. & Santostefano, G. (1985). Light-enhanced dark respiration in phytoplankton. *Verhandlungen/Internationale Vereinigung Limnologie*, Vol.22, pp. 2830-2833

Grobbelaar, J.U. (2006). Photosynthetic response and acclimation of microalgae to light fluctuations In: *Algal Cultures Analogues of Blooms and Applications*, D.V. Subba Rao (Ed.), 671-683, Science Publishers, Enfield, NH, Plymouth, UK

Grobbelaar, J.U. & Soeder, C.J. (1985). Respiration losses in planktonic green algae cultivated in raceway ponds. *Journal of Plankton Research*, Vol.7, No.4, pp. 497-506, ISSN 0142-7873

Grobbelaar, J.U., Nedbal, L. & Tichý, V. (1996). Influence of high frequency light/dark fluctuations on photosynthetic characteristics of microalgae photoacclimated to different light intensities and implications for mass algal cultivation. *Journal of Applied Phycology*, Vol.8, No.4-5, pp. 335-343, ISSN 0921-8971

Grosser, K., Hieronymi, M., Macke, A., Wahl, M., Eden, C., Zimmer, M. & Croot, P. (2008). The role of light fluctuations on marine processes. *Future Ocean: Cluster Retreat*, 17.-18.03, Salzau

Hieronymi, M. & Macke, A. (2010). Monte Carlo radiative transfer simulations on the influence of surface waves on underwater light fields. *20th Ocean Optics Conference, Anchorage, Alaska*

Hüner, N.P.A., Öquist, G. & Sarhan, F. (1998). Energy balance and acclimation to light and cold. *Trends in Plant Science*, Vol.3, No.6, pp. 224-230, 1360-1385

Jensen, A. (1993). Present and future needs for algae and algal products. *Hydrobiologia*, Vol.261, pp. 15-23, ISSN 0018-8158

Jewson, D.H. & Wood, R.B. (1975). Some effects on integral photosynthesis of artificial circulation of phytoplankton through light gradients. *Verhandlungen des Internationalen Verein Limnologie*, Vol.19, pp. 1037-1044

Kirk, J.T.O. (1994). *Light and Photosynthesis in Aquatic Ecosystems*, 2nd edition, Cambridge University Press, London, New York

Knapp, A.K. & Smith, W.K. (1987). Stomatal and photosynthetic responses during sun shade transitions in sub-alpine plants - Influence on water use efficiency. *Oecologia*, Vol.74, No.1, pp. 62-67, ISSN 0029-8549

Kok, B. (1953). Experiments on photosynthesis by *Chlorella* in flashing light. In: *Algal Culture from Laboratory to Pilot Plant*, J.S. Burlew (Ed.), 63-75, Carnegie Institution of Washington #600, Washington, DC

Kok, B., Forbush, B. & McGloin, M. (1970). Cooperation of charges in photosynthetic O_2 evolution. 1. A linear 4-step mechanism *Photochemistry and Photobiology*, Vol.11, No.6, pp. 467-475

Kübler, J.E. & Raven, J.A. (1996). Nonequilibrium rates of photosynthesis and respiration under dynamic light supply. *Journal of Phycology*, Vol.32, No.6, pp. 963-969, ISSN 0022-3646

Laws, E.A., Terry, K.L., Wickman, J. & Chalup, M.S. (1983). A simple algal production system designed to utilize the flashing light effect. *Biotechnology and Bioengineering*, Vol.25, No.10, pp. 2319-2335, ISSN 0006-3592

Leakey, A.D.B., Press, M.C., Scholes, J.D. & Watling, J.R. (2002). Relative enhancement of photosynthesis and growth at elevated CO_2 is greater under sunflecks than uniform irradiance in a tropical rain forest tree seedling. *Plant Cell and Environment*, Vol.25, No.12, pp. 1701-1714, ISSN 0140-7791

Leakey, A.D.B., Scholes, J.D. & Press, M.C. (2005). Physiological and ecological significance of sunflecks for dipterocarp seedlings. *Journal of Experimental Botany*, Vol.56, No.411, pp. 469-482, ISSN 0022-0957

Lee, C.-G. (1999). Calculation of light penetration depth in photobioreactors. *Biotechnology and Bioprocess Engineering*, Vol.4, pp. 78-81

Lee, Y.K. (2001). Microalgal mass culture systems and methods: Their limitation and potential. *Journal of Applied Phycology*, Vol.13, No.4, pp. 307-315, 0921-8971

Lidholm, J., Gustafsson, P. & Öquist, G. (1987). Photoinhibition of photosynthesis and its recovery in the green alga *Chlamydomonas reinhardii. Plant and Cell Physiology*, Vol.28, No.6, pp. 1133-1140, ISSN 0032-0781

Long, S.P., Humphries, S. & Falkowski, P.G. (1994). Photoinhibition of photosynthesis in nature. *Annual Review of Plant Physiology and Plant Molecular Biology*, Vol.45, pp. 633-662

Marra, J. (1978). Effect of short-term variations in light-intensity on photosynthesis of a marine phytoplankter - Laboratory simulation study. *Marine Biology*, Vol.46, No.3, pp. 191-202, ISSN 0025-3162

Marra, J. & Heinemann, K. (1982). Photosynthesis response by phytoplankton to sunlight variability. *Limnology and Oceanography*, Vol.27, No.6, pp. 1141-1153, ISSN 0024-3590

Matthijs, H.C.P., Balke, H., VanHes, U.M., Kroon, B.M.A., Mur, L.R. & Binot, R.A. (1996). Application of light-emitting diodes in bioreactors: Flashing light effects and energy economy in algal culture (Chlorella pyrenoidosa). *Biotechnology and Bioengineering*, Vol.50, No.1, pp. 98-107, ISSN 0006-3592

Munro, M.H.G., Blunt, J.W., Dumdei, E.J., Hickford, S.J.H., Lill, R.E., Li, S.X., Battershill, C.N. & Duckworth, A.R. (1999). The discovery and development of marine compounds with pharmaceutical potential. *Journal of Biotechnology*, Vol.70, No.1-3, pp. 15-25, ISSN 0168-1656

Myers, J. (1953). Growth characteristics of algae in relation to the problems of mass culture. In: *Algal culture from laboratory to pilot plant*, J.S. Burlew (Ed.), Carnegie Institution of Washington #600, Washington, DC

Nedbal, L., Tichy, V., Xiong, F.H. & Grobbelaar, J.U. (1996). Microscopic green algae and cyanobacteria in high-frequency intermittent light. *Journal of Applied Phycology*, Vol.8, No.4-5, pp. 325-333, ISSN 0921-8971

Osmond, C.B. (1994). What is photoinhibition? Some insights from comparisons of sun and shade plants. In: *Photoinhibition of Photosynthesis: From Molecular Mechanisms to the Field*, N.R. Baker and J.R. Bowyer (Eds.), 1-24, BIOS Scientific Publishers, Oxford

Park, K.-H. & Lee, C.-G. (2000). Optimization of algal photobioreactors using flashing lights. *Biotechnology and Bioprocess Engineering*, Vol.5, pp. 186-190

Park, K.H., Kim, D.I. & Lee, C.G. (2000). Effect of flashing light on oxygen production rates in high-density algal cultures. *Journal of Microbiology and Biotechnology*, Vol.10, No.6, pp. 817-822, ISSN 1017-7825

Patent 6509188 (2001). Photobioreactor with improved supply of light by surface enlargement, wavelength shifter bars or light transport (United States patent)

Phillips, J.N. & Myers, J. (1954). Growth rate of *Chlorella* in flashing light. *Plant Physiology*, Vol.29, No.2, pp. 152-61, ISSN 0032-0889

Pinchasov, Y., Kotliarevsky, D., Dubinsky, Z., Mauzerall, D.C. & Feitelson, J. (2005). Photoacoustics as a diagnostic tool for probing the physiological status of phytoplankton. *Israel Journal of Plant Sciences*, Vol.53, No.1, pp. 1-10

Pons, T.L. & Pearcy, R.W. (1992). Photosynthesis in flashing light in soybean leaves grown in different conditions .2. Lightfleck utilization efficiency. *Plant Cell and Environment*, Vol.15, No.5, pp. 577-584, ISSN 0140-7791

Powles, S.B. (1984). Photoinhibition of photosynthesis induced by visible light. *Annual Review of Plant Physiology* Vol.35, pp. 15-44, ISSN0066-4294

Qiang, H. & Richmond, A. (1996). Productivity and photosynthetic efficiency of *Spirulina platensis* as affected by light intensity, algal density and rate of mixing in a flat plate photobioreactor. *Journal of Applied Phycology*, Vol.8, No.2, pp. 139-145, ISSN 0921-8971

Richmond, A. (2000). Microalgal biotechnology at the turn of the millennium: A personal view. *Journal of Applied Phycology*, Vol.12, No.3-5, pp. 441-451, ISSN 0921-8971

Richmond, A. (2004). Principles for attaining maximal microalgal productivity in photobioreactors: an overview. *Hydrobiologia*, Vol.512, No.1-3, pp. 33-37, ISSN 0018-8158

Richmond, A. & Vonshak, A. (1978). *Spirulina* culture in Israel. *Arch fur Hydroiologie (Beih. Ergebn. Limnol)*. Vol.11, pp. 274-280

Richmond, A., Zhang, C.W. & Zarmi, Y. (2003). Efficient use of strong light for high photosynthetic productivity: interrelationships between the optical path, the optimal population density and cell-growth inhibition. *Biomolecular Engineering*, Vol.20, No.4-6, pp. 229-236, ISSN 1389-0344

Sato, T., Yamada, D. & Hirabayashi, S. (2010). Development of virtual photobioreactor for microalgae culture considering turbulent flow and flashing light effect. *Energy Conversion and Management*, Vol.51, No.6, pp. 1196-1201, ISSN 0196-8904

Savidge, G. (1986). Growth and photosynthetic rates of *Phaeodactylum* tricornutum Bohlin in a cyclical light field. *Journal of Experimental Marine Biology and Ecology*, Vol.100, No.1-3, pp. 147-164, ISSN 0022-0981

Schubert, H., Sagert, S. & Forster, R.M. (2001). Evaluation of the different levels of variability in the underwater light field of a shallow estuary. *Helgoland Marine Research*, Vol.55, No.1, pp. 12-22, ISSN 1438-387X

Stramski, D. (1986). Fluctuations of solar irradiance induced by surface waes in the Baltic. *Bulletin of the Polish Academy of Sciences - Earth Sciences*, Vol.34, pp. 333-344

Terry, K.L. (1986). Photosynthesis in modulated light - Quantitative dependence of photosynthetic enhancement on flashing rate. *Biotechnology and Bioengineering*, Vol.28, No.7, pp. 988-995, ISSN 0006-3592

Tjus, S.E., Moller, B.M. & Scheller, H.V. (1999). Photoinhibition of Photosystem I damages both reaction centre proteins PSI-A and PSI-B and acceptor-side located small Photosystem I polypeptides. *Photosynthesis Research*, Vol.60, No.1, pp. 75-86, ISSN 0166-8595

Ugwu, C.U., Aoyagi, H. & Uchiyama, H. (2008). Photobioreactors for mass cultivation of algae. *Bioresource Technology*, Vol.99, No.10, pp. 4021-4028, ISSN 0960-8524

Veal, C.J., Carmi, M., Dishon, G., Sharon, Y., Michael, K., Tchernov, D., Hoegh-Guldberg, O. & Fine, M. (2010). Shallow-water wave lensing in coral reefs: a physical and biological case study. *Journal of Experimental Biology*, Vol.213, No.24, pp. 4304-4312, ISSN 0022-0949

Walsh, P. & Legendre, L. (1983). Photosynthesis of natural phytoplankton under high-frequency light fluctuations simulating those induced by sea-surface waves. *Limnology and Oceanography*, Vol.28, No.4, pp. 688-697, ISSN 0024-3590

Yamasaki, H. & Nakamura, T. (2008). Flicker light effects on photosynthesis of symbiotic algae in the reef-building coral *Acropora digitifera* (Cnidaria:Anthozoa:Scleractinia). *Pacific Science*, Vol.62, No.3, pp. 341-350, ISSN 0030-8870

Yoshimoto, N., Sato, T. and Kondo, Y. (2005). Dynamic discrete model of flashing light effect in photosynthesis of microalgae. *Journal of Applied Phycology*, Vol.17, No.3, pp. 207-214, ISSN 0921-8971

7

An Approach Based on Synthetic Organic Chemistry Toward Elucidation of Highly Efficient Energy Transfer Ability of Peridinin in Photosynthesis

Takayuki Kajikawa and Shigeo Katsumura
Kwansei Gakuin University
Japan

1. Introduction

Photosynthesis has driven the development of life which is powered by the efficient capture and conversion of sunlight. Carotenoids are naturally occurring pigments that absorb sunlight in the spectral region in which the sun irradiates maximally. These molecules transfer the absorbed energy to chlorophylls and the primary photochemical events of photosynthesis are initiated. More than a half of photosynthesis is performed in the ocean, although the oceanic photosynthesis is relatively less studied. Marine carotenoid, peridinin, has been known as the main light-harvesting pigment in photosynthesis in the sea and forms the unique water soluble peridinin-chlorophyll a (Chl a)–protein (PCP) complex. The crystal structure of the main form of the PCP trimer from *Amphidinium carterae* was determined by X-ray crystallography as shown in Fig. 1 (A) (Hoffman et al., 1996). Each of the polypeptides binds eight peridinin molecules and two Chl a molecules, and the allene function of peridinin exsists in the center of the PCP. In this complex, a so-called antenna pigment, peridinin exhibits exceptionally high (> 95%) energy transfer efficiencies to Chl a (Song et al., 1976; Mimuro et al., 1993). This energy transfer efficiency is thought to be related to the unique structure of peridinin, which possesses allene and ylidenebutenolide functions and the unusual C37 carbon skeleton referred to as a 'nor-carotenoid' (Fig. 1 (B)) (Stain et al., 1971). There are, however, no studies on the relationship between the structural features of peridinin and its super ability for the energy transfer in the PCP complex.

In order to clear this efficient energy transfer mechanism, there are many and hot discussions in spectroscopic fields. In particular, the presence of an intramolecular charge transfer (ICT) excited state of peridinin has been proposed. It has been anticipated that the highly efficient energy transfer is caused through this key energy level, and a detailed discussion on this is described in the later chapters. This particular excited state is thought to be related to the intricate structure of peridinin. However, the precise nature of the ICT excited state and its role in light-harvesting have not yet been entirely clear, and there are no studies on the relation between the structural features of peridinin and its super ability for the energy transfer in the PCP complex. This is because the synthesis of various kinds of desired peridinin derivatives are not easy. Then, we started the research work to clear the subjects of why peridinin possesses a unique allene group, a ylidenebutneolide ring and an irregular C37 skeleton, and how these functions play a role in the exceptionally high energy transfer and in the special excited state, ICT state.

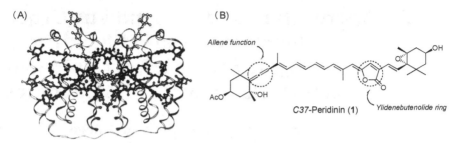

Fig. 1. (A) Crystal structure of PCP complex and (B) the structure of peridinin

2. Syntheses and stereochemical characteristics of peridinin derivatives

2.1 Design of peridinin derivatives and synthetic strategy

We have focused on the subjects of why peridinin possesses a unique allene group, a ylidenebutenolide ring and the irregular C37 carbon not to be usual C40 skeleton, how these functional groups play a role in the exceptionally high energy transfer, and how they affect the ICT state. In order to solve these questions, we designed and began to synthesize allene-modified, ylidenebutenolide-modified and conjugated chain-modified derivatives of peridinin (Fig. 2). For example, in order to understand the exact roles of the allene group, we designed following three peridinin derivatives. Acetylene derivative **2** possesses an epoxy-acetylene, olefin derivative **3** has an epoxy-olefin, and diolefin derivative **4** has a conjugating olefin group instead of the hydroxy-allene group. Next, in order to understand why peridinin possesses the irregular C37 skeleton, we designed three peridinin derivatives as a series of different π–electron chain length compounds. These are C33 derivative **5** which has two fewer double bonds than peridinin, C35 derivative **6** which has one less double bond, and C39 derivative **7** which has one more double bond. On the other hand, in order to understand the role of the γ-ylidenebutenolide group, we designed the open-ring peridinin derivatives **8** and **9**. Derivative **8** possesses a triple bond and a methyl ester group, and derivative **9** has a double bond and also a methyl ester group instead of the γ-ylidenebutenolide group. These derivatives would provide useful information on the roles of these unique functional groups by comparing their data on the spectroscopies and energy transfer efficiencies.

According to the stereocontrolled synthesis of peridinin, which we previously established and the strategy is shown in Fig. 3 (Furuichi et al., 2002, 2004), we planned to synthesize these peridinin derivatives by a coupling between the pairs of C17-allenic segment **10** and the corresponding ylidenebutenolide-modified half-segments **15-17** or C20-ylidenebutenolide half-segment **14** and the corresponding allene-modified half-segments **11-13** using the modified Julia olefination reaction (Baudin et al., 1991, 1993) (Fig. 3). Namely, we planned to synthesize the allene modified derivatives **2** and **3** by a coupling between the corresponding allene modified half-segment **12** and **13** and C20-ylidenebutenolide half-segment **14**. Next, the syntheses of both C35 and C39 peridinin derivatives **6** and **7** would be possibly synthesized by utilizing the modified Julia olefination of the appropriate allenic

An Approach Based on Synthetic Organic Chemistry Toward Elucidation of Highly Efficient Energy Transfer
Ability of Peridinin in Photosynthesis

137

half segments such as **10** and **11** with the suitable γ-ylidenebutenolide half segments such as **14** and **15**, respectively. Thus, a coupling between C15-allenic segment **11** and C20-ylidenebutenolide segment **14** would produce C35 peridinin derivative **6**. Meanwhile, applying the same method to the coupling between C17-allenic segment **10** and C22-ylidenebutenolide segment **15** might produce the desired C39 peridinin derivative **7**, if the coupling product would be enough stable to be handled. On the other hand, we planned to synthesize acetylene ester derivative **8** and olefin ester derivative **9** by a coupling between the C17-allenic half-segment **10** and the corresponding ylidenebutenolide-modified half-segments **16** and **17**. We have really achieved the synthesis of these complex peridinin derivatives **2-9** by this efficient strategy.

Fig. 2. Structure of peridinin and its derivatives

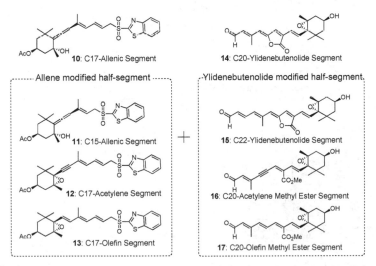

Fig. 3. Synthetic strategy

2.2 Syntheses of allene modified derivaitves

First, the synthetic studies of the allene modified derivatives **2-4** are described. The synthesis of the longer conjugated half-segments was more difficult and needed the milder reaction conditions than those of the peridinin synthesis. Under stereospecific manner for the construction of the desired conjugated chains, palladium catalyzed sp-sp^2 and sp^2-sp^2 couplings were very effective. Meanwhile, the synthesis of the allenic functional group has already been established as follows: acetylene derivative **19** was prepared starting from (-)-actinol **18** as shown in Fig. 4. The Sonogashira cross-coupling between **19** and vinyl iodide **20** in the presence of catalytic amounts of $Pd(PPh_3)_4$ and CuI in diisopropylamine produced the desired ester **21** in 84% yield. In the case of using organic solvents, such as THF and CH_2Cl_2, the yield was lower. The conjugated diene ester **21** thus obtained was transformed into allenic triol **23** by the stereospecific hydride redution in 80% yield, whose method was already established and generally used for the synthesis of the allenic carotenoids. The stereochemistry of the obtained allenic triol **23** was well explained by the following

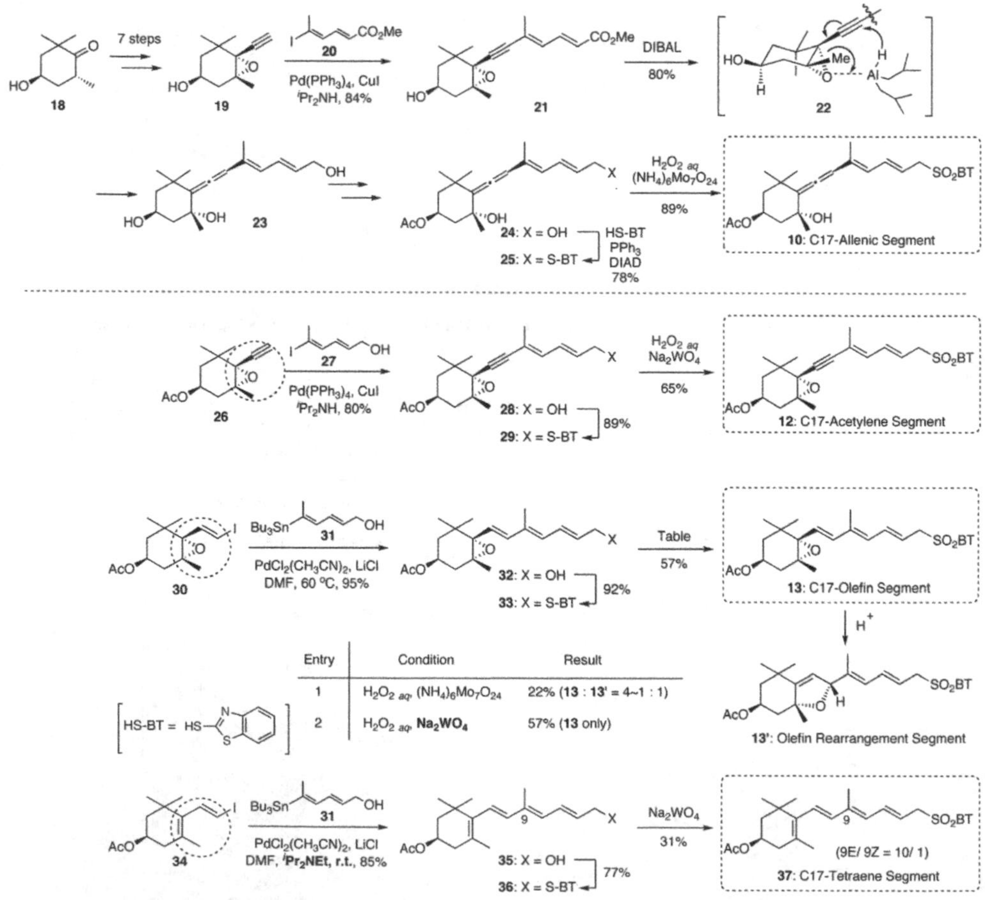

Entry	Condition	Result
1	H_2O_2 aq, $(NH_4)_6Mo_7O_{24}$	22% (**13** : **13'** = 4~1 : 1)
2	H_2O_2 aq, Na_2WO_4	57% (**13** only)

Fig. 4. Synthesis of allenic and allene modified half-segments

An Approach Based on Synthetic Organic Chemistry Toward Elucidation of Highly Efficient Energy Transfer
Ability of Peridinin in Photosynthesis
139

consideration; this reaction proceeded by intramolecular S_N2' hydride reduction resulting from the coordination between the oxygen atom of the epoxide in **21** and the aluminum atom of DIBAL, as shown in **22**. The obtained acetyl diol **24** was transformed into the C17-allenic segment **10** using the Mitsunobu reaction with 2-mercaptobenzothiazole, followed by oxidation of the resulting sulfide **25** with aqueous 30% H_2O_2 and $(NH_4)_6Mo_7O_{24}$.

The terminal segments **26**, **30** and **34** were led to the each half-segment **12**, **13** and **37** shown in Fig. 4, respectively. The Sonogashira cross-coupling between **26** and vinyl iodide **27** under the same reaction condition produced the desired alcohol **28** in 80% yield. The obtained alcohol **28** was transformed into the acetylene segment **12** using the Mitsunobu reaction with 2-mercaptobenzothiazole, followed by oxidation of the resulting sulfide with aqueous 30% H_2O_2 and $Na_2WO_4 \cdot 2H_2O$, which was milder than $(NH_4)_6Mo_7O_{24}$ (Schulz et al., 1963).

On the other hand, the olefin segment **13** would be obtained by a coupling between vinyl iodide **30** and vinylstannane **31**. The Stille cross-coupling reaction of **30** with vinyl stannane **31** in the presence of $PdCl_2(CH_3CN)_2$ and LiCl gave the desired alcohol **32** in exellent yield as a single isomer. The Stille cross-coupling of the opposite conbination between the corresponding stannane and iodide did not afford the desired result. The alcohol **32** was transformed into sulfide **33** by the same procedure. Oxidation of **32** under the same reagent as that for the preparation of **12** gave the desired **13**. However, the use of 30 % H_2O_2 and $(NH_4)_6Mo_7O_{24}$, which is a little strict condition, gave a mixture of the desired **13** and the isomerized **13'** in low yield, and the ratio of **13** and **13'** was not reproducible (1: 4 to 1: 1). It is noteworthy that sulfone **13** was easily isomerized to **13'** by a trace amount of hydrochloric acid in $CDCl_3$.

Next was the synthesis of diolefin segment **37**. The Stille cross-coupling of **34** with vinyl stannane **31**, which was used in the synthesis of **13**, afforded tetraene alcohol **35** as a single isomer. In this coupling, the reaction smoothly proceeded at room temperature, and when $^iPr_2NEt$ was not used, a mixture of **35** and its 9Z-isomer was obtained in a ratio of eight to one by NMR. The amount of 9Z-isomer seemed to increase at higher reaction temperature, for instance, 9E/9Z = 3/1 at 60 °C, which is the same condition to that of the synthesis of **13**. The desired sulfone **37** was obtained from **35** by the Mitsunobu reaction with 2-mercaptobenzothiazole, followed by oxidation of the resulting sulfide with aqueous 30% H_2O_2 and $Na_2WO_4 \cdot 2H_2O$, as a mixture of 9E/9Z = 10/1 in 31% yield. The use of 30% H_2O_2 with $(NH_4)_6Mo_7O_{24}$ and mCPBA gave a complex mixture. Oxidation of the allylic sulfide to the corresponding sulfone in longer conjugated polyenes was still problematic.

We chose the modified Julia olefination as the final C-C coupling reaction, because most of this olefination proceeded even at -78 °C. Such low temperature reaction was well suitable for the construction of the poly functionalized polyene chain such as peridinin. For instance, the crucial modified Julia olefination was explored as the final key step in the synthesis of acetylene derivative **2**. The reaction of an anion derived from **12** with **14** at -78 °C smoothly proceeded within 5 min in the dark to produce the peridinin derivatives in 42% amount as a mixture of stereoisomers. Due to the previous experiments in our carotenoid syntheses and the reports of the Brückner's and de Lera's groups that the modified Julia olefination of polyene compounds generally produced the Z-isomer at the connected double bond (Bruckner et al., 2005; Vaz et al., 2005), we tried to isomerize the connected double bond monitoring by HPLC as shown in Fig. 5. The resulting mixture was allowed to stand in benzene at room temperature under fluorescent light in an argon atomosphere. The isomerization under fluorescent light was faster than that in the dark. After 2 days, we

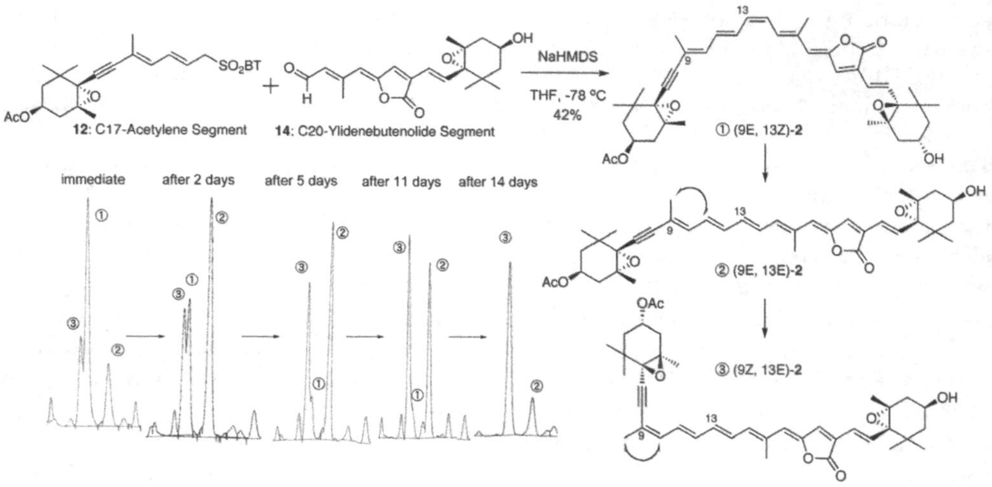

Fig. 5. Isomerization and structure of acetylene derivative

observed that the initially generated major peak (peak 1 in the immediate situation) changed to another major peak (peak 2). After 11 days, while the peak 2 gradually decreased, the peak 3 increased. After 14 days, the peak 2 became the major peak in an equilibrium state. We isolated all peaks by both the mobile-phase and the reverse-phase HPLC, and elucidated their structures by NMR (400 and 750 MHz). Thus, we clarified that the peak 1 was (9E, 13Z)-isomer 2, the peak 2 was (9E, 13E)-all-*trans* acetylene derivatve 2, and peak 3 was (9Z, 13E)-isomer 2. All-*trans* derivative 2 did not isomerize to the 9Z-isomer at –20 °C but gradually isomerized at room temperature in the dark. Obviously, all-*trans* isomer was unstable at room temperature and easily isomerized to the 9Z-isomer (Vaz et al., 2006), which was the most stable isomer. In addition, olefin derivative 3 and diolefin derivative 4 were synthesized by the same procedure (Kajikawa et al., 2009a).

2.3 Syntheses of polyene-chain modified derivaitves

Next, we synthesized polyene chain modified peridinin derivatives by using a stereocontorolled domino one-pot formation of the ylidenebutenolide as a key step. First, dibromide 40 was obtained by a sequence of the Wittig reaction, $^1O_2$ oxygenaiton followed by a treatment with diisopropylethylamine in the presence of allyl bromide, and the Corey's dibromination from aldehyde 38, which was prepared from (-)-actinol in 53% for 4 steps. Dibromide 40 was successfully transformed into alkyne 41 by the treatment with TBAF in 81% yield (Tanaka et al., 1980). Next was the key stereocontrolled preparation of the ylidenebutenolide segment 14 from alkyne 41 (Fig. 6). Thus, alkyne 41 was treated with vinyl iodide 42 and cuprous iodide in triethylamine for 1 h followed by an addition of formic acid after confirming the consumption of the starting 41 by TLC analysis. The mixture was then further stirred at room temperature for overnight to produce the desired ylidenebutenolide 45 in 49% yield under the stereocontrolled fashion in one-pot. This three-step domino one-pot reaction to prepare the ylidenebutenolide 45 could be explained in detail by the possible mechanism shown in Fig. 6. At first, Sonogashira coupling of 41 and iodide 42 proceeded to afford the desired coupling product, in which the π-allylpalladium generated from the allylester group was formed and coordinated to the alkyne such as 43.

An Approach Based on Synthetic Organic Chemistry Toward Elucidation of Highly Efficient Energy Transfer
Ability of Peridinin in Photosynthesis
141

Next, the intermediary **43** underwent π-allylalkenylpalladium(II)-assisted regio- and stereoselective intramolecular cyclization to form the π-allylalkenylpalladium lactone intermediate **44**. In the final step, the π-allylalkenylpalladium moiety was removed by hydrogenolysis with formic acid to give the desired ylidenebutenolide **45**. MnO_2 oxidation of **45** gave the stereocontrolled ylidenebutenolide segment **14**.

The crucial one-pot ylidenebutenolide formation from **41** and vinyl iodide **46**, which was previously synthesized by us, was explored as the key step in the synthesis of C33 peridinin derivative **5**. Thus, a mixture of **41** and **46** was stirred in the presence of catalytic amounts of $Pd(PPh_3)_4$ and cuprous iodide in triethylamine at 45 °C for 10 min. After the complete consumption of **41** was ascertained by TLC, formic acid was added to the reaction mixture and then the mixture was stirred at 45 °C for 10 min to produce the desired C33 peridinin derivative in 35% yield as a mixture of stereoisomers in one-pot. The undesired 11Z-isomer **46** resulted in the undesired 11Z-isomer of the compound **5**. The resulting mixture was then allowed to isomerize in benzene at room temperature under fluorescent light in an argon atmosphere to successfully produce the desired **5** as a mixture of stereoisomers (Fig. 7).

Next, the stereocontrolled preparation of C22-ylidenebutenolide segment **15** from alkyne **41** was fortunately successful by the same procedure; a mixture of **41** and vinyl iodide **47** was stirred at 45 °C for 10 min to produce the desired ylidenebutenolide compound **15** in 40% yield as the 13'E/ 13'Z mixture (10/ 1). The reaction with the corresponding hydroxy derivative of **47** did not give the desired result because of its instability.

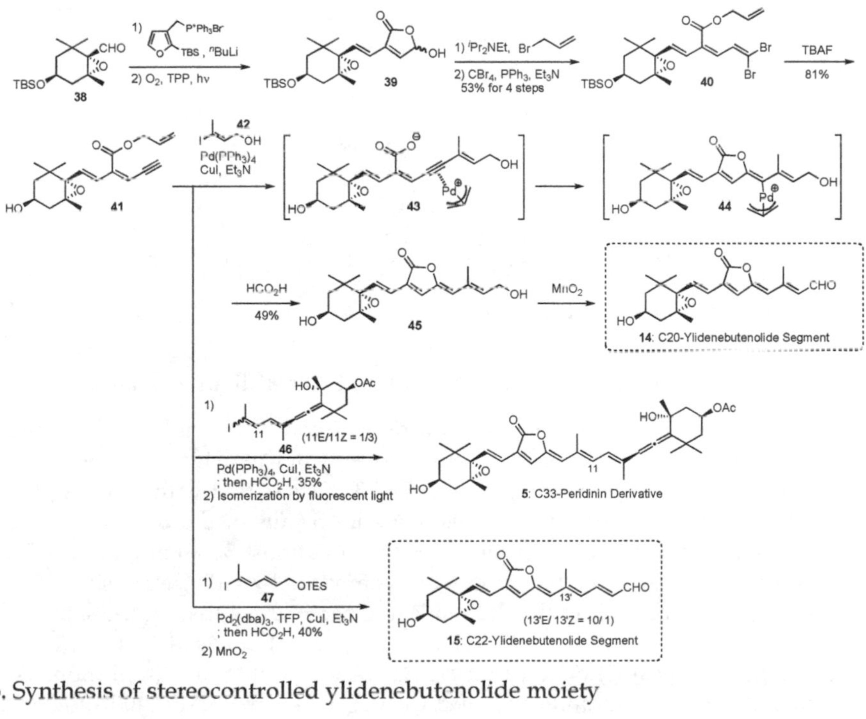

Fig. 6. Synthesis of stereocontrolled ylidenebutenolide moiety

We thus successfully synthesized the C20- and C22-ylidenebutenolide half-segments **14** and **15**, and C33 peridinin derivative **5** by the same way. The isomerization of C33 peridinin derivative was shown in Fig. 7 (A). The resulting mixture was then allowed to isomerize in benzene at room temperature under fluorescent light in an argon atmosphere. After 2 days, we observed that the initially generated major peak (peak 2) in Fig. 7 (A) changed into another major peak (peak 1) in the HPLC. In addition, peak 3 became larger after 2 days, when the situation would be an equilibrium state. We isolated all peaks by the mobile-phase HPLC and elucidated their structures by NMR (400 MHz), and we elucidated that peak 1 was fortunately (11E, 11'Z)-all-*trans* C33 peridinin **5**, peak 2 was (11Z, 11'Z)-isomer **5'** and peak 3 was (11E, 11'E)-isomer **5''**, respectively. Interestingly, (11E, 11'E)-isomer **5''** was the secondarily larger isomer in the equilibrium state.

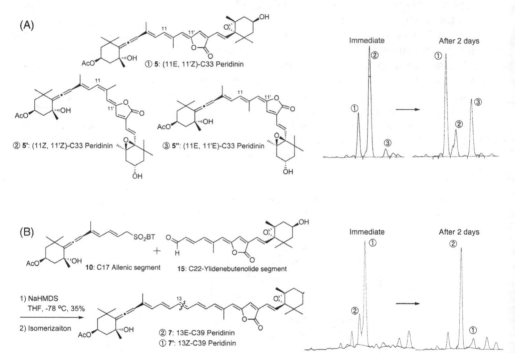

Fig. 7. Structure and HPLC analysis of (A) C33 and (B) C39 peridinin derivatives

Furthermore, relatively unstable C39 peridinin derivative **7** was synthesized by the same protocol as shown in Fig. 7 (B). Thus, the anion derived from **10**, which was the allenic half-segment of the established peridinin synthesis (Fig. 3), was stirred with **15** under the same condition. Fortunately, the reaction completed within 5 min in the dark to produce the coupling products as a mixture of the stereoisomers in almost 35% amount, in which the 13Z-isomer (peak 1) was estimated to be 48% of the mixture by HPLC analysis (13E-isomer: peak 2 was 19%). Isomerization to the desired **7** was again attempted by the same method. After 2 days, a large amount of the 13Z-isomer **7'** (peak 1) changed to the all-*trans* C39 peridinin derivative **7** (peak 2) (57% based on HPLC analysis) in an equilibrium state. We then isolated both compounds, and confirmed their structures by NMR (400 and 750 MHz). The synthesized all-*trans* C39 peridinin derivative gradually decomposed within one month under an argon gas atmosphere at around –20 ºC. This instability was in good contrast to the case of peridinin, which could be stored without any remarkable decomposition under the same conditions. Meanwhile, C35 peridinin derivative was also synthesized by the same procedure (Kajikawa et al., 2009b).

2.4 Syntheses of ylidenebutenolide modified derivatives and the stereochemical and spectral characteristics of the synthesized derivatives

Finally, the synthesis of ylidenebutenolide modified peridinin derivatives is described. The acetylene ester derivative was synthesized by the same protocol. Namely, the coupling between C17-allenic segment **10** and C20-acetylene methyl ester segment **16** by using the modified Julia olefination, which was followed by the isomerization gave all-*trans* **8** (Fig. 8 (A)) (Kajikawa et al., 2010).

On the other hand, there were some difficulties in the synthesis of the olefin ester derivative (Fig. 8). We obtained the only 9'E-olefin ester segment **17** by the similar synthetic process, which was not the desired 9'Z half-segment resulting from the contribution of the carbonyl group of the methyl ester. We then tried to connect the segments **10** and **17** by the modified Julia olefination and to obtain all-*trans* **9-3** by the isomerization. The anion derived from **10** was stirred with a mixture of stereoisomers of **17** under the same conditions used for the coupling of the previous peridinin derivatives. The reaction was completed within 5 min in the dark to produce the coupling products as a mixture of the stereoisomers in 46% amount, whose HPLC is shown in Fig. 8 (B). The major peak (peak 1) was estimated to be 45% of the mixture by HPLC analysis (other isomers were 16%, 14%, 5%, 5%, and others). Isomerization to the desired all-*trans* **9-3** was attempted under the same conditions previously used. After 5 days, the initially generated major peak (peak 1) changed to another peak (peak 2; 44% based on HPLC analysis) in an equilibrium state. We then isolated both compounds and elucidated their structures by NMR (400 and 750 MHz), and clarified that peak 1 was (13Z, 9'E)-isomer **9-1** and peak 2 was (13E, 9'E)-isomer **9-2**. Unfortunately, we could not obtain the desired all-*trans* (13E, 9'Z)- isomer **9-3**.

We investigated the stability of the synthesized ring opened derivatives **8** and **9**, and found the isolated all-*trans* acetylene ester derivative **8-1** was more labile than **8-2** (13Z-isomer). For instance, the isolated all-*trans* derivative **8-1** (13E-isomer) isomerized to the dihydrofuran derivative **8'** by a trace amount of hydrogen chloride in CDCl$_3$, but the corresponding isomerization of **8-2** (13Z-isomer) was not observed (Fig. 8). In addition, the all-*trans* derivative **8-1** rapidly isomerized to give a mixture of Z-isomers upon illumination. This might occur due to the contribution of the carbonyl group of the methyl ester similar to the case of the 9'E-olefin ester derivative **9-2**.

In the PCP complex, peridinin exhibits an exceptionally high efficiency of energy transfer to Chl a. In order to make clear the effect of the ylidenebutenolide, we needed to measure the energy transfer efficiencies in peridinin derivatives. Futhermore, it was tried to construct the corresponding PCP derivatives using the synthesized peridinin derivatives **8** and **9** to compare with the energy transfer efficiencies of peridinin (private information from Dr. H. A. Frank). First, it was attempted to reconstitute the PCP apoprotein using the 9'E-olefin ester derivative **9-2** under the same conditions that were successful for peridinin (Ilagan et al., 2006), but the reconstitution was not observed. It was also tried to reconstitute the PCP apoprotein using the 13Z-isomer **8-2**, but it did not bind the protein either. The reason might be that these compounds were bent into a *cis* configuration, and hence they might not fit properly into the protein binding site. These results apparently showed that the γ-ylidenbutenolide of peridinin at least contributes to the stereochemical stability of the compound and would keep the all-*trans* conformer suitable for incorporation into the protein to form the PCP complex.

The maximum absorption wavelengths (λ_{max}) in the electronic spectra of peridinin (**1**) and the synthesized derivatives **2~7**, **8-1** and **9-2** in hexane were measured and are summerized in Fig. 9. Evidently, the diolefin derivative **4** and C39 peridinin derivative **7** showed the longer λ_{max} than that of peridinin. The λ_{max} value in polyene chain modified derivatives **5-7** increased almost 20 nm per one olefin unit added to the conjugated polyene. On the other hand, the olefin derivative **3**, having eight conjugated carbon-carbon double bonds like peridinin, showed the shorter λ_{max} than that of peridinin. The open-ring derivative also displayed a shorter λ_{max} than peridinin (**1**) due to shorter effective π-electron conjugated chain length, although the 9′E-olefin ester derivative **9-2** had the same conjugated carbon-carbon double bonds compared with peridinin. These results show that the allene and ylidenebutenolide group at least contribute to giving rise to the λ_{max} value desirable for the marine organism to absorb light in the blue-green region of the visible spectrum.

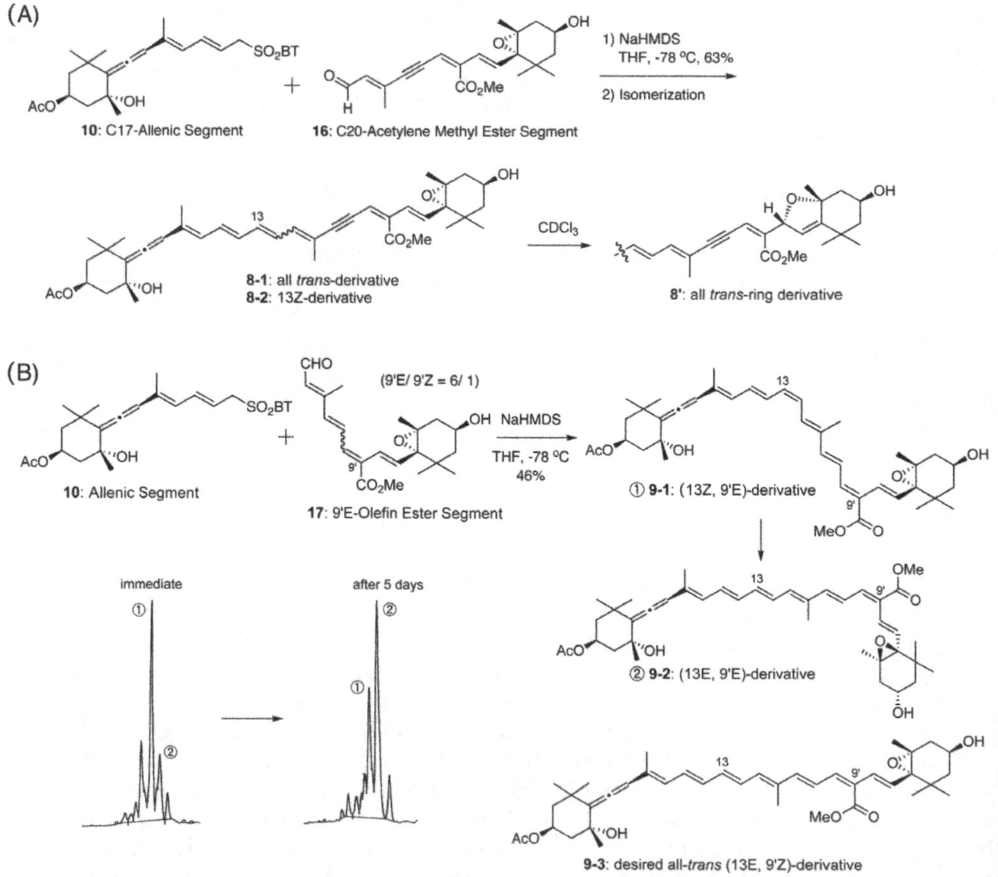

Fig. 8. Stereochemical characteristics of ylidenebutenolide modified derivatives

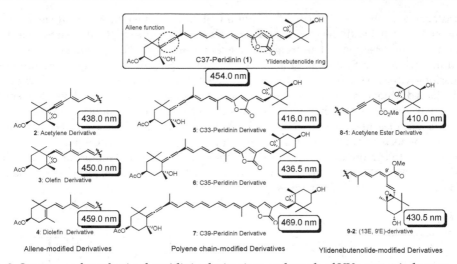

Fig. 9. Structure of synthesized peridinin derivatives and result of UV spectra in hexane

3. Relationships between the unique structure and the special exited state of peridinin

3.1 Measurement of ultrafast time-resolved optical absorption spectra

From studies on peridinin and other carotenoids, it has been known that there are at least two important low-lying excited singlet states, denoted S_1 and S_2, which are related to the highly efficient energy transfer from peridinin to chlorophyll a (Fig. 10(A)). To elucidate the mechanism of this efficient energy transfer, it is important that we make clear the characteristics of these excited states and the energy transfer pathways such as those from S_2 to Q_X and/or from S_1 to Q_Y. Recently, many researchers have tried to examine this particular mechanism. The conjugated double bonds of most carotenoids are symmetry and these double bonds can be regarded as polyenes described in terms of the idealized C_{2h} point group in the spectroscopic fields (Hudson et al., 1973). The lowest excited singlet (S_1) state is assigned to the $2^1A_g^-$ state, and the second lowest singlet (S_2) state is assigned to the $1^1B_u^+$ state. The excitation to S_1 from the ground state is symmetry forbidden and is not directly accessible by one-photon processes in contrast to the allowed absorption to S_2 state (Polivka et al., 2004). On the other hand, the conjugated double bond of peridinin and other carbonyl-containing carotenoids are asymmetric due to the presence of the conjugated carbonyl group, and these oxygenated carotenoids display a pronounced solvent dependence of its lowest excited singlet state lifetime (S_1 lifetime). Namely, it has been proposed that the findings are consistent with the presence of an intramolecular charge transfer (ICT) state, which is uniquely formed in carotenoids containing the carbonyl group in conjugation with the π-electron system of double bonds (Frank et al., 2000; Zigmantas et al., 2004). It has also been argued that changes in the position of the ICT state related to the S_1 state rationalize the dependence on solvent polarity concerning S_1 lifetime. In the case of peridinin, the relationship of these energy levels is well discussed based on the detailed experimental works. The ICT in the excited state manifold of peridinin is shown to be higher in energy than the S_1 state in nonpolar solvents and shifts below S_1 with increasing solvent polarity

(Bautista et al., 1999) (Fig.10 (B)). In addition, it is suggested that the efficient energy transfers are related with this ICT state. Proposals for the nature of the ICT state include its being a separate electronic state from S_1 (Vaswani et al., 2003; Papagiannakis et al., 2005), quantum mixed with S_1 (Shima et al., 2003) or simply S_1 itself. Although there are many discussions with experiments and calculations, the precise nature of the ICT state remains to be elucidated. Under these back-ground on the proposed attractive energy level, ICT energy level, a new approach, that the synthesis of a series of peridinin analogues followed by their spectroscopic measurements are investigated, has been started as a collaboration work between Connecticut University, Osaka City University and Kwansei Gakuin University of Hyogo. Thus, to explore the nature of the ICT state in carbonyl-containing carotenoids, both steady-state and ultrafast time resolved optical spectroscopy have been performed on peridinin and its synthetic derivatives.

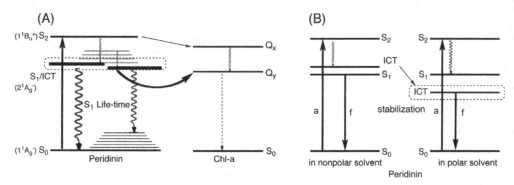

Fig. 10. (A) Enegy transfer from peridinin to chl-a and (B) the nature of ICT state

The ultrafast time resolved optical absorption experiments of peridinin (1) and many other carbonyl-containing natural carotenoids such as fucoxanthin and spheroidenone were measured, and their S_1 lifetimes were reported by the analysis of their ultrafast time resolved optical absorption (Frank et al., 2000). The lifetime of three natural carotenoid was reported to depend on the polarity of the measured solvents, and this effect is attributed to the presence of an intramolecular charge transfer (ICT) state in the manifold of the excited states of these molecules. We then measured the lifetime of the lowest excited single state of the four compounds, which are C33, C35 and C39 synthesized derivatives along with peridinin, and the results are listed in Fig. 11 (A). The data listed in Fig. 11 (A) show that the lifetime is shorter in the polar solvent, methanol, and is longer in a non-polar solvent, n-hexane. This means that the ICT states in the excited state manifold of peridinin and its three derivatives are higher energy than the S_1 state in nonpolar solvents, and they shift to a lower energy than the S_1 state in polar solvents. These experiments on peridinin and its derivatives revealed an increasing solvent effect with the decreasing π-electron chain length. This result agrees with the experimental results carried out on conjugated apo-carotenoids (Ehlers et al., 2007). The lifetime of the lowest excited singlet state of C33 peridinin derivative 5 is the one most strongly dependent on the solvent polarity. In fact, this is the strongest solvent dependence on the lifetime of the carotenoid excited state so far yet reported. Moreover, the most striking observation in the data is that the lifetime of the ICT state converges to a value of 10 ± 1 ps in the polar solvent, methanol, for all the peridinin analogues regardless of the extent of π-electron conjugation. Potential energy level diagrams for four molecules in polar and nonpolar solvents are described as shown in Fig. 11 (B). Based on the results of S_1 lifetime, althought S_1 state gradually drops as longer polyene chain in hexane, the ICT state exists in the same position in methanol. We dramatically observed that the behavior of ICT states were obviously different from that of S_1 states in the series of our synthesized peridinin derivatives including peridinin itself. These results strongly support the idea that the S_1 and ICT states act as independent states. We can not, however, conclude clearly whether ICT state is separate or mixed energy level from S_1 state. The unexpected phenomena, that the ICT state exists in the same position in methanol, is quite intereisting. We can presume that this nature of the ICT state is very important for energy transfers because the environment in methanol is considered to be the nearly same to that of PCP complex (Akimoto et al., 1996).

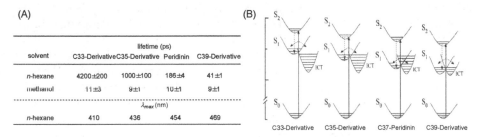

(A)

solvent	lifetime (ps)			
	C33-Derivative	C35-Derivative	Peridinin	C39-Derivative
n-hexane	4200±200	1000±100	186±4	41±1
methanol	11±3	9±1	10±1	9±1
	λ_{max} (nm)			
n-hexane	410	436	454	469

Fig. 11. (A) Result of S_1 lifetime and (B) proposed energy level diagram

3.2 Measurement of Stark spectra

The precise relationship between S_1 and ICT energy levels and also the nature of ICT are described in the previous chapter. On the other hand, there has also been a suggestion of the relationship between S_2 and ICT states. The measurement of electroabsorption spectroscopy (Stark spectra) of peridinin has been reported (Premvardhan et al., 2005). Stark spectra can determine the change in electrostatic properties and estimate the change of the static dipole moment ($|\Delta\mu|$) between in the ground state and in the excited state. Thus, the value of this change represents by $|\Delta\mu|$. Based on the observation, it is found that the absorption band from S_0 to S_2 showes large static dipolemoment change. In addition, it is suggested that in PCP complex there may be strong dipole-dipole coupling between peridinin and chlorophyll a. The large dipole moment would allow for strong dipolar interaction between peridinin and Chl a in PCP, and would contribute to high energy transfer. It has also been recently proposed that the presence of the ICT excited state promotes dipolar interactions with Chl a in the PCP complex and facilitates energy transfer via a dipole mechanism (Zigmantas et al., 2002). Although the magnitude of the static dipole moment is suggested to be very important, the relationship between the structural features of peridinin and the dipole moment has not been made clear. We then measured the Stark absorption spectra of perdinin along with its allene modified and polyene chain modified derivatives. Stark spectra is particularly suitable for peridinin and its derivatives, because the presence of the ICT state would be directly discernible.

The Stark spectra and the maximum absorption of the electronic spectra of peridinin (**1**) and the synthesized derivatives (**2-4, 6** and **7**) are summarized in Fig. 12. The Stark spectra of peridinin, allene modified and polyene chain modified derivatives were recorded in methyl methacrylate polymer at 77 K. The $|\Delta\mu|$ values were corresponding to the CT absorption band. As the results of peridinin and allene modified derivatives, peridinin showed the largest $|\Delta\mu|$ value among all of them. Namely, peridinin yielded a $|\Delta\mu|$ value of 5.42 (x 10^{-29} C · m), acetylene derivative **2** showed 2.47, olefin derivative **3** showed 4.22, and diolefin derivative **4** showed 4.25. A $|\Delta\mu|$ value of peridinin was in agreement with a reported value (Premvardhan et al., 2005). The $|\Delta\mu|$ value generally shows a larger number with the increasing π-electron chain length theoretically. Although peridinin possesses fewer conjugating double bonds and shows a shorter λ_{max} rather than that of diolefin derivative **4**, the $|\Delta\mu|$ value of peridinin was the largest among the four compounds. The difference in the $|\Delta\mu|$ value is evidently attributable to the difference in the functional groups. Thus, we have understood that the unique allene group contributes to production of the large dipole moment in the molecule. These results strongly suggest that the allene group of peridinin is essential for formation of the effective ICT state, which would allow the quantitative energy transfer to Chl a in the PCP complex. This is the first experimental evidence that shows the allene group in peridinin enhances the ICT character (Kusumoto et al., 2010).

In addition, as the results of peridinin and polyene chain modified derivatives, peridinin (**1**) also showed the largest $|\Delta\mu|$ value among all of them. Namely, peridinin (**1**) yielded a $|\Delta\mu|$ value of 5.42 (x 10^{-29} C · m), C35 peridinin derivative **6** showed 4.25, and C39 peridinin derivative **7** did 5.29. The $|\Delta\mu|$ value generally shows a larger number with the increasing π-electron chain length (Kajikawa et al., 2009b). Although peridinin possesses fewer conjugated double bonds and shows a rather shorter λ_{max} than that of C39 peridinin derivative **7**, the $|\Delta\mu|$ value of peridinin (**1**) was the largest among the three compounds. Thus, the C37 skeleton of peridinin (**1**) would also contribute to the large dipole moment of the molecule in the exited state to facilitate energy transfer. This would be at least a partial answer to the question of why peridinin (**1**) possesses the irregular C37 skeleton.

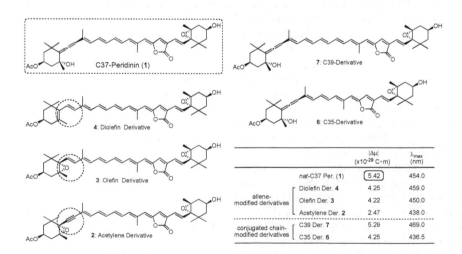

		$|\Delta\mu|$ (x10^{-29} C·m)	λ_{max} (nm)
	nat-C37 Per. (1)	5.42	454.0
allene-modified derivatives	Diolefin Der. 4	4.25	459.0
	Olefin Der. 3	4.22	450.0
	Acetylene Der. 2	2.47	438.0
conjugated chain-modified derivatives	C39 Der. 7	5.29	469.0
	C35 Der. 6	4.25	436.5

Fig. 12. Structure of Peridinin and its derivatives and the result of Stark spectra

4. Conclusion

The research on the efficient energy transfer mechanism of the primary photochemical events of photosynthesis have been exactly developed by the new approach based on the synthetic organic chemistry. Namely, the relationship between the characteristic structure of peridinin and the nature of its ICT state has been gradually made clear by synthesizing a series of peridinin analogues and comparing their characteristics to those of peridinin.

Stark spectra of peridinin, allene modified, and polyene chain modified derivatives have showed that the dipole moment of the excited state ($|\Delta\mu|$ value) of peridinin is the largest among the series of six compounds, although the increasing π-electron chain length generally shows a larger value of $|\Delta\mu|$. These results apparently show that the allene group and C37 skeleton of peridinin effectively contributes to production of the large dipole moment in the molecule in excited state, which would result in the high energy transfer efficiencies to Chl a in the PCP complex. This is an answer of why peridinin possesses the unique allene bond and the irregular C37 skeleton.

In addition, the ultrafast time resolved optical absorption spectra of polyene chain modified derivatives including peridinin show that the lifetime of the lowest excited singlet state of C33 peridinin derivative has the strongest solvent dependence so far yet reported. Furthermore, the data reveal the striking observation that the lifetime of the ICT state converges to a value of 10 ± 1 ps in methanol for all peridinin analogues regardless of the extent of π-electron conjugation. These data strongly support the notion that the S_1 and ICT states behave independently.

On the other hand, comparing the stereochemical stability and spectral characteristics of the synthesized ylidenebutenolide modified analogues to those of peridinin has resulted in the conclusion that this particular functional group at least contributes to maintaining the stereochemistry of the conjugated double bonds in the all-*trans* configuration and giving rise to a λ_{max} value desirable for the marine organism to absorb light in the blue-green region of the visible spectrum.

These inherent characteristics of peridinin are important clues for elucidating the energy transfer mechanism from the light-harvesting carotenoids to chlorophylls. The studies to measure the energy transfer efficiencies of peridinin derivatives are currently in progress to further understand the exact role of these unique functional groups.

5. Acknowledgment

We thank Dr. Thomas Netscher of DSM Nutritional Products, Ltd., for the donation of (-)-actinol **18**. We also would like to thank Prof. H. A. Frank (University of Connecticut) with ultrafast experiments and Prof. H. Hashimoto (Osaka City University) with Stark spectra measurement. This work was supported by a Grant-in-Aid for Science Research on Priority Areas 16073222 from the Ministry of Education, Culture, Sports, Science and Technology, and Matching Fund Subsidy for a Private University, Japan. T. K. is also grateful for recieving the Scholar Ship of JSPS.

6. References

Akimoto, S.; Takaichi, S.; Ogata, T.; Nishiyama, Y.; Yamazaki, I. & Mimuro, M. (1996). Excitation energy transfer in carotenoid chlorophyll protein complexes probed by femtosecond fluorescence decays. *Chemical Physics Letters*, 260, 1-2 (September 1996), pp. 147-152 ISSN 0009-2614

Baudin, J. B.; Hareau, G.; Julia, S. A. & Ruel, O. (1991). A direct synthesis of olefin by reaction of carbonyl-compounds with lithioderivatives of 2-[alkyl-sulfonyl or (2'-alkenyl)-sulfonyl or benzyl- sulfonyl]-benzothiazoles. *Tetrahedron Letters*, 32, 9 (February 1991), pp. 1175-1178 ISSN 0040-4039

Baudin, J. B.; Hareau, G.; Julia, S. A. & Ruel, O. (1993). Stereochemistry of direct olefin formation from carbonyl-compounds and lithiated heterocyclic sulfones. *Bulletin de la Societe Chimique de France*, 130, 6 (February 1993), pp. 856-878 ISSN 0037-8968

Bautista, J. A.; Connors, R. E.; Raju, B. B.; Hiller, R. G.; Sharples, F. P.; Gosztola, D.; Wasielewski, M. R. & Frank, H. A. (1999). Excited state properties of peridinin: observation of a solvent dependence of the lowest excited singlet state lifetime and spectral behavior unique among carotenoids. *Journal of Physical Chemistry B*, 103, 41 (October 1999), pp. 8751-8758 ISSN 1089-5647

Bruckner, R. & Sorg, A. (2005). Unexpected cis-selectivity in (Sylvestre) Julia olefination with Bu_3Sn-containing allyl benzothiazolyl sulfones: Stereoselective synthesis of 1,3-butadienyl- and 1,3,5-hexatrienylstannanes *Synlett*, 2, (February 2005), pp. 289-293 ISSN 0936-5214

Ehlers, F; Wild, D. A.; Lenzer, T. & Oum, K. (2007). Investigation of the $S_{1/ICT}$ to S_0 internal conversion lifetime of 4'-apo-β-caroten-4'-al and 8'-apo-β-caroten-8'-al: dependence on conjugation length and solvent polarity. *Journal of Physical Chemistry A*, 111, 12 (March 2007), pp. 2257-2265 ISSN 1089-5639

Frank, H. A.; Bautista, J. A.; Josue, J.; Pendon, Z.; Hiller, R. G.; Sharples, F. P.; Gosztola, D. & Wasielewski, M. R. (2000). Effect of the solvent environment on the spectroscopic properties and dynamics of the lowest excited states of carotenoids. *Journal of Physical Chemistry B*, 104, 18 (May 2000), pp. 4569-4577 ISSN 1089-5647

Furuichi, N.; Hara, H.; Osaki, T.; Mori, H. & Katsumura, S. (2002). Highly efficient stereocontrolled total synthesis of the polyfunctional carotenoid peridinin. *Angewandte Chemie-International Edition*, 41, 6, (November 2001), pp. 1023-1026, ISSN 1433-7851

Furuichi, N.; Hara, H.; Osaki, T.; Nakano, M.; Mori, H. & Katsumura, S. (2002). Stereocontrolled total synthesis of a polyfunctional carotenoid, peridinin. *Journal of Organic Chemistry*, 69,(July 2004), pp-7949-7959, ISSN 0022-3263

Hofmann, E.; Wrench, P. M.; Sharples, F. P.; Hiller, R. G.; Welte, W. & Diederichs, K. (1996). Structural basis of light harvesting by carotenoids: Peridinin-chlorophyll-protein from Amphidinium carterae. *Science*, 272, 5269, (June 1996), pp. 1788-1791, ISSN 0036-8075

Hudson, B. S. & Kohler, B. E. (1973). Polyene spectroscopy: The lowest energy excited singlet state of diphenyloctatetraene and other linear polyenes. *The Journal of Chemical Physics*, 59, 9, (April 1973), pp. 4984-5002

Ilagan, R. P.; Chapp, T. W.; Hiller, R. G.; Sharples, F. P.; Polivka T. & Frank, H. A. (2006). Optical spectroscopic studies of light-harvesting by pigment-reconstituted peridinin-chlorophyll-proteins at cryogenic temperatures. *Photosynthesis Research*, 90, 1, pp. 5-15 ISSN 0166-8595

Kajikawa T.; Aoki K.; Singh R. S.; Iwashita T.; Kusumoto T.; Frank H. A.; Hashimoto H. & Ilagan, Katsumura S. (2009). Syntheses of allene-modified derivatives of peridinin

toward elucidation of the effective role of the allene function in high energy
transfer efficiencies in photosynthesis. *Organic & Biomolecular Chemistry*, 7, 18, pp.
3723-3733 ISSN 1477-0520

Kajikawa T.; Hasegawa S.; Iwashita T.; Kusumoto T.; Hashimoto H.; Niedzwiedzki D. M.;
Frank H. A. & Katsumura S. (2009). Syntheses of of C33-, C35- and C39-Peridinin
and their Spectral Characteristics. *Organic Letters*, 11, 21, (November 2009), pp.5006-
5009 ISSN 1523-7060

Kajikawa T.; Aoki K.; Iwashita T.; Niedzwiedzki D. M.; Frank H. A. & Katsumura S. (2010).
Syntheses of ylidenbutenolide-modified derivatives of peridinin and their
stereochemical and spectral characteristics. *Organic & Biomolecular Chemistry*, 8, 11,
pp. 2513-2516 ISSN 1477-0520

Kusumoto T.; Horibe T.; Kajikawa T.; Hasegawa S.; Iwashita T.; Cogdell R. J.; Birge R. R.;
Frank H. A.; Katsumura S. & Hashimoto H. (2010). Stark Absorption Spectroscopy
of Peridinin and Allene-Modified Analogues. *Chemical Physics*, 373, 1-2, (July 2010),
pp. 71-79 ISSN 0301-0104

Mimuro, M.; Nishimura, Y.; Takaichi, S.; Yamano, Y.; Ito, M.; Nagaoka, S. Yamazaki, I.
Katoh, T. & Nagashima, U. (1993). The effect of molecular structure on the
relaxation processes of carotenoids containing a carbonyl group, *Physics letters*, 213,
5-6, (October 1993), pp. 576-580 ISSN 0009-2614

Papagiannakis, E.; Larsen, D. S.; van Stokkum, I. H. M.; Vengris, M.; Hiller, R. G. & van
Grondelle, R. Resolving the excited state equilibrium of peridinin in solution.
Biochemistry, 43, 49, (December 2004), pp.15303-15309 ISSN 0006-2960

Polivka, T. & Sundstrom, V. (2004). Ultrafast dynamics of carotenoid excited states: From
solution to natural and artificial systems. *Chemical Reviews*, 104, 4, (April 2004), pp.
2021-2071 ISSN 0009-2665

Premvardhan, L.; Papagiannakis, E.; Hiller, R. G. & Grondelle, R. V. (2005). The charge-
transfer character of the S_0 to S_2 transition in the carotenoid peridinin is revealed by
stark spectroscopy. *Journal of Physical Chemistry B*, 109, 32, (August 2005), pp. 15589-
15597 ISSN 1520-6106

Schulz, H. S.; Freyermuth, H. B. & Buc, S. R. (1963). New catalysts for the oxidation of
sulfides to sulfones with hydrogen peroxide, *Journal of Organic Chemistry*, 28, (April
1960), pp. 1140-1142

Shima, S.; Ilagan, R. P.; Gillespie, N.; Sommer, B. J.; Hiller, R. G.; Sharples, F. P.; Frank, H. A.
& Birge, R. R. (2003). Two-photon and fluorescence spectroscopy and the effect of
environment on the photochemical properties of peridinin in solution and in the
peridinin-chlorophyll-protein from Amphidinium carterae, *Journal of Physical
Chemistry A*, 107, 40, (Octobor 2003), pp. 8052-8066 ISSN 1089-5639

Song, P. S.; Koba, P.; Prezelin, B. B. & Haxo, F. T. (1976). Molecular topology of
photosynthetic light-harvesting pigment complex, peridinin-chlorophyll-a-protein,
from marine dinoflagellates, *Biochemistry*, 15, 20, pp. 4422-4427 ISSN 0006-2960

Stain H.; Svec, W. A.; Aitzetmuller, K.; Grandolfo, M.C.; Katz, J. J.; Kjosen, H.; Norgard, S.;
Liaaen-Jensen, S.; Haxo, F. H.; Wegfahrt, P. & Rapoport, H. (1971). Structure of
peridinin, the characteristics dinoflagellate carotenoid. *Journal of the American
Chemical Society*. 93, 7, (April 1971), pp. 1823-1825

Tanaka, R.; Zheng, S. Q.; Kawaguchi, K.; Tanaka, T. (1980). Nucleophilic-attak on halogeno(phenyl)acetylenes by halide-ion. *Journal of the Chemical Society Perkin Transactions 2*, 11, pp.1714-1720 ISSN 0300-9580

Vaz, B.; Alvarez, R.; Souto, J. A. & de Lera, A. R. (2005). γ-Allenyl allyl benzothiazole sulfonyl anions undergo cis-selective (Sylvestre) Julia olefinations, *Synlett*, 2, (February 2005), pp. 294-298 ISSN 0936-5214

Vaz, B.; Dominguez, M.; Alvarez, R. & de Lera, A. R. (2006). The Stille reaction in the synthesis of the C37-norcarotenoid butenolide pyrrhoxanthin. Scope and limitations. *Journal of Organic Chemistry*, 71, 16, (August 2006), pp. 5914-5920 ISSN 0022-3263

Zigmantas, D.; Hiller, R. G.; Sundstroem, V. & Polivka, T. (2002). Carotenoid to chlorophyll energy transfer in the peridinin-chlorophyll a-protein complex involves an intramolecular charge transfer state. *Proceedings of the National Academy of Sciences of the United States of America*. 99, 26, (December 2002), pp. 16760-16765 ISSN 0027-8424

Zigmantas, D.; Hiller, R. G.; Sharples, F. P.; Frank, H. A.; Sundstrom, V. & Polivka, T. Effect of a conjugated carbonyl group on the photophysical properties of carotenoids. *Physical Chemistry Chemical Physics*. 6, 11, (June 2004), pp. 3009-3016 ISSN 1463-9076

Vaswani, H. M.; Hsu, C. P.; Head-Gordon, M. & Fleming, G. R. Quantum chemical evidence for an intramolecular charge-transfer state in the carotenoid peridinin of peridinin-chlorophyll-protein. *Journal of Physical Chemistry B*. 107, 31, (August 2003), pp. 7940-7946 ISSN 1520-6106

Part 3

Artificial Photosynthesis Applications

8

Using Chlorophyll Fluorescence Imaging for Early Assessment of Photosynthesis Tolerance to Drought, Heat and High Illumination

María José Quiles, Helena Ibáñez and Romualdo Muñoz
Department of Plant Biology, Faculty of Biology, University of Murcia, Murcia
Spain

1. Introduction

Plants are frequently exposed to environmental stress both in natural and agricultural conditions and it is common for more than one abiotic stress to occur at a given time; for example, drought, heat and high illumination. The concept of stress is intimately associated with that of stress tolerance, which is the plant's ability to cope with an unfavourable environment. Plants exhibit great variations in their tolerance to stress. Some plants show sufficient developmental plasticity to respond to a range of light regimes, growing as sun plants in sunny areas and as shade plants in shady habitats. However, other species of plants are adapted to sunny environments or to shaded environments, and they show different levels of tolerance to high illumination. Generally, sun plants support exposure to high light better than shade plants, which experience photoinhibition (Bray et al., 2000; Levitt, 1980; Osmond, 1994; Saura & Quiles, 2008; Wentworth et al., 2006).

Abiotic stress limits crop productivity and plays a major role in determining the distribution of plant species across different types of environments. Thus, understanding the physiological processes that underlie stress injury and the tolerance mechanisms of plants to environmental stress is of immense importance for both agriculture and the environment. Tolerance to environmental stresses results from integrated events occurring at all organization levels, from the anatomical and morphological to the cellular, biochemical and molecular level. At the biochemical level, plants alter their metabolism in various ways to accommodate environmental stress, with photosynthesis being the most important of these ways.

The photosynthetic apparatus in the plants absorbs large amounts of light energy and processes it into chemical energy. The absorption of photons excites the pigment molecules and this excitation energy is used in the photochemical reactions of photosynthesis, though part of the excitation energy is dissipated by fluorescence (emission of photons by chlorophyll molecules) and heat emission, principally in the antenna system. These three processes (photochemistry, fluorescence and thermal energy dissipation) compete in the dissipation of the excitation energy, while the total energy dissipated is the sum of all three (Long et al., 1994). Estimation of these processes in different conditions will permit us compare the competition that exists among the three of them and to evaluate possible alterations in the functioning of the photosynthetic apparatus. When plants are exposed for long periods of time to more light than they can use, photosynthesis is inhibited in a phenomenon known as photoinhibition. If the excess of absorbed light energy is not dissipated safely toxic species may be produced which can damage the photosynthetic apparatus. Photosynthetic organisms therefore contain a complex set of regulatory and repair mechanisms to avoid this situation. However, even with all these protective mechanisms, the photosynthetic apparatus is still sometimes damaged. PS II is the most sensitive site to photoinhibition, whereas PS I is more stable, probably because it plays a photoprotective role through cyclic electron flow (Quiles and López, 2004). In high-light conditions, the xanthophylls cycle operates, of which violaxanthin together with antheraxanthin and zeaxanthin are components (Demmig-Adams & Adams, 1993; Lichtenthaler et al., 1992; Schindler & Lichtenthaler, 1994, 1996). The xanthophylls cycle is essential to prevent the rapid photoinhibition of PS II (Havaux & Gruszecki, 1993; Lichtenthaler & Babani, 2004; Ruban & Horton, 1995). Sun plants accumulate zeaxanthin during high-light stress of several hours to photoprotect their photosynthetic apparatus against photoinhibition and photooxidation, whereas shade plants do not possess zeaxanthin but only its oxidized form violaxanthin with some traces of antheraxanthin, and these plants are more sensitive to photoinhibition (Lichtenthaler & Babani, 2004).

The classic methods used to determine the damage induced by adverse factors in leaves (such as the measurement of transpiration, respiration and photosynthesis rates; stomatal conductance; water potential; the concentration of photosynthetic pigments, stress metabolites and heat shock proteins) are all quite slow and require considerable effort. Moreover, many of these methods only provide one datum per leaf and measurement, and involve the destruction of tissues, so that subsequent measurements are not possible in the same leaf. In many cases, these methods are effective only for assessing the damage caused by stress to the plant in advanced situations, when the damage is visible, but do not allow early detection of alterations caused by adverse conditions, before the damage becomes visible. This is unfortunate because early detection is important and, in many cases, would make it possible to prevent the onset of irreversible damages. For these reasons, it would be of great interest to develop rapid, non-destructive and quantitative techniques for the early detection of stress in plants. One non-intrusive method for monitoring photosynthetic events and for judging the physiological state of the plant is to measure the chlorophyll fluorescence emitted by intact plant leaves, using a fluorometer (Sayed, 2003). Based on pulse amplitude modulation (PAM) and the saturation pulse method (Schreiber, 2004), chlorophyll fluorometry provides quantitative information concerning the maximal quantum yield of PS II in dark-adapted leaves, the fluorescence yield, the effective PSII quantum yield or photochemical efficiency and the non-photochemic quenching of fluorescence, which represents the heat dissipation in the antenna system (Müller et al., 2001). Three major components of non-photochemical quenching have been identified in plants, namely, energy-dependent quenching, photoinhibitory quenching and state-transition quenching, which are related to trans-thylakoid proton gradient, photoinhibition and energy redistribution, respectively (Allen, 1992; Krause, 1988; Krause & Weis, 1991).

The rate of fluorescence emission depends on the absorbed light flux and on all the competing reactions that result in a return of the excited chlorophyll molecule to the ground state. The most important of those reactions are the photochemical reactions, thermal deactivation and the excitation energy transfer. In the PS II reaction center, the primary

Using Chlorophyll Fluorescence Imaging for Early Assessment of Photosynthesis Tolerance to Drought, Heat and High Illumination

157

photochemical reaction is the transfer of one electron from pigment P_{680} in the first excited singlet state (P_{680}^*) to pheophytin a. From there, the electron is transferred to the primary quinone-type acceptor, Q_A. When all the reaction centers are in an active state, with the quinones totally or partially oxidized ("open" state), the fluorescence yield is minimal (F_0). However, when Q_A is fully reduced, the excitation of P_{680} cannot result in stable charge separation and all the reaction centers are in a "closed" state; in this situation, maximum fluorescence yield (F_M) is obtained. The variable fluorescence emission (F_V), is the difference between F_M and F_0 ($F_V = F_M - F_0$). In a dark-adapted leaf, the plastoquinone pool is fully oxidized, the reaction centers are open and the fluorescence emitted under a weak measuring light is minimal (F_0). When a saturating pulse of white light is given the plastoquinone pool is reduced, the rate of Q_A reduction being faster than the rate of reoxidation, the reaction centers are closed and F_M is reached; at that moment the maximal quantum yield of PS II can be estimated as F_V / F_M. This ratio is an important and easily measurable parameter of the physiological state of the photosynthetic apparatus in intact plant leaves. Additionally, the kinetic of the increase in fluorescence during the saturation pulse can be displayed as the fluorescence induction curve. Most fluorescence is emitted by the PS II antenna, and PS I only contributes around 1-2 % of the total fluorescence; for this reason, the changes in this radiation reflect the state of PS II (Krause and Weis, 1991).

In recent years, the versatility of the chlorophyll fluorescence measurement technique has increased significantly with the development of fluorescence imaging systems, these provides a powerful tool for investigating leaf photosynthesis in a variety of conditions and reveal a wide range of internal leaf characteristics, including spatial variations due to differences in physiology and development (for a review see Papageorgiou & Govindjee, 2004). This technique may also represent a simple and effective tool for the early detection of the effects caused by adverse factors (Oxborough, 2004a), which affect photosynthesis and cause an imbalance in the processes of excitation energy dissipation. Fluorescence imaging permits us to compare the variation in these processes and to study any damage caused in the same leaf as time progresses. However, not all fluorescence parameters are suitable for the early detection of plant stress. Usually, changes in the maximum quantum yield of PS II are used as an indicator of the functional state of the photosynthetic apparatus (Barbagallo et al., 2003; Oxborough, 2004b), since this parameter, which has a value between 0.70 and 0.85 in unstressed leaves, falls under the influence of adverse factors (Ehlert & Hincha, 2008; Havaux & Lannoye, 1985; Joshi & Mohantly, 2004; Quiles & López, 2004; Teicher et al., 2000). However, in the present paper we show that in both sun (*Chrysanthemum morifolium*) and shade (*Spathiphyllum wallisii*) plants exposed to drought, high illumination and heat and showing no visible damage, the images of the maximal quantum yield of PS II (F_v/F_m) in dark-adapted leaves vary little from those in control plants, and, in all cases, the values are quite normal (above 0.74). Therefore, other fluorescence parameters are required to assess the tolerance of plants to those adverse factors. In this respect, we show that images of fluorescence yield, the effective PSII quantum yield and the non-photochemical quenching of fluorescence in illuminated leaves clearly showed variations in the energy dissipation processes between sun and shade plants exposed to drought, high illumination and heat. As a consequence, the measurement of these fluorescence parameters can be considered a better and earlier indicator of functional alterations of the photosynthetic apparatus than maximal quantum yield of PS II (F_v/F_m), which, as we have mentioned, shows only small variations in response to these adverse conditions.

2. Materials and methods

2.1 Plant material and growth conditions

Chrysanthemum morifolium (sun plant) and *Spathiphyllum wallisii* (shade plant) were grown in 500 mL pots at 22-25 °C in the greenhouse with a natural photoperiod, under daytime irradiation maxima of around 800 and 200 µmol·m^{-2}·s^{-1} PPFD (sun and shade plants, respectively) and controlled watering to avoid drought stress (control conditions). To simulate stress conditions, adult plants were transferred to cultivation chambers and exposed to 18 h photoperiods of high light intensity (1060 µmol·m^{-2}·s^{-1} PPFD) supplied by a 100 W Flood Osram (Augsburg, Germany) white light lamp, at 35 °C, followed by 6h night-periods at 24 °C, decreasing the irrigation to 50 mL/day, which was applied after the start the night period.

2.2 Plant water status and pigments measurements

Plant water status was estimated by measuring the relative water content of leaves (RWC). The leaves were collected and immediately weighed to determine fresh weight (FW). They were then re-hydrated for 24 h at 4 °C in darkness to determine the turgid weight (TW), and subsequently oven-dried for 24 h at 85 °C to determine the dry weight (DW). The RWC was determined as 100 x (FW-DW)/ (TW-DW).

Total Chlorophyll and carotenoids were determined by Lichtenthaler & Wellburn's method using 80 % (v/v) acetone as solvent.

2.3 Chlorophyll fluorescence measurements

Chlorophyll fluorescence was imaged, using the MINI-version of the Imaging-PAM (Heinz Walz GmbH, Effeltrich, Germany), in selected leaves attached to plants for the control and stress conditions and the measurements were made after the last night-period. The fluorometer used employs the same blue LEDs for the pulse modulated measuring light, continuous actinic illumination and saturation pulses. The minimal fluorescence yield (F_0) and the maximal fluorescence yield (F_m), were measured in dark-adapted samples. F_0 was measured at a low frequency of pulse modulated measuring light, while F_m was measured with the help of a saturation pulse. The maximal quantum yield of PS II was calculated as $F_v/F_m = (F_m-F_0)/F_m$.

Light response curves were realised illuminating the selected leaves with actinic light of different intensities (20, 41, 76, 134, 205, 249, 300, 371, 456, 581,726 µmol·m^{-2}·s^{-1} PPFD), with 2 min illumination periods at each intensity. After each illumination periods a saturation pulse was applied to determine the relative electron transport rate, the effective PS II quantum yield of illuminated samples ((F_m'-F)/F_m') and non-photochemical quenching (Nq)), all of which were automatically calculated by the ImagingWin software. Results are shown as color-coded images of the maximal quantum yield of PS II (F_v/F_m), the fluorescence yield (F), the effective PS II quantum yield of illuminated samples ((F_m'-F)/F_m') and non-photochemical quenching (Nq)).

3. Results

The relative water content (RWC) was measured in the leaves from *Chrysanthemum morifolium* and *Spathiphyllum wallisii* plants at the start the experiment (control) and immediately after exposure to each stress photoperiod (Figure 1). The RWC values decreased in both species to around 60 % after the two stress photoperiods with low watering, indicating that the plants were subjected to moderate water deficit.

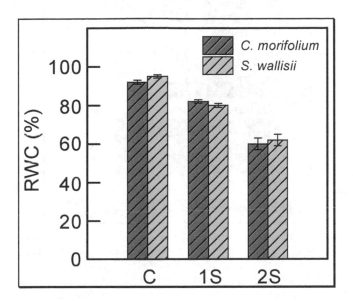

Fig. 1. The relative water content (RWC) of leaves from *Chrysanthemum morifolium* and *Spathiphyllum wallisii* plants in control conditions (C) and after exposure to one (1S) and two (2S) stress photoperiods (18 h, 1060 μmol·m^{-2} s^{-1} PPFD, 35 °C and low watering). The stress photoperiods were separated by 6 h night-periods at 24 °C. The values are means ± SE from four independent experiments.

The fluorescence imaging technique was used to assess photosynthesis in sun and shade plants. Figure 2 shows the images of the maximal quantum yield of PS II (F_v/F_m) from a typical leaf, and the means values ± SE of *C. morifolium* and *S. wallisii* plants in control conditions and exposed to one and two stress photoperiods. The results are shown as colour-coded images according to the pattern shown below the images. All the leaves provided similar images with a homogeneous colour throughout the leaf. The mean F_v/F_m values in all cases were higher than 0.74, indicating that maximal quantum yield of PS II in leaves from *C. morifolium* and *S. wallisii* plants, in control conditions and exposed to one and two stress photoperiods, was quite normal (Krause & Weis, 1991; Schereiber et al., 1997) and that the maximal photosynthetic capacity of PS II in these species was probably unaffected by the stress condition used here; furthermore it seems that the photosynthetic apparatus is protected by mechanisms that dissipate excess excitation energy.

Figure 3 shows the amounts of total chlorophylls and carotenoids in leaves from *C. morifolium* and *S. wallisii* plants in control conditions and exposed to one and two stress photoperiods. No significant difference was observed between the control plants and those exposed to stress photoperiods in either sun or shade plants.

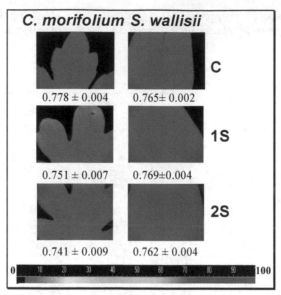

Fig. 2. Images and values of the maximal quantum yield of PS II (F_v/F_m) from typical leaves of *Chrysanthemum morifolium* and *Spathiphyllum wallisii* plants in control conditions (C) and exposed to one (1S) and two (2S) stress photoperiods (18 h, 1060 µmol m^{-2} s^{-1} PPFD, 35 °C and low watering). The stress photoperiods were separated by 6 h night-periods at 24 °C. Images are colour coded according to the pattern (0 to 1 x 100 range) shown below the images. The figure shows representative images from four independent experiments and the values are means ± SE from four different entire leaves.

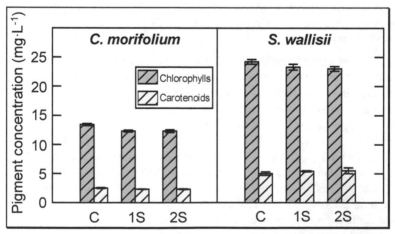

Fig. 3. Total chlorophyll and carotenoids of leaves from *Chrysanthemum morifolium* and *Spathiphyllum wallisii* plants in control conditions (C) and exposed to one (1S) and two (2S) stress photoperiods (18 h, 1060 µmol m^{-2} s^{-1} PPFD, 35 °C and low watering). The stress photoperiods were separated by 6 h night-periods at 24 °C. The values are means ± SE from four independent experiments.

Figure 4 shows the light response curves for the relative electron transport rate in leaves from *C. morifolium* and *S. wallisii* plants in control conditions and exposed to one and two stress photoperiods. In both species, when light was not excessive, the relationship between the relative electron transport rate and the light intensity was linear (optimum line, Fig. 4). When the light became excessive, the relative electron transport rate decreased below the values predicted by the optimum line. Finally, when the photonic flux density was increased, a satured rate was reached, which represents the capacity of photosynthetic electron transport (Schreiber et al., 1997). In low light intensity of less than 100 μmol $\cdot$ m^{-2} s^{-1}, the relative electron transport rate was similar in control and stress-exposed *C. morifolium* plants, but not in *S. wallisii*, where the values in plants exposed to stress photoperiods were lower than those predicted by the optimum line. the capacity of photosynthetic electron transport was greater in *C. morifolium* than in *S. wallisii* control plants and decreased in the plants of both species exposed to stress photoperiods.

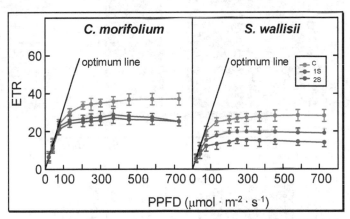

Fig. 4. Light response curves for the relative electron transport rate (ETR) in intact dark-adapted leaves of *Chrysanthemum morifolium* and *Spathiphyllum wallisii* plants in control conditions (green graphics) and exposed to one (blue graphics) and two (red graphics) stress photoperiods (18 h, 1060 μmol $\cdot$ m^{-2} s^{-1} PPFD, 35 °C and low watering). The stress photoperiods were separated by 6 h night-periods at 24 °C. The values are means $\pm$ SE from six independent experiments.

Figure 5 shows the images obtained at two light intensities (20 and 300 μmol $\cdot$ m^{-2} s^{-1}) of the effective PS II quantum yield (Y(II)), the fluorescence yield (F) and non-photochemical quenching (Nq)) from a typical leaf of *C. morifolium* and *S. wallisii* plants in control conditions and exposed to one and two stress photoperiods. For comparison purposes, the data from the analysed entire leaves were also averaged and the medium values $\pm$ SE are shown in the histograms.

Leaves from the control plants and those exposed to stress photoperiods showed changes in the images of the fluorescence parameters in both sun and shade species illuminated with low and high light intensity. With low illumination (20 μmol $\cdot$ m^{-2} s^{-1}) the photochemical efficiency of control plants was approximately 0.5 and the leaves provided Y(II) images of a green-blue colour in both species; the fluorescence emission of control plants was higher in sun (0.32) than in shade species (0.21), while Nq was lower in sun plants (0.16, yellow

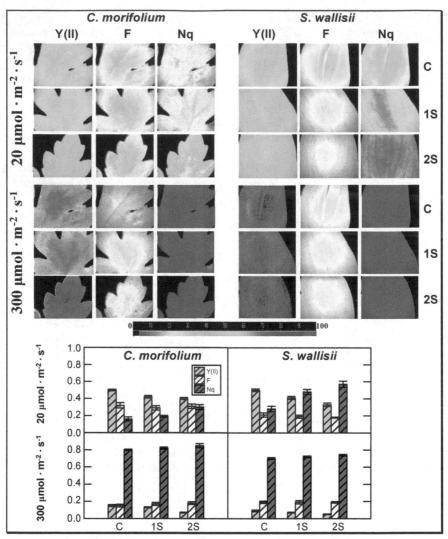

Fig. 5. Images at 20 and 300 µmol ·m⁻² ·s⁻¹ PPFD of the effective PS II quantum yield (Y(II)), the fluorescence yield (F) and non-photochemical quenching (Nq) from a typical leaf attached to *Chrysanthemum morifolium* and *Spathiphyllum wallisii* plants, in control conditions (C) and exposed to one (1S) and two (2S) stress photoperiods (18 h, 1060 µmol ·m⁻² ·s⁻¹ PPFD, 35 °C and low watering). The stress photoperiods were separated by 6 h night-periods at 24 °C. Images are colour coded according to the pattern (0 to 1 x 100 range) shown below the images. The histograms show the means ± SE of parameters calculated from variable chlorophyll fluorescence measurements in six entire leaves.

images) than in shade plants (0.28, green images). After one and two stress photoperiods, the effective PSII quantum yield and the fluorescence emission decreased, moreso in shade than in sun plants, while Nq increased, moreso in shade (blue images) than in sun plants (yellow-green images). When the same leaves were illuminated with 300 μmol·m^{-2}·s^{-1}, the Y (II) decreased in both species, although values were higher in sun than in shade species, and the *C. morifolium* images showed orange-red colour in control leaves, which changed to red after two stress photoperiods, while *S. wallisii* images showed red colour in all cases.

The F emission decreased in sun plants and the images showed orange and orange-green colours, whereas in shade plants the F emission decreased in leaves of the control plants, but after stress photoperiods the leaf images showed only slight differences from those illuminated with 20 μmol·m^{-2}·s^{-1}. The Nq increased significantly in both species, although the values were higher in sun than in shade species.

4. Discussion

Fluorescence imaging represents a simple and non-invasive tool for the early detection of effects caused by adverse factors, which affect photosynthesis causing an imbalance in the processes of excitation energy dissipation (Long et al., 1994). This technique permits us to compare, by means of imagines, the variation in these processes and to study any damage caused in the same leaf as time progresses. Usually, changes in F_v/F_m of leaves adapted to dark, which represents the maximal quantum yield of PS II (Krause & Weis, 1991), are used as an indicator of the functional state of the photosynthetic apparatus (Barbagallo et al., 2003; Krause & Jahns, 2004; Oxborough, 2004b), since this parameter, which has a value of between 0.70 and 0.85 in unstressed leaves, falls under the influence of adverse factors (Ehlert & Hincha, 2008; Havaux & Lannoye, 1985; Joshi & Mohantly, 2004; Quiles & López, 2004; Teicher et al., 2000).

Sun plants (*C. morifolium*) and shade plants (*S. wallisii*) were exposed to photoperiods with low watering, high illumination and heat. Even after two stress photoperiods no visible damage was observed in either plant species (not shown). Neither did the concentration of photosynthetic pigments or the F_v/F_m values show any significant decrease after the stress photoperiods, suggesting that chloroplasts are protected by mechanisms that dissipate excess excitation energy to prevent damage to the photosynthetic apparatus under adverse conditions. In this respect, we have reported that chlororespiration and cyclic electron flow pathways are involved in the tolerance to adverse factors in both sun and shade species (Díaz et al., 2007; Gamboa et al., 2009; Ibañez et al., 2010; Quiles, 2006; Tallón & Quiles, 2007). However, when the light response curves for the relative electron transport rate were depicted, differences were observed between control plants and those exposed to stress photoperiods, the capacity of photosynthetic electron transport being lower in plants exposed to stress photoperiods in both species. In *C. morifolium* after one or two stress photoperiods, the values were similar and the capacity of photosynthetic electron transport was approximately 22% lower than in control plants. However, in *S. wallisii* differences between plants exposed to one and two stress photoperiods were observed and the capacity of photosynthetic electron transport after one and two stress photoperiods was approximately 27 and 44%, respectively, lower than that of control plants.

The images of the fluorescence yield, the effective PS II quantum yield or photochemical efficiency and the non-photochemic quenching of fluorescence, which represents the heat dissipation in the antenna system (Müller et al., 2001), also showed significant differences, indicating that plants exposed to stress photoperiods behaved differently as regards the processes of dissipation of excitation energy, in each species. At low illumination (20 μmol·m^{-2}·s^{-1}), fluorescence emission predominates over heat dissipation in the sun species, while the contrary occurs in the shade species, heat dissipation predominates over fluorescence emission. However, at high illumination (300 $\mu mol \, m^{-2} \, s^{-1}$), when the photochemical efficiency significantly decreased because of the light saturation, the sun species was more efficient in dissipating excess energy in the form of heat.

5. Conclusions

We conclude that fluorescence imaging is a useful method for the early assessment of photosynthesis tolerance to adverse conditions, such as drought, high light and heat, when there is still no visible damage to plants. However, not all fluorescence parameters are effective, and analysis of the maximum quantum yield in leaves adapted to darkness was unable to detect significant differences between control plants and plants exposed to stress photoperiods. In contrast, the analysis in illuminated leaves of the relative electron transport rate and the fluorescence parameters, Y(II), F and Nq, which are representative of the three processes of excitation energy dissipation (photochemistry, fluorescence and thermal dissipation, respectively) showed significant differences in the two species studied, indicating that sun species (C. morifolium) had greater tolerance to drought, heat and high illumination than the shade species (S. wallisii).

6. Acknowledgment

This work was supported by the Spanish Ministry of Science and Innovation (grant BFU2008-00331).

7. List of abbreviations

DW, dry weight;
F, fluorescence yield;
F_m, maximal fluorescence yield in the dark adapted state;
F_0, minimal fluorescence yield in the dark adapted state;
F'_m, maximal fluorescence yield in the light adapted state;
F_v, variable fluorescence;
FW, fresh weight;
LED, light-emitting diode;
Nq, non-photochemical quenching;
PAM, pulse amplitude modulation;
PPFD, photosynthetic photon flux density;
PS, photosystem;
RWC, relative water content;
TW, turgid weight;
Y(II), effective PS II quantum yield.

8. References

Allen, J.F. (1992). Protein phosphorylation in regulation of photosynthesis. *Biochim Biophys Acta.* 1098: 275-335.

Barbagallo, R.P.; Oxborough, K.; Pallet, K.E. & Baker N.R. (2003). Rapid, non-invasive screening for perturbations of metabolism and plant growth using chlorophyll fluorescence imaging. *Plant Physiol.* 132:485-493.

Using Chlorophyll Fluorescence Imaging for Early Assessment of Photosynthesis Tolerance to Drought, Heat and High Illumination

165

Bray, E.A.; Bailey-Serres, J. & Weretilnyk, E. (2000). Responses to abiotic stresses. In *Biochemistry and Molecular Biology of Plants*, W. Buchanan, W. Gruissem & R. Jones (Eds.), pp1158-1203. American Society of Plant Physiologists, ISBN 0-943088-37-2, Rockville, MD, USA.

Demmig-Adams, B. & Adams, W.W. (1993). Photoprotection and other responses of plants to high light stress. *Annu Rev Plant Physiol Plant Mol Biol* 43: 599-626.

Díaz, M.; De Haro, V.; Muñoz, R. & Quiles, M.J. (2007). Chlororespiration is involved in the adaptation of *Brassica* plants to heat and high light intensity. *Pant Cell Environ* 30:1578-1585.

Ehlert, B. & Hincha K. (2008). Chlorophyll fluorescence imaging accurately quantifies freezing damage and cold acclimation responses in Arabidopsis leaves. *Plant Methods* 4:12 doi:10.1.186/1746-4811-4-12.

Gamboa, J., Muñoz, R. & Quiles, M.J. (2009). Effects of antimycin A and n-propyl gallate on photosynthesis in sun and shade plants. *Plant Sci* 177:643-647.

Havaux, M. & Gruszecki, W.I. (1993). Heat and light-induced Chl a fluorescence changes in potato leaves containing high or low levels of the carotenoid zeaxanthin, indicators of a regulatory effect of zeaxanthin on thylakoid membrane fluidity. *Photochem Photobiol* 58: 607-614.

Havaux, M. & Lannoye, R. (1985). In vivo chlorophyll fluorescence and delayed light emissions as rapid screening techniques for stress tolerance in crop plants. Z. *Pflanzenzucht.* 95: 1-13.

Ibáñez, H.; Ballester, A.; Muñoz, R. & Quiles, M.J. (2010). Chlororespiration and tolerance to drought, heat and high illumination. *J Plant Physiol.* 167:732-738.

Joshi, M.K. & Mohanty, P. (2004). Chlorophyll *a* fluorescence as a probe of heavy metal ion toxicity in plants. In: *Chlorophyll a fluorescence. A signature of photosynthesis*, G. Papageorgiou & Govindjee Eds.), pp. 637-661. Springer, ISBN 1-4020-3217-X, Dordrecht. The Netherlands

Krause, G.H. (1988). Photoinhibition of photosynthesis. An evaluation of damaging and protective mechanisms. *Physiol Plant.* 74: 566-574.

Krause, G.H. & Jahns, P. (2004). Non-photochemical energy dissipation determined by chlorophyll fluorescence quenching: characterization and function. In: *Chlorophyll a fluorescence. A signature of photosynthesis,* G. Papageorgiou & Govindjee Eds.), pp. 463-495. Springer, ISBN 1-4020-3217-X, Dordrecht. The Netherlands.

Krause, G.H. & Weis, E. (1991). Chlorophyll fluorescence and photosynthesis. The basics. *Ann Rev Plant Physiol Plant Mol Biol* 42: 313-349.

Levitt, J. (1980). *Responses of plants to enviromental stresses*, Vol. 1. Academic Press, ISBN 0124455018, New York, USA.

Lichtenthaler, H.K. & Babani, F. (2004). Light adaptation and senescence of the photosynthetic apparatus. Changes in pigment composition, chlorophyll fluorescence parameters and photosynthetic activity. In: *Chlorophyll a fluorescence. A signature of photosynthesis,* G. Papageorgiou & Govindjee Eds.), pp. 713-736. Springer, ISBN 1-4020-3217-X, Dordrecht. The Netherlands.

Lichtenthaler, H.K.; Burkart, S.; Schindler, C. & Stober, F. (1992). Changes in photosynthetic pigments and in vivo chlorophyll fluorescence parameters under photoinhibitory growth conditions. *Photosynthetica*, 27: 343-353.

Lichtenthaler, H.K. & Wellburn, A.R. (1983). Determinations of total carotenoids and chlorophylls a and b of leaf extracts in different solvents. *Biochemical Society Transactions* 11: 591-592.

Long, S.P.; Humphries, S. & Falkowski P.G. (1994). Photoinhibition of photosynthesis in nature. *Ann Rev Plant Physiol Plant Mol Biol.* 45: 633-662.

Müller, P.; Li, X.P. & Niyogi, K.K. (2001). Non-photochemical quenching: a response to excess light energy. *Plant Physiol.* 125:1558-1566.

Osmond, C.B. (1994). What is photoinhibition? Some insights from comparisons of shade and sun plants. In: *Photoinhibition of Photosynthesis: From Molecular Mechanisms to the Field.* N.R. Baker & J.R.Bowyer (Eds.), pp. 1-24. BIOS Scientific Publishers, ISBN 1-872748-03-1, Oxford. United Kingdom.

Oxborough, K. (2004a). Using chlorophyll a fluorescence imaging to monitor photosynthetic performance. In: *Chlorophyll a fluorescence. A signature of photosynthesis,* G. Papageorgiou & Govindjee Eds.), pp. 409-428. Springer, ISBN 1-4020-3217-X, Dordrecht. The Netherlands.

Oxborough, K. (2004b). Imaging of chlorophyll a fluorescence: theoretical and practical aspects of an emerging technique for the monitoring of photosynthetic performance. *J Exp Bot.* 55:1195-1205.

Papageorgiou, G. & Govindjee. (2004). *Chlorophyll a fluorescence. A signature of photosynthesis.* Springer, ISBN 1-4020-3217-X, Dordrecht. The Netherlands

Quiles, M.J. (2006). Stimulation of chlororespiration by heat and high light intensity in oat plants. *Plant Cell Environ.* 29:1463-1470.

Quiles, M.J. & López N.I. (2004) Photoinhibition of photosystems I and II induced by exposure to high light intensity during oat plant growth. Effects on the chloroplast NADH dehydrogenase complex. *Plant Sci.* 166: 815-823.

Ruban, A.V. & Horton, P. (1995). Regulation of non-photochemical quenching of chlorophyll fluorescence in plants. *Aus J Plant Physiol.* 22: 221-230.

Sayed, O.H. (2003). Chlorophyll fluorescence as a tool in cereal crop research. *Photosynthetica,* 41: 321-30.

Saura, P. & Quiles, M.J. (2009). Assessment of Photosynthesis Tolerance to Herbicides, Heat and High Illumination by Fluorescence Imaging. *Open Plant Sci J.* 3: 7-13.

Schindler, C. & Lichtenthaler, H.K. (1994). Is there a correlation between light-induced zeaxanthin accumulation and quenching of variable chlorophyll a fluorescence? *Plant Physiol Bioch.* 32: 813-823.

Schindler, C. & Lichtenthaler, H.K. (1996). Photosynthetic CO2 assimilation, chlorophyll fluorescence and zeaxanthin accumulation in field-grown maple trees in the course of a sunny and cloudy day. *J Plant Physiol.* 148: 399-412.

Schreiber U. (2004). Pulse-Amplitude (PAM) fluorometry and saturation pulse method. In *Chlorophyll a fluorescence. A signature of photosynthesis.* Edited by Papageorgiou G, Govindjee. Dordrecht: Springer; 279-319.

Schreiber, U.; Gademann, R.; Ralph, P.J. & Larkum, A.W.D. (1997). Assessment of photosynthetic performance of *Prochloron* in *Lissoclinum patella* in hospite by chlorophyll fluorescence measurements. *Plant Cell Physiol.* 38:945-951.

Tallón, C. & Quiles, M.J. (2007). Acclimation to heat and high light intensity during the development of oat leaves increases the NADH DH complex and PTOX levels in chloroplasts. *Plant Sci.* 173: 438-445.

Teicher, H.B.; Moller, B.L. & Scheller, H.V. (2000). Photoinhibition of photosystem I in field-grown barley (*Hordeum vulgare* L.): induction, recovery and acclimation. *Photosynth Res.* 64: 53-61.

Wentworth, M.; Murchie, E.H.; Gray, J.E.; Villegas, D. & Pastenes, C. (2006) Differential adaptation of two varieties of common bean to abiotic stress. *J Exp Bot.* 57: 699-709.

9

Friend or Foe? Exploring the Factors that Determine the Difference Between Positive and Negative Effects on Photosynthesis in Response to Insect Herbivory

John Paul Délano Frier[1],
Carla Vanessa Sánchez Hernández[2] and Axel Tiessen[1]
[1]Unidad de Biotecnología e Ingeniería Genética de Plantas, Cinvestav -
Irapuato, Guanajuato,
[2]Centro Universitario de Ciencias Biológicas y Agropecuarias,
Universidad de Guadalajara, Zapopan, Jalisco,
México

1. Introduction

Photosynthesis is a central process for the survival of plants, providing essential elements for growth and reproduction. As its name implies, the excitation of chlorophyll by photons, initiates an electron current that results in the generation of NADPH and ATP, which are subsequently used by the plant to sustain growth and development. However, precise and challenging energy equilibrium between the synthesis of NADPH and ATP and the downstream reactions that consume them must be maintained to ensure that the photosynthetic process runs efficiently. This balancing act is not trivial considering the necessity to integrate extremely rapid, temperature-independent photochemical reactions with relatively slow, temperature-dependent biochemical reactions. In addition, the equilibrium can be perturbed with relative ease when plants face adverse ambient conditions, including biotic stresses, leading to negative effects on the photosynthetic apparatus and concomitant disruption of the orderly transfer of electrons to their proper final acceptors. To avoid or minimize damage, plants have developed several strategies to maintain stability, the details of which will be explored in this chapter.

Together with many other biotic stresses, insect herbivory is known to alter photosynthetic activity and/or photosynthetic gene expression levels in affected plants. The effects are predominantly negative, although compensatory responses are not uncommon. Both positive and negative effects can be detected in injured and adjacent, intact, plant tissue, which may include, in some species, photosynthetically active stems in addition to leaves. Indirect effects on photosynthesis have been clearly demonstrated with the use of fluorescence and thermal imaging system techniques. The negative insect herbivory-derived effects on photosynthesis are frequently explained by a theoretical framework constructed on the argument that the resource-rich photosynthetic apparatus is sacrificed to provide the metabolic precursors and energy needed for the proper deployment of the resource-demanding defense responses, although the concept of a down-regulated photosynthetic apparatus as a protective measure against oxidative damage has also been proposed. Jasmonic acid (JA), frequently interacting with ethylene, is recognized as the primary regulator of the defense response against defoliating insects and is strongly associated with the down-regulation of photosynthesis genes. Conversely, the mechanism(s) responsible for the compensatory photosynthetic activity that allows increased plant growth or fitness after insect herbivory are not well understood. Experimental evidence gathered to date suggests that the onset of compensatory photosynthesis may be dependent on many factors. The timing of herbivory injury, which also influences the plant's source/sink relationships along its phenology, is important. Also influential are the type of tissue damaged, the type and extension of damage, which in turn may be influenced by highly specific factors such as the composition of the herbivore's saliva and the type of endosimbiotic bacteria colonizing the insect's gut, the development stage of the herbivore, its feeding guild and even differences within guilds. Finally, the tolerance capacity of the plant to injury, the environmental conditions surrounding the plant, and the type of defense produced, if any, in response to herbivory, will also define whether a given assault on the plant leads to compensatory responses. The suggested role for phytochrome as a regulator of resource allocation between plant growth and anti-herbivore defense also implies its participation as a signaling element in herbivory-dependent changes in photosynthesis. Moreover, emerging data suggest that stress acclimation and chloroplast-to-nucleus signaling is mediated by phytohormones acting via AP2/EREBP transcription factors, which are believed to play a major and diversified role in environmental signal integration.

The overall perspective on the varying factors that influence the effect that insect herbivory may have on photosynthesis is that of complexity. This chapter will also concentrate on the description of these multiple factors, including emerging data that have shed light on the poorly understood mechanisms that regulate photosynthesis-related gene expression or that define how herbivore damage is interpreted by the plant, either as a "friendly" jolt to increase photosynthesis and stimulate growth and promote fitness or as an "act of war" leading to austerity measures to privilege a defense effort.

2. Photosynthesis under stress: How a vital process copes with a permanently changing environment

Photosynthesis constitutes a highly integrated process involving four multi-subunit membrane-protein complexes: photosystem II (PSII), photosystem I (PSI), cytochrome $b6f$ and F-ATPase (Nelson & Yocum, 2006) that is exquisitely designed to funnel the electron current initiated by photon absorption to the reducing reactions that generate NADPH from $NADP^+$. The process of electron transport also generates an electrochemical proton gradient that powers ATP synthesis. The reductive power and ATP generated are then employed to reduce inorganic sources of carbon, nitrogen and sulfur needed for the synthesis of carbohydrates, amino acids and proteins used to maintain cellular homeostasis and growth respiration for cell division and expansion (Paul & Foyer, 2001). The need to balance the light energy absorbed by the photosystems with the energy consumed by metabolic sinks of the plant renders the photosynthetic process highly sensitive to any disturbing change in environmental conditions, such as fluctuating illumination, limitation of CO_2 fixation by low temperatures, salinity or low nutrient or water availability, and biotic stress. Several molecular short-term and long-term acclimation mechanisms are deployed by photosynthetic organisms (predominantly green algae and land plants) to maintain or restore photosynthetic efficiency under adverse conditions and counteract stresses (Öquist & Huner, 2003; Ensminger et al., 2006). On a time scale of minutes, organisms can reduce the efficiency of energy transfer to PSII either by redistributing light energy to PSI at the expense of PSII through state transitions or by dissipating excess energy as heat by non-photochemical quenching associated with the light-harvesting complex (LHC) antenna (Hüner et al., 1998; Szyszka et al., 2007). Alternatively, it has also been proposed that quenching of excess energy may occur at the reaction centers, in addition to zeaxanthin-dependent antenna quenching (Krause & Weis, 1991; Bukhov et al., 2001; Matsubara & Chow, 2004; Ivanov et al., 2006). Long-term acclimation responses include alterations in light harvesting antenna size and adjustments of PSI: PSII stoichiometry that balance the excitation light energy absorbed by the two photosystems (Yamazaki et al., 2005; Ozaki et al., 2007, Solanke & Sharma, 2008; Muramatsu et al., 2009). A significant body of experimental evidence indicates that redox signals from photosynthetic electron transport and reactive oxygen species (ROS) or ROS-scavenging molecules play a central role in the regulation of acclimation and stress responses (Foyer & Noctor, 2009; Pfannschmidt et al., 2009; Mubarakshina et al., 2010).

2.1 Short-term acclimation mechanisms
2.1.1 State transitions
State transitions represent a short-term response, occurring within a time-frame of seconds to minutes, required to balance the light excitation energy between the antennae systems of PSII and PSI which preferentially absorb 650 and 700 nm light, respectively. Because of these differences in light absorption properties, changes in light conditions, such as those happening under shaded or light-limiting conditions, or as a consequence of shifts in the spectral filtering properties of leaf canopies, can lead to unequal excitation of the two photosystems (Allen & Forsberg, 2001; Haldrup et al., 2001; Wollman, 2001; Dietzel et al., 2008; Pesaresi et al., 2010). Preferential excitation of PSI leads to the oxidation of the plastoquinone (PQ) pool and to state 1. In state 1 the mobile light-harvesting antenna is bound to PSII and the photosynthetic electron transport chain acts mostly in a linear mode generating NADPH and ATP. Preferential excitation of PSII relative to PSI leads to a reduced state of the PQ pool and thus to the docking of plastoquinol to the Qo site of the cytochrome *b6f* complex. It is the PQ redox state only, acting independently of photoreceptors (Fey et al., 2005), that then activates the so-called redox-sensitive thylakoid LHCII kinase needed to phosphorylate the peripheral LHCII "mobile pool", which then migrates laterally, as a consequence of charge repulsion, from PSII to PSI (state 2) (Lunde et al., 2000). Chemical cross-linking and RNA interference approaches performed in Arabidopsis plants maintained in state 2 have provided evidence for an association of LHCII polypeptides to a specific PSI docking domain composed of subunits PsaL, PsaH and PsaO (Lunde et al., 2000; Zhang et al., 2004; Pesaresi et al., 2010), whereas a gentle mechanical fractionation of the thylakoid membranes showed that the lateral movement of phosphorylated LHCII might be confined to a very limited portion of the thylakoid membranes, more precisely to the grana margins (Tikkanen et al., 2008). Regardless of the mechanism, the re-distribution of the light-harvesting chlorophyll to PSI at the expense of

PSII results in a balanced excitation of PSII and PSI to ensure optimal quantum efficiency for photosynthetic electron transport. Under PQ oxidizing conditions the LHCII kinase is inactive, LHCII is or becomes dephosphorylated and is relocated to PSII (state I). The identity of the redox-independent and constitutively active protein phosphatase that presumably dephosphorylates LHCII during the transition from State 2 to 1 was the subject of intense research and remained unknown until recently. The search finally yielded fruit with the identification of a LHCII-specific phosphatase, called PPH1/TAP38. This enzyme dephosphorylates LHCII upon a transition from state 2 to state 1 (Pribil et al., 2010; Shapiguzov et al., 2010), by specifically dephosphorylating the major trimeric Lhcb1 and Lhcb2 proteins. It is a chloroplast protein that is mainly associated with the stromal membranes of the thylakoid membranes and belongs to the family of monomeric PP2C type phosphatases. Its regulatory role was demonstrated in experiments where the loss of PPHI/TAP38 gave rise to an increase in the antenna size of PSI and strongly impaired state transitions.

2.1.2 The nature of the redox-sensitive thylakoid LHCII kinase

LHCII kinase activity was first reported in 1977 (Bennett, 1977, 1979), a finding that also triggered an intensive search for its identity. A first approximation of its nature came from a screening for proteins capable of interacting with the N-terminal region of the light-harvesting proteins known to contain the amino-acid targets for phosphorylation during states 1–2 transition. This approach led to the identification of a small family of three kinases, called TAK kinases (for thylakoid associated kinases) in *Arabidopsis thaliana* (Snyders & Kohorn, 1999), whose exact role in state transitions regulation remains undefined until now, notwithstanding biochemical and genetic experimental evidence showing that TAKs do indeed participate in LHCII phosphorylation (Snyders & Kohorn, 2001). Moreover, the failure to identify TAK orthologs in the green motile unicellular alga *Chlamydomonas reinhardtii* either suggests that the TAK kinases perform a role which is specific to land plants or that a considerable diversion of these kinases happened as a result of evolutionary divergence between green algae and plants.

Later, advantage was taken of the large chlorophyll fluorescence changes that occur during a transition from states 1 to 2 in *C. reinhardtii* (that can be efficiently measured with a fluorescence video imaging system) to screen insertional mutants that could lead to the identification of the LHCII kinase and other factors of the signal transduction pathway leading to state transitions (Fleischmann et al., 1999; Kruse et al., 1999). This strategy permitted the isolation of the *state transition-deficient mutant 7* (*stt7*), which was found to encode a thylakoid-associated Ser-Thr protein kinase, and of another protein kinase of unknown function but related to Stt7, called Stl1. Orthologs of these two proteins, called STN7 and STN8, respectively, were subsequently found in Arabidopsis, rice and in marine algae (Depége et al., 2003). Utilization of Arabidopsis T-DNA insertion lines with disruptions in the *STN7* or *STN8* genes helped elucidate the function of these proteins. Thus, STN7 was found to be required for state transitions and for the specific phosphorylation, under state 2 conditions, of several LHCII proteins that did not include the major thylakoid proteins CP43, D1 and D2 (Bellafiore et al., 2005). An additional site directed mutagenesis approach provided conclusive evidence demonstrating that the kinase activity of STN7 is essential for state transitions. It is now known that Stt7 and STN7 are both structurally and functionally related.

Friend or Foe? Exploring the Factors that Determine the Difference Between Positive and Negative
Effects on Photosynthesis in Response to Insect Herbivory

171

The recent characterization of Stt7 in *C. reinhardtii* revealed a structural organization in which a transmembrane helix separates its stroma-exposed catalytic domain from its lumen-located N-terminal end. This organization permits the co-localization of the catalytic site with the target sites on the LHCII proteins. It also identified two conserved cysteine residues that are critical for its activity (Lemeille et al., 2009). In addition, co-immunoprecipitation assays have shown that Stt7 interacts with Cyt *b6f*, PSI and LHCII, suggesting that all these protein complexes might be clustered together, possibly in very restricted areas of thylakoid membranes, such as the grana margins (Tikkanen et al., 2008; Lemeille et al., 2009; see above). The curious fact that the Qo site of the cytochrome *b6f* complex, which is critical for the activation of the kinase, is on the lumen side necessarily implies that a signal for kinase activation needs to be transported across the membrane. A mobile Rieske protein (Zhang et al., 1998; Breyton, 2000) and subunit V (also called PetO) from the cytochrome *b6f* complex, which is the only protein of the complex capable of under-going reversible phosphorylation during state transitions (Hamel et al., 2000), have been proposed as possible signal transducing candidates in a pathway model that suggests that sensing of the structural changes of the Rieske protein by the lumenal domain of PetO is transmitted through its trans-membrane region to its stromal domain in order to allow its interaction with the kinase. Another interesting feature found in Stt7 and STN7 involves the presence of two conserved Cys residues near the N-terminal end which could be the targets of thioredoxin (Rintamaki et al., 2000). The loss of state transitions and LHCII phosphorylation as the result of site-directed mutagenesis of either of the conserved Cys residues in both Stt7 and STN7 strongly suggests that these residues play an important role in the activation of the kinase. It is also likely that the high-light-induced reduction of this bond may occur through a trans-thylakoid thiol-reducing pathway driven by the ferredoxin-thioredoxin system which is also required for cytochrome *b6f* assembly and heme biogenesis (Lemeille et al., 2009). Therefore, it appears likely that STN7 kinase activity is regulated not by PQ alone, but by a complex network involving co-operative redox control by PQ and the Cyt *b6f* complex, as well as by the ferredoxin/ thioredoxin system in the stroma of the chloroplasts (Rintamaki et al., 2000; Lemeille et al., 2009)

Although it is clear that the Stt7/STN7 kinase is required for LHCII phosphorylation and for state transitions, it is not yet known whether it acts in a kinase cascade or recognizes LHCII as its direct substrate. Some of these uncertainties were dispelled by the findings of a recent study that compared the thylakoid phosphoproteome of the wild-type strain and the *stt7* mutant of *C. reinhardtii* under state 1 and state 2 conditions (Lemeille et al., 2010). The study revealed that under state 2 conditions several Stt7-dependent phosphorylations occur in the Lhcbm1/Lhcbm10, Lhcbm4/Lhcbm6/Lhcbm8/Lhcbm9, Lhcbm3, Lhcbm5, and CP29 proteins located at the interface between PSII and its light-harvesting system. One of the two Stt7-dependent phosphorylation sites detected specifically in CP29 under state 2 was proposed to play a crucial role in the dissociation of CP29 from PSII and/or in its association to PSI where it serves as a docking site for LHCII in state 2. Moreover, the Stt7-dependent phosphorylation of the thylakoid protein kinase Stl1 under state 2 conditions, suggested the existence of a thylakoid protein kinase cascade. Curiously, the auto-phosphorylation of Stt7 in state 2, was found not to be required for state transitions. Additional findings included the identification of redox (or state 2)-dependent but Stt7-independent, and redox-independent phosphorylation sites.

The existence of the conserved STN7/ STN8 and Stt7/Stl1 kinase couples in Arabidopsis and *Chlamydomonas* also suggests a possible functional interaction between STN7/Stt7 and STN8/Stl1. This is in accordance with data proposing that these proteins appear to act synergistically, since the de-phosphorylation phenotype of LHCII and PSII core proteins in the double mutant *stn7/stn8* is more pronounced than than those observed in the two single mutants (Bonardi et al., 2005; Vainonen et al., 2005). Moreover, field tests revealed that fitness, as measured by seed production, was significantly decreased in the double mutant whereas it was decreased to a smaller extent in *stn7* and not significantly affected in *stn8* mutants, respectively (Frenkel et al., 2007).

2.1.3 The importance of state transitions in flowering plants
The magnitude of state transitions is much larger in *C. reinhardtii* than in flowering plants, where displacements have been reported to involve up to 85% of the LHCII antenna from PSII in State 2. In contrast, only 20-30% of the total LHCII is mobile in green plants. Moreover, state transitions in green algae represent a unique adaptive mechanism that allows the organism to switch between linear (State 1) and cyclic (State 2) electron flow through PSI (Finazzi et al., 2001), whereas *C. reinhardtii* mutants unable to undergo state transitions, such as *stt7*, exhibit altered photosynthetic performance and a marked decrease in growth rate (Depége et al., 2003). Conversely, plant development and fitness under laboratory and field conditions have been found to be only marginally affected in Arabidopsis mutants impaired in state transitions (Lunde et al., 2000; Bonardi et al., 2005; Bellafiore et al., 2005; Frenkel et al., 2007). However, a marked decrease in growth rate relative to the parental single mutants, which was accompanied by a consistent drop in the effective quantum yield of PSII and an increase in the reduction state of the PQ pool, was detected in double Arabidopsis mutants affected both in the linear electron transport leading to an increased pool of reduced PQ (i.e. *psad1-1* and *psae1-3*), and state transitions (i.e. *stn7-1 or psal-1*) (Lunde et al., 2000; Pesaresi et al., 2009). This behavior implied that state transitions become critical for plant performance when linear electron flow is perturbed. Further spectroscopic analyses performed on the different genotypes led to the conclusion that, in flowering plants as in green algae, state transitions play an important role in balancing energy distribution between photosystems.

2.2 Long Term acclimation Responses (LTR)
2.2.1 LTR mechanisms
Besides inducing short term acclimation processes such as state transitions, changes in light conditions are known to lead to long term responses (LTR) characterized by changes in the amounts of the antenna proteins of PSII and PSI and in photosystem stoichiometry. These changes are implemented over periods lasting hours or days (Dietzel et al., 2008). This process is achieved through a signaling network involving coordinate gene expression in the nucleus and chloroplast (Pfannschmidt, 2003; Pfannschmidt et al., 2009). Most experimental evidence gathered to date indicates that STN7 is also required for triggering LTR (Allen & Pfannschmidt, 2000; Bonardi et al., 2005; Tikkanen et al., 2006), suggesting a dual role for STN7, acting as a common redox sensor and/or signal transducer for both state transitions and LTR responses. However, experimental evidence obtained with mutant or silenced Arabidopsis lines affected in various components required for state transitions (including the novel TSP9 protein, suggested to function in the signaling pathway due to its partial

Friend or Foe? Exploring the Factors that Determine the Difference Between Positive and Negative
Effects on Photosynthesis in Response to Insect Herbivory
173

dissociation from the thylakoid membrane upon phosphorylation (Pesaresi et al., 2009), showed that neither LHCII phosphorylation, nor the conformational changes in the thylakoid associated with state transitions themselves, appear to play any role in LTR (Carlberg et al., 2003; Pesaresi et al., 2009). This argued against the possibility that the signal pathways leading to state transitions and LTR were part of a hierarchically organized signaling cascade, with changes in PQ redox state first triggering state transitions and then LTR, via a STN7-dependent phosphorylation cascade. In order to conciliate the above data, an alternative hypothesis proposing that the PQ redox state must reach a still undefined threshold value to be able to induce the specific, and reversible, STN7-dependent phosphorylation steps that trigger the signaling events leading to LTR was considered (Pesaresi et al., 2010).

In most species investigated, the re-adjustment of photosystem stoichiometry involves an enhanced expression of the PSI reaction-center genes *psaA* and *psaB* (which encode the P700 apoproteins) upon active reduction of the PQ pool or repression of its oxidation. LTRs are also known to involve the regulated expression of the PSII reaction-center gene *psbA* (encoding the D1 protein) (Pfannschmidt, 2003) and changes in several other physiological and molecular parameters, including the chlorophyll *a/b* ratio, steady state chlorophyll fluorescence and structural modifications of the thylakoid membrane system (Bonardi et al., 2005; Tikkanen et al., 2006). Several proteins were recognized as possible regulators of photosystem stoichiometry in the cyanobacteria *Synechocystis* sp., including photomixotrophic growth-related and CO_2-concentrating-mechanism proteins, a probable esterase, cytochrome c oxidase subunits II and III and a hypothetical protein with a von Willebrand factor type A domain. The latter suggested a role for protein-protein interactions in the regulation of photosystem stoichiometry in these organisms (Ozaki et al., 2007). In addition, the depletion of the vesicle inducing protein in plastids 1 (Vipp1), believed to be essential for thylakoid membrane formation in Arabidopsis and cyanobacteria, was found to negatively affect photosystem stoichiometry in *Synechocystis* sp. This effect was associated with a concerted decrease in the number of thylakoid layers and associated photosystem I (PSI) complexes in individual cyanobacterial cells, and an enrichment of PSI monomeric species resulting from of PSI trimer destabilization (Fuhrmann et al., 2009).

More recently, photosystem stoichiometry adjustment in plants and algae, was found to be governed by a modified two-component sensor kinase of cyanobacterial origin, known as chloroplast sensor kinase (CSK) (Puthiyaveetil et al., 2011), acting together with chloroplast sigma factor 1 (SIG1) and a plastid transcription kinase (PTK). These findings confirmed previous data implicating CSK as a control of chloroplast gene expression (Puthiyaveetil et al., 2008), via its role as a sensor of the PQ redox state. Moreover, they confirmed the concept assigning different signaling pathways to state transitions and photosystem stoichiometry adjustments, with the two pathways sensing PQ redox state independently of each other (i.e the reduced and oxidized forms of the quinone recognized for state transitions and photosystem stoichiometry, respectively).

The LTR is also accompanied by dynamic changes in metabolite pools that depend to the prevailing illumination (Bräutigam et al., 2009). For instance, the propagation of plants under PSI-specific light is known to cause a lower accumulation of transitory starch. Moreover, contrasting light conditions have been observed to exert different co-regulation effects on biosynthetic pathways for organic acids and several amino acids linked to secondary metabolism in plants. Thus, the LTR appears to contribute also to the adaptation of plant primary productivity to environmental conditions (Pesaresi et al., 2010).

All evidence gathered to data indicates that short- and long-term photosynthetic acclimation responses are triggered by changes in the redox state of the PQ pool and require the modulated activity of the kinase STN7. Due to its dual regulatory role, STN7 initiates a phosphorylation cascade that induces state transitions by phosphorylating LHCII and promotes the LTR process via the phosphorylation of as yet unknown chloroplast proteins. Beyond this point, the LTR signaling pathway is divided into two main branches: one is responsible for transcriptional regulation of chloroplast gene expression, while the other controls the expression of nuclear photosynthesis-related genes at transcriptional and post-transcriptional levels (see Figure 1) (Pesaresi et al., 2010).

2.3 Acclimation responses under high or excess excitation pressure

State transitions and LTR are acclimation responses that typically occur under low-light conditions and are controlled via redox signals. Under conditions resulting in high or excess excitation pressure other acclimation responses are activated, such as non-photochemical dependent antenna quenching, the D1 repair cycle or various other stress-response programs. These responses are also controlled via redox signals originating from the photosynthetic process (i.e. the PQ redox state and signals from the PSI acceptor side), but may also involve the participation of ROS such as hydrogen peroxide (H_2O_2) or singlet oxygen (Pfannschmidt et al., 2009).

2.3.1 Non-photochemical quenching

Non-photochemical quenching (NPQ) is a rapid de-excitation photo-protective quenching mechanism (qE) that involves dissipation of excess energy occurring upon short-term high light exposure. In addition, NPQ is considered to act as a "light intensity counter," providing the photosynthetic membrane with a "memory" of the light-exposure history of the leaf (Horton et al., 2008; Foyer & Noctor, 2009). qE is induced by a low thylakoid lumen pH (i.e. a high ΔpH) generated by photosynthetic electron transport in excess light and involves the harmless thermal dissipation of excess energy in the chlorophyll (Chl) singlet excited states ($^1Chl^*$) in photosystem II (PSII) of green plants and algae. qE is designed to minimize alternative reaction pathways that generate toxic photo-oxidative intermediates. It functions by activating a lumen-localized violaxanthin de-epoxidase enzyme that catalyses the conversion of violaxanthin to zeaxanthin via the intermediate antheraxanthin, in what is known as the xanthophyll cycle (which is completed by the conversion of zeaxanthin back to violaxanthin by means of a zeaxanthin epoxidase activated under limiting light conditions). The process also requires the protonation of PsbS, a PSII subunit that plays a role in the regulation of photosynthetic light harvesting and is also necessary for qE *in vivo* (possibly by establishing a binding site for zeaxanthin that facilitates the de-excitation of singlet excited chlorophyll via energy or electron transfer). Experimental evidence has shown that energy transfer from chlorophyll molecules to a chlorophyll-zeaxanthin heterodimer that undergoes charge separation is the main mechanism for excess energy dissipation during feedback de-excitation (Horton et al., 1999; Külheim et al., 2002; Niyogi et al., 2005; Holt et al., 2004, 2005). However, quenching of excess energy can also occur independently of zeaxanthin via a reversible inactivation of a fraction of photosystem II

(PSII) centers (Ivanov et al., 2006, 2008) or through conformational changes within the PSII antenna, as recently reported (Johnson et al., 2009).

2.3.2 D1 repair cycle

In addition to the D1 and D2 proteins that conform its core reaction center, PSII contains a and β subunits of cytochrome $b559$, the $psbI$ gene product and a few low molecular weight polypeptides. The D1-D2 heterodimer within PSII binds all the electron carriers and cofactors necessary for electron transport (Nanba & Satoh, 1987; Mattoo et al., 1989). The reaction center protein D1 of PSII is also the primary target of photo-inhibition (Mattoo et al., 1984; Prasil et al., 1992). Due to its intrinsic vulnerability, the short-lived D1 protein must be constantly replaced by new copies via a complicated and evolutionary conserved process known as the PSII or D1 repair cycle, whose significance remains elusive (Mattoo et al., 1989; Andersson & Aro, 2001; Baena-Gonzales & Aro, 2002; Yokthongwattana & Melis, 2006; Edelman & Mattoo, 2008). Nevertheless, the process has an undoubted physiological importance considering that an accumulation of photo-inactivated PSII centers, leading to a decreased photochemical efficiency and the consequent photo-damage, occurs whenever its repair capacity is surpassed (Melis, 1999; Andersson & Aro, 2001).

D1 is a target of at least five post-translational modifications during its life cycle, including N-acetylation, palmitoylation and phosphorylation (Edelman & Mattoo, 2008). One or more of these post-translational modifications could potentially alter protein degradation kinetics, although the use of nitric oxide donors to inhibit *in vivo* phosphorylation of the D1 protein suggested that redox-dependent phosphorylation and D1 degradation in plants are not linked events (Booij-James et al., 2009). The DegP and FtsH proteases have been shown to be involved in D1 degradation *in vitro* (Haussuhl et al., 2001; Kanervo et al., 2003; Lindahl et al., 2000). The physiological significance of these specific proteases was demonstrated in the Arabidopsis *var2* (for *yellow variegated2*) or *var1* mutants, lacking the FtsH2 or FtsH5 membrane-bound metalloproteases, respectively, and the *fu-gaeri1* (*fug1*) mutant that suppresses interfering leaf variegation in *var1* and *var2*, all of which led to an inefficient degradation of the D1 protein and a concomitant increase in ROS levels that was connected to an enhanced susceptibility to photoinhibition (Bailey et al., 2002; Kato et al., 2009). Similar results were reported in cyanobacteria, where impaired D1 protein turnover was detected in an FtsH inactivation mutant (Silva et al., 2003). Moreover, experiments in which an increased transcription of two FtsH-coding genes and of FtsH protease activity was found to be induced upon transfer of cyanobacteria to high light, demonstrated that FtsH proteolysis is a light regulated process (Hihara et al., 2001; Singh et al., 2005).

2.3.3 Other stress-response programs

The accumulation of excitation energy produced when the rate of absorption of photons exceeds the rate of utilization of excitation energy in photosynthetic electron transport leads to an accumulation of reduced electron acceptors that eventually produce excited states of chlorophyll (i.e. triplet state). This process is presumed to predominantly occur in the PSII reaction center where quenching by carotenoids is less effective. Triplet state chlorophyll readily reacts with oxygen to give rise to singlet oxygen, a highly destructive excited oxygen species causing photo-oxidations (Triantaphylidés & Havaux, 2009) (Figure 1). Superoxide, H_2O_2 (produced via reduction or dismutation of superoxide) and hydroxyl radicals, all of which are more reactive than ground state triplet O_2, can also be produced by numerous pathways in photosynthetic cells. It is generally accepted that PSI is the major site of superoxide generation in the photosynthetic electron transport (PET) chain (Asada, 2006) (Figure 1). In addition to production linked to PET and respiratory electron transport (RET) chains, the photorespiration pathway is a major producer of H_2O_2 (Peterhansel et al., 2010). Photorespiration is due to the oxygenase activity of ribulose-1, 5-bisphosphate carboxylase-oxygenase (Rubisco), which produces 2-phosphoglycolate. This small molecule is metabolized through a sequence of reactions that includes H_2O_2 production by glycolate oxidase. Thus, the implementation of protective/metabolizing systems to prevent the deleterious effects of ROS accumulation in plants, are essential to maintain the process of photosynthesis in the oxygen-rich atmosphere of this planet.

The perturbation of the equilibrium between ROS production and scavenging that is frequently produced in plants under stress, can result in a transient increase in ROS levels that is closely associated with the emergence of various disorders such as cell death, disease, and aging (Neill et al., 2002; Overmyer et al., 2003). ROS exert this effect either by reacting with, and irreversibly damaging, a large variety of bio-molecules or by altering the expression of genes that affect signal transduction pathways in a highly selective, specific, and sometimes antagonistic, manner (Apel & Hirt, 2004; Danon et al., 2005; Gadjev et al., 2006; Laloi et al., 2007; Lee et al., 2007). Strong evidence suggesting that H_2O_2 either directly or indirectly antagonizes singlet-oxygen-mediated signaling was obtained recently using a ingenious experimental approach in which Arabidopsis *flu* mutant plants, which generate singlet oxygen in plastids during a dark-to-light transition (see below), were found to produce a more intense stress responses when the H_2O_2 concentration was reduced non-invasively by the over-expression of a thylakoid ascorbate peroxidase (Murgia et al., 2004; Laloi et al., 2007). In addition, low molecular weight antioxidants (e.g., ascorbate, glutathione) serve not only to limit the lifetime of the ROS signals but also participate in an extensive range of other redox signaling and regulatory functions (Foyer & Noctor, 2009).

Considering the above, ROS are considered to be primary diffusible and reactive mediators of signaling linked to electron transport status. For instance, singlet oxygen was considered for many years as a highly toxic molecule with very limited diffusion. However, the utilization of specific probes capable of detecting singlet oxygen in the aqueous phase of isolated thylakoid suspensions and the cytoplasm of high light stressed cells of *C. reinhardtii*, strongly suggested that singlet oxygen can diffuse significant distances from its site of production to activate specific gene expression, such as the nuclear-encoded glutathione peroxidase homolog GPXH (Fisher et al., 2007). However, the physiological relevance of these findings remains questionable considering that the fraction of mobile singlet oxygen was probably small, was detectable only at very high light intensities and has been observed only in this species. H_2O_2 is also recognized as an important signaling molecule. Its role as a signal conveyor was reinforced by data, generated using spin trapping electron paramagnetic resonance spectroscopy and H_2O_2-sensitive fluorescence dyes, that showed that up to 5% of the total H_2O_2 produced inside the chloroplasts was able to diffuse out of the chloroplasts, and in the process evade the effective antioxidant systems located inside this organelle. Moreover, H_2O_2 diffusion was shown to increase concomitantly with light intensity and time of illumination (Mubarakshina et al., 2010). Additional observations have suggested that glutathione, whose synthesis is affected by changes in photosynthesis, may also act as a plastid signal that controls expression of stress defense genes in the nucleus (Wachter et al., 2005; Mullineaux & Rausch, 2005; Rausch et al., 2007).

Friend or Foe? Exploring the Factors that Determine the Difference Between Positive and Negative
Effects on Photosynthesis in Response to Insect Herbivory

177

Identifying ROS specific signaling pathways leading to changes in nuclear gene expression is hampered by the fact that several chemically distinct ROS are generated simultaneously during stress within the plastid compartment. This problem was partially solved by the generation of the *flu* mutant of Arabidopsis that generates singlet oxygen in plastids in a controlled and non-invasive manner (Meskauskiene et al., 2001; op den Camp et al., 2003). Thus, this mutant accumulates excess protochlorophyllide in the dark that, upon illumination, acts as a photosensitizer capable of generating singlet oxygen (Flors & Nonell, 2006; Hideg et al., 2006; op den Camp et al., 2003). Light induced generation of singlet oxygen has revealed a rapid change in nuclear gene expression that differs substantially from nuclear gene expression profiles activated by superoxide or H_2O_2, further supporting the proposal that superoxide/H_2O_2- and singlet oxygen-dependent signaling occur via distinct pathways (Laloi et al., 2006; op den Camp et al., 2003).

The high reactivity of singlet oxygen, together with its unlikely ability to leave the plastid compartment, suggested that its physiological impact depended on the generation of more stable second messengers within the plastid, which were assumed to activate a signaling cascade outside of the plastid compartment. Two components of singlet oxygen signaling, the EXECUTER 1 and 2 proteins localized in the chloroplast, were recently identified as additional signaling components of singlet oxygen (Wagner et al., 2004; Lee et al., 2007) (Figure 1). However, the mechanisms involved in singlet oxygen sensing and signal transduction to the nucleus remain to be characterized, although experimental evidence supporting a positive role for abscisic acid (ABA), ethylene-, salicylic acid (SA)- and JA-dependent signaling pathways in the singlet oxygen- induced response was recently reported (Danon et al., 2005; Ochsenbein et al., 2006). In addition, approximately 50 genes encoding putative transcription factors have been identified to be rapidly and strongly induced within 30 min of the release of singlet oxygen. These include ethylene responsive factors, WRKY transcription factors, zinc-finger proteins and several DNA-binding proteins. Many genes involved in putative signal-transduction pathways and calcium regulation, such as protein kinases, calcium and calmodulin-binding proteins, were also identified (Danon et al., 2005; Laloi et al., 2006; Lee et al., 2007; op den Camp et al., 2003).

As mentioned above, H_2O_2 activates a response program different from singlet oxygen which is also more stable. Its experimentally proven capacity to diffuse across the chloroplast envelope is believed to be a pivotal step in a model that involves an H_2O_2-dependent activation of a mitogen-activated protein kinase (MAPK) cascade in the cytosol that subsequently affects gene expression in the nucleus (Kovtun et al., 2000; Vranova et al., 2002; Apel & Hirt, 2004; Mittler et al., 2004) (Figure 1). However, the polar nature of H_2O_2, which would be expected to limit its capacity to diffuse across hydrophobic membranes unaided, has been a strong argument used to question this model. Alternatively, it has been proposed that H_2O_2 transport is mediated by aquaporin channels (Bienert et al., 2007; Dynowski et al., 2008). This proposal, is supported by the hypersensitivity to H_2O_2 observed in yeast cells expressing Arabidopsis aquaporins in the plasma membrane, but has yet to be demonstrated in plants. Another nebulous aspect of the model is the mechanism by which H_2O_2, produced in multiple cell sites and in response to various different stresses and stimuli, acquires the specificity needed to act as a reliable signal conveying information on the state of chloroplasts to the nucleus.

Further experimental support for the presence of independent redox signaling pathways acting via differentiated signaling cascades came from the characterization of the so called *redox-imbalanced* (*rimb*) mutants, which were detected using an Arabidopsis reporter gene line expressing luciferase under control of the redox-sensitive 2-cysteine peroxiredoxin A (2CPA) promoter (Heiber et al., 2007). Valuable information shedding light on the nature of redox signaling should be expected when the identity of the RIMB genes and their biochemical function is determined.

2.3.4 Tetrapyrrole and metabolite signaling

Pioneering experiments designed to understand plastid signaling were based on the application of norflurazon. This bleaching herbicide was found to be a potent experimental tool by its ability to profoundly disrupt chloroplast function, due predominantly by its strong inhibition of carotenoid biosynthesis. It was also shown to trigger the release of ROS upon illumination and prevent the light-dependent induction of nuclear photosynthesis-related genes (Oelmuller & Mohr, 1986). The isolation and characterization of the so called *gun* mutants (for *genomes uncoupled*), most of which coded for proteins that are involved in tetrapyrrole biosynthesis, led some workers to suggest that tetrapyrrole intermediates serve as a plastid signal to regulate the expression of nuclear genes for photosynthetic proteins (Mochizuki et al., 2001; Larkin et al., 2003; Strand, 2003; Koussevitzky et al., 2007). However, the conclusive results derived from subsequent studies strongly suggested that tetrapyrrole pathway intermediaries are not directly linked to plastid signaling (Mochizuki et al., 2008; Moulin et al., 2008).

Conversely, the concept of metabolic signaling arose from the unlikeness that ROS or redox compounds themselves act as signaling molecules that traverse the cytosol (see above). In this context, messengers that are metabolically more inert and less readily inactivated during diffusion through the cell represent much more promising signaling candidates (Baier & Dietz, 2005). This alternative signaling pathway is justified on the informative nature regarding the metabolic state of the chloroplast that is contained in the relatively large exchange of primary metabolites, such as carbohydrates, or of the xanthophyll derivative abscisic acid (ABA) phytohormone, with the rest of the cell. In this regard, sugar-signaling pathways have been considered as likely candidates for the regulation of photosynthetic acclimation under various stress conditions. Moreover, the expression of nuclear encoded photosynthetic genes (e.g. *CAB2* and *rbcS*) was found to be inversely correlated with intercellular soluble sugar levels (Oswald et al., 2001).

Thus, one proposed scenario envisions that metabolite concentration changes could be sensed by cytosolic or nuclear receptors to regulate nuclear gene expression. One of these sensors could be the cytosolic hexokinase, known to be crucial for sensing and responding to glucose signals intra-cellularly (Figure 1). Alternatively, the transport of carbohydrates across the chloroplast membrane might directly communicate information on the redox state of the chloroplast by means of so called 'redox valves'. Two well-studied examples of carbohydrate shuttles that export reducing power from the chloroplast are the malate/oxaloacetate and triose-phosphate shuttles (Heineke et al., 1991). The 'malate valve' comprises the malate/oxaloacetate translocator (Taniguchi et al., 2002) together with the chloroplast and cytosolic isoforms of NAD(P)H malate dehydrogenase, and is thought to constitute the central mechanism for the export of excess reducing power from the chloroplast. The dihydroxyacetonephosphate/3-phosphoglycerate shuttle involves the triosephosphate translocator (TPT) (Flügge et al., 1989) that functions primarily as a dihydroxyl acetonephosphate/phosphate exchanger to maintain sucrose synthesis in the cytosol, but that could also export both ATP and NADPH from chloroplasts into the cytosol (Figure 1). Interestingly, the altered nuclear gene expression detected in the *tpt* (Biehl et al., 2005) and *cue1* mutants, the latter affected in the phosphoenolpyruvate/phosphate translocator (PPT) in the inner chloroplast envelope (Streatfield et al., 1999) supports the proposed role played by metabolite exchange in redox signaling .

Finally, the role of ABA as a signal relying information on the chloroplast status to the nucleus can be explained by evoking the multiple effects that photosynthetic activity rates have on the biosynthetic ABA pathway that is partially localized in the chloroplast (Baier & Dietz, 2005). For instance, oxidative stress conditions leading to increased ABA through inductive effects on the synthesis of its xanthophyll precursor in the chloroplast, could provide a link between the redox state and ROS levels in the plastid and gene expression in the nucleus. In this respect, it has been speculated that the repressed photosynthetic gene expression produced by norflurazon treatment might be associated with reduced levels of ABA resulting from a depressed carotenoid biosynthesis (Kleine et al., 2009).

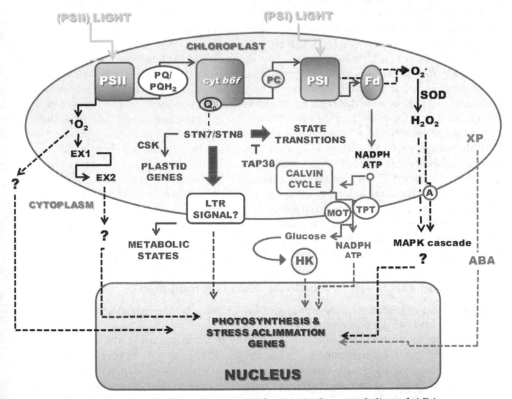

Fig. 1. Proposed plastid signal transduction pathways (redox, *metabolic* and ABA-dependent) involved in the regulation of acclimation responses to stress, including state transitions, long-term responses (LTR) and/or the activation/ repression of specific target genes in the chloroplast and nucleus. Redox signals are generated within the electron transport chain (purple) or by generation of reactive oxygen species (ROS) (black). The plastoquinone pool (PQ/PQH$_2$, in purple letters) is the origin for at least two redox

signaling pathways that are active under low or high light. These can lead to rapid state transitions, involving reversible association of the mobile pool of LHCII to PSI or PSII, or LTR. Qo (red oval) represents the docking site where plastoquinol binds to the cytochrome *b6f* complex. Both processes are dependent on the redox-regulated activity of the STN7 kinase (possibly in conjunction with STN8; purple letters). The TAP38 phosphatase (in red letters) regulates state transitions by specifically dephosphorylating LHCII. On the other hand, LTR involves chloroplast and nuclear gene expression (purple arrows). LTR-dependent plastid gene expression is believed to be regulated by the *Chloroplast Sensor Kinase* (CSK, in purple letters), while nuclear gene expression might require a putative LTR signal (in purple box). LTR-dependent changes in gene expression and protein accumulation can also lead to the establishment of two distinct metabolic states (purple letters, in *cytoplasm*) derived from the exposure to either PSI or PSII light (yellow letters and arrows). ROS are generated by transfer of electrons from PSI or reduced ferredoxin (Fd) to oxygen-generating superoxide (O_2^-, in black letters). This is detoxified by superoxide dismutase (SOD, in black letters) to hydrogen peroxide (H_2O_2; in black letters). Un-scavenged H_2O_2 might able to diffuse freely across the chloroplast envelope or through water channels (or aquaporins, A; blue letter in blue circle) and is thought to start MAP kinase cascades in the cytosol. Singlet oxygen ($^1O_2$; in black letters), is generated at PSII. Its high reactivity and short half-life require additional signaling components, such as Executer 1 and 2 (EX1, EX2, in black letters), although evidence in green algae suggests that $^1O_2$ might be able to diffuse out of the chloroplast. *Metabolic* plastid signaling has been proposed to require the activity of the malate/oxaloacetate (MOT; in green letters) and triosephosphate (TPT; in green letters) translocators needed for the export of excess reducing power and ATP from the chloroplast. Alternatively, metabolite concentration changes (e.g., glucose originated from the Calvin cycle; green letters) could be sensed by cytosolic or nuclear receptors (i.e. hexokinase; HK, green letters in yellow circle) to modify nuclear gene expression. Abscisic acid (ABA, in orange letters) whose synthesis is derived from ROS-sensitive xanthophyll precursors (XP, in orange letters) in the chloroplast, has also been proposed to act as a signal relying information on the chloroplast status to the nucleus. Dotted arrows represent pathways mediated by unknown components that have not been entirely elucidated. Tetrapyrrole and ROS-scavenging-associated signaling pathways are not shown (Adapted from Kleine et al., 2009, Pfannschmidt et al., 2009, and Pesaresi et al., 2010).

3. Photosynthesis under biotic stress: How is it regulated?

3.1 Integration of metabolic, hormonal and environmental signals in stress acclimation and retrograde signaling

Plants are able to integrate and evaluate a diversity of input signals to optimize acclimation responses to stressful environmental growth conditions and to ensure plant survival. Frequently observed adaptation responses are growth retardation, reduced metabolism and photosynthesis, reallocation of metabolic resources and increased antioxidant capacity. Cumulative evidence showing strong stress-related effects on ROS and auxin levels, coupled with the stress-induced morphogenetic changes often produced during adaptation, indicate that these physiologically active metabolites play a prominent role in the integration of the stress-regulatory networks, acting through mechanisms that remain poorly understood (Tognetti et al., 2011). The elaborate ROS signaling network is also known to act in concert with other hormonal networks and with plastid signaling to regulate developmental processes, in addition to abiotic and biotic stress tolerance responses (Kleine et al., 2009; Tognetti et al., 2011). For instance, the local and systemic acclimation in Arabidopsis leaves in response to excess excitation energy associated with cell death and regulated by specific redox changes of the PQ pool, also caused a rapid decrease of stomatal conductance, global induction of genes involved in ROS scavenging and pathogen resistance, increased ROS production and enhanced ethylene signaling. In addition, evidence was provided which showed that multiple hormonal/ROS signaling pathways not only regulate the plant's response to excess excitation energy, but also control induced systemic acquired resistance and basal defenses to virulent bacterial pathogens. The balanced activity of the disease resistance and signaling-related proteins coded by the *LSD1, EDS1, PAD4,* and *EIN2* genes was found to be necessary to regulate the steps leading to programmed cell death, light acclimation, and defense responses that are initiated, at least in part, by redox changes of the PQ pool (Mühlenbock et al., 2008). Further evidence coupling chloroplast-controlled disease resistance with ROS accumulation was obtained with the Arabidopsis mutant *rph1* (for *resistance to Phytophthora 1*), which was found to be susceptible to the pathogen *Phytophthora brassicae* as a consequence of a reduced oxidative burst, a runaway cell-death response, and failure to properly activate the expression of defense-related genes. The finding that the *RPH1* gene encodes an evolutionary highly conserved chloroplast protein was in accordance with a prominent chloroplast-dependent role in the activation of immune responses to *Phytophthora,* not only in Arabidopsis but in potato, as well (Belhaj et al., 2009).

In addition, signal integration at the level of transcription factor (TF) activation appears to be majorly controlled by the family of APETALA 2/ ethylene response element binding protein (AP2/EREBP) TFs, which are abundantly represented in Arabidopsis, poplar and rice (Dietz et al., 2010). By dint of their activation of different innervating pathways, or their ability to bind to multiple target elements, AP2/EREBP TFs are known to integrate several signaling inputs. A couple of examples are the ERF1 TF, which is controlled by ethylene and JA (Lorenzo et al., 2004), and the dehydration-responsive element binding TF TINY, that connects abiotic stress signaling via DRE-dependent regulation to biotic stress signaling via ethylene response elements (Sun et al., 2008). It is considered that combinatorial target gene regulation by different signals may involve different mechanisms, including: (i) cross-talk in the signaling pathways; (ii) stimuli-dependent TF activation, e.g. by homo- or heterotypic dimer and oligomer formation, respectively; (iii) competition for the same or binding to different cis elements; and (iv) amplification cascades that can be modulated by interfering signals (Dietz et al., 2010).

3.2 The negative effect of abiotic and biotic stress on photosynthetic gene expression

The down-regulation of photosynthetic gene transcription is frequently observed in plants subjected to stress. Thereby, environmental stresses, including drought, salinity and low temperatures can negatively affect photosynthetic gene expression in addition to an induction of compensating physiological and biochemical alterations (Saibo et al., 2009; Chaves et al., 2009). Similarly, a whole gamut of biotic insults caused by arthropods, fungi, bacteria and viral pathogens triggers a uniform and apparently regulated reduction in transcription of nuclear genes coding for the major components of photosynthesis, regardless of the plant host. The widespread negative effect on photosynthesis caused by biotic stressors was recently evidenced by a meta-genomic analysis in which the transcriptome data from microarray experiments representing twenty two different forms of biotic damage on eight different plant species (predominantly Arabidopsis, five other herbaceous species and two tree species) was performed. In this study, transcript levels of photosynthesis light reaction, carbon reduction cycle and pigment synthesis genes were decreased regardless of the type of biotic attack. Interestingly, down-regulation of photosynthesis-related genes contrasted with the corresponding up-regulation of genes coding for the synthesis of JA and those involved in the responses to SA and ethylene. This clear difference in expression patterns suggested that the up-regulation of defense-related biosynthetic genes could be part of the overall defense response responsible for re-allocating resources from growth to defense (see below; Bilgin et al., 2010).

Apparently, insect herbivory caused by chewing insects, had the weakest negative effect on photosynthetic gene expression as compared to other biotic stressors, whereas a general down-regulation of photosynthesis genes was observed in plants infested by aphids and the whitefly *Bemisia tabaci* (Bt) (Heidel & Baldwin, 2004; Zhu-Salzman et al., 2004; Voelckel et al., 2004; Qubbaj et al., 2005; Yuan et al., 2005; Kempema et al., 2007; Bilgin et al., 2010). Additional data pertaining modifications in gene expression produced by Bt larval feeding in tomato plants at different stages of development was generated using a suppression-subtractive-hybridization (SSH) approach (Estrada-Hernández et al., 2009). In addition to the identification of several genes whose expression was differentially modified at different larval phases during the infestation process, the study showed a down-regulation of photosynthetic gene expression which was in accordance to the general negative trend associated with biotic-stress. However, upon closer examination, it became apparent that a more or less defined phase-dependent change in photosynthetic gene expression occurred in response to Bt infestation, which favored an up-regulation of photosystem II genes in the late two phases of Bt development in detriment of genes coding for components of other photosynthetic complexes, and also of the oxygen evolving complex and the Calvin cycle. A similar behavior was observed in Bt-infested tomatillo (*Physalis* spp.) plants (C. Sánchez-Hernández, personal communication). This led to the proposal that the contrasting pattern of gene expression, which occurred concomitantly with an up-regulation of oxidative stress genes leading to tissue senescence, could represent an additional strategy used by Bt, besides their reported ability avoid plant defenses, to favor infestation, (Walling, 2008; Estrada-Hernández et al., 2009; Délano-Frier et al., 2009). Support for this proposal was recently provided by a study showing that the application of the bacterial toxin coronatine to tomato seedlings, led to a reduced expression of photosynthesis related genes and a 1.5- to 2-fold reduction in maximum quantum efficiency of PS II, which occurred concomitantly with ROS generation and necrotic cell death (Ishiga et al., 2009)

Mention should be made, however, that given the long functional lifetime of most photosynthesis-related proteins (the highly labile D1 protein is a notable exception, see above), reduced gene photosynthetic gene expression does not necessarily translate into an immediate loss of function. Such behavior is believed to permit reallocation of nitrogen to the defense response, due to repressed transcription, without significantly affecting the rate of carbon assimilation (Bilgin et al., 2010).

3.3 The role of JA in on photosynthetic gene expression and growth regulation

Jasmonates play a central role in regulating plant defense responses to herbivores (Howe & Jander, 2008; Spoel & Dong, 2008) and also inhibit growth and photosynthesis by participating in the down-regulation of photosynthesis-related genes (Creelman & Mullet,

Friend or Foe? Exploring the Factors that Determine the Difference Between Positive and Negative
Effects on Photosynthesis in Response to Insect Herbivory

183

1997; Wasternack & Parthier, 1997; Hui et al., 2003; Reymond et al., 2004; Giri et al., 2006; Zavala & Baldwin, 2006; Yan et al., 2007). Ample evidence demonstrating the direct and indirect negative effect on plant growth and/or photosynthesis exerted by JA is available. Early reports described that JA treatment of barley leaves inhibits activity of PSII electron transport (Maslenkova et al., 1999), whereas barley plants treated with methyl jasmonate (MeJA) suffered a repressed translation of transcripts coding for Rubisco small subunit, chlorophyll a/b binding protein and photosystem II proteins (Roloff et al., 1994). Moreover, plants treated with MeJA or genetically manipulated to accumulate higher JA concentrations were found to develop shorter petioles or have a reduced total seed production (Cipollini, 2007; Bonaventure et al., 2007), or suffer reduced root growth (Henkes et al., 2008). Also, induction by diverse types of herbivores of the lipoxygenase pathway, which represents the initial step of JA biosynthesis and jasmonate signaling, was found to be associated with reduced photosynthesis and vegetative growth (Heidel & Baldwin, 2004; De Vos et al., 2005; Kempema et al., 2007; Bilgin et al., 2010), similarly to experiments showing that herbivore-induced JA signaling suppressed re-growth and contributed to apical dominance (Zavala & Baldwin, 2006). It has been proposed that the above effects on plant growth are modulated by the gene *JASMONATE-ASSOCIATED1* (*JAS1*) (Yan et al., 2007). Moreover, a cross-talk between ABA- and JA-responsive gene expression in response to insect herbivory, mediated by the action of MYC and MYB TFs, has been proposed as a mechanism to coordinate the expression of defensive and dehydration-responsive genes (Yamaguchi-Shinozaki & Shinozaki, 2006). Such interaction is deemed to be needed by the plant in order to deal with the increased leaf dehydration and accompanying senescence produced by defoliating herbivores (Aldea et al., 2005; Lim et al., 2007; see below). The observed influence of light quality and perception by phytochromes on JA induced defense responses and resource allocation was also indicative of an indirect connection between JA and photosynthesis (Ballare, 2009; Moreno et al., 2009).

3.4 Main photosynthetic genes targeted by biotic stress

Curiously, the gene coding for the Rubisco enzyme, an absolutely vital component of the carbon assimilation machinery in plants, was found to be one of the primary photosynthetic genes targeted by herbivore attack, in addition to genes coding for the components of the antenna complexes in both photosystems (Logemann et al., 1995; Ehness et al., 1997; Hermsmeier et al., 2001; Hahlbrock et al., 2003; Hui et al., 2003; Montesano et al., 2004; Zou et al., 2005). Proteomic investigations provided further evidence of the vulnerability of the CO_2 fixing process to insect attack by showing that herbivory, or even the application of the insect's saliva, also reduced the abundance of Rubisco activase (RCA) in *Nicotiana attenuata* and Arabidopsis (Giri et al., 2006; Thivierge et al., 2010). RCA is a key regulatory enzyme of photosynthetic carbon assimilation that modulates the activity of Rubisco by facilitating the removal of inhibiting sugar phosphates from its active site (Portis, 1995). Additional findings derived from the proteomic experimental approaches indicated that Rubisco large subunit and RCA, in addition to PS I P700 apoprotein A1 suffered several caterpillar-specific modifications, including the conversion of Cys192 in Rubisco to the the thiolate anion, which may lead to decreased enzyme activity and protein degradation, and adverse modifications, in RCA, of the protein domains involved in ATP binding (Thivierge et al., 2010).

A valid explanation for the slower growth and down-regulation of photosynthetic-related genes elicited by herbivore damage is that it may be required to liberate resources, e.g. the high proportion of leaf N that is invested in photosynthetic proteins, primarily Rubisco, for defense-related processes (Baldwin, 2001). Another possibility is that these changes could represent a variety of the *scorched earth* strategies used by plants to buffer the impact of herbivory, based on the premise that a reduction in growth and nutrient availability, resulting from the combined effects of decreased photosynthesis, inhibited nitrate assimilation and diminished levels of amino acids and of the main dietary protein (i.e Rubisco) in leaves, will reduce the nutritional quality of the plant to the feeding insect, and consequently, the degree of attractiveness for subsequent damage (Hermsmeier et al., 2001; Hahlbrock et al., 2003; Schwachtje et al., 2006).

Silencing of either *RCA* or *Rubisco* in *N. attenuata* suppressed photosynthetic capacity, as expected, but uncovered further refinements regarding the defensive role played by a reduced expression of these genes in plants subjected to insect damage (Giri et al., 2006; Mitra et al., 2008). Surprisingly, insect performance of both specialist and generalist insect pests was increased in RCA-silenced plants, a result attributed mostly to an impairment of the JA-Ile/leucine signaling pathway required for the expression of defense-related genes coding for trypsin protease inhibitors or for enzymes needed for biosynthesis of defensive metabolites such as diterpene glycosides (Mitra et al., 2008). The negative effect exerted by RCA silencing on JA-Ile/leucine signaling and related herbivore resistance traits was hypothesized to occur in response to reduced ATP levels produced in carbon and energy depleted plants having decreased photosynthetic rates, since ATP is needed for the adenylation of JA, the first step in the amino acid conjugation process needed to regulate this hormone's activity (Staswick et al., 2002). On the other hand, no negative effects on JA signaling were detected in Rubisco-silenced plants, which nevertheless suffered greater damage when confronted by larvae of a specialist insect. These were proposed to have a higher tolerance than generalist insects to the protein deficiency resulting from Rubisco silencing due to their improved ability to detoxify plant defenses (Green et al., 2001). One of the principal conclusions reached from the results of the above series of experiments was that the nature of the photosynthetic genes affected as a consequence of insect herbivory will have important repercussions in their relation to plant defense.

4. Plant responses to herbivory from the phostosynthetic perspective

4.1 The profound impact of insect herbivory on plants

Herbivory can negatively affect ecosystems by decreasing photosynthesis and net primary production. Estimates of global crop production losses caused by foliage-feeding insects typically range from 5% to 30%, with losses estimated to exceed 50% were it not for the widespread application of pesticides (Mattson & Addy, 1975; Oerke & Dehne, 1997). Additional losses range from 2 to 15% in forests and from 4 to 24% in old-fields and grasslands, while insect outbreaks have been known to reduce net primary productivity by 70% to 100% in some terrestrial ecosystems (Cyr & Pace, 1993). Insect herbivory reduces leaf area or depletes leaf fluids by mining and cell content feeding. It can also be selective, targeting other tissues such the phloem, xylem (Haile et al., 1999; Macedo et al., 2003a, b; Heng-Moss et al., 2006), or the stems (Macedo et al., 2005, 2007). Insect feeding typically reduces photosynthesis, although evidence showing positive or neutral effects (i.e. tolerance or compensatory responses; see below) on photosynthesis has also been reported. Most of the time, this variability stems from characteristic factors of a given plant insect interaction, including damage intensity (e.g. total vs. partial defoliation; dispersed vs. concentrated damage; phloem feeding vs. defoliation) and location (e.g. proximity to veins), type of tissue that is preferentially damaged (as mentioned above) and the way tissue is damaged (e.g. chewing vs. scraping; crushing vs. piercing, etc.). Another important factor that influences the outcome of herbivory on photosynthesis has to do whether insect damage induces the accumulation of autotoxic defensive allelochemicals (see below).

4.2 Positive or neutral effects of insect herbivory on photosynthesis

Resistance and tolerance represent two general strategies of plant defense against herbivores, although interactions between these two strategies are assumed to occur under certain conditions, i.e. when the resources available for defense are limited or when both defensive strategies are physiologically costly (Leimu & Koricheva, 2006). Resistance involves the reduction of the amount of herbivore damage whereas tolerance leads to a reduction of the impact of herbivory on plant fitness (Rausher et al., 1993; Stowe et al., 2000). Resistance traits include mechanical and chemical characters that reduce herbivore performance (antibiosis) or preference (antixenosis). Conversely, proposed mechanisms for tolerance/compensation are re-growth stimulation, elevated rates of photosynthesis in remaining leaves of partially defoliated plants, increased branching through the release of apical dominance, alteration of phenology or plant architecture, production of new leaf area, utilization of high pre-herbivory stored carbon resources or the ability to reallocate them to less vulnerable tissues, resorption of nutrients from senescent/damaged leaves, especially nitrogen (N) and phosphorus (P), alteration of the external light environment and higher reproductive efficiency through increased percentage of fruit set (Mabry & Wayne, 1997; Hjalten et al., 1993; Strauss & Agrawal, 1999; Hochwender et al., 2000; Tiffin, 2000; Anten et al., 2003; Silla & Escudero, 2003; Leimu & Koricheva, 2006; Schwachtje et al., 2006). Compensatory ability in plants varies widely across species, and the degree in which it is manifested depends on the amount of leaf lost, with complete rather than fragmented defoliation usually being more conductive to an increased rate of net photosynthesis in the remaining or newly formed leaves (Welter, 1989). The mode of herbivore damage and herbivore type may also determine whether the overall effect on photosynthesis in the plant. This was elegantly evidenced in a recent report showing that herbivory on *N. attenuata* by *Tupiocoris notatus*, a cell-content feeder, (or by application on wounded plants of its salivary secretions), induced an elevated photosynthetic activity, and consequent CO_2 assimilation, that appeared to compensate for lost tissue and for the fitness costs associated with the deployment of direct and indirect defenses. This compensatory effect was shown to be specific for this insect, since feeding by chewing *Manduca sexta* larvae resulted in a strong down-regulation of photosynthesis (Halitschke et al., 2011).

Environmental conditions and the timing of the herbivory event are also influential factors. Thus, compensation to damage in terms of timing of herbivory is usually more effective when required early in the growing season or before the reproductive phase has started. For example, a study performed in Lebanese cucumber (*Cucumis sativus*) to compare the ability to compensate for foliar herbivory at both the pre-flowering and flowering stages found that damage produced before flowering allowed plants to compensate more efficiently, in terms of vegetative biomass and fruit production, for leaf losses that sometimes reached 80% of the total leaf area in the plant groups examined. Higher compensation was correlated with a higher photosynthetic efficiency and capacity, and with less dissipation of light energy as heat, leading to the proposal that herbivore-damaged plants may be induced to use a greater proportion of the absorbed light energy for photosynthesis as a result of altered carbohydrate source-sink relationships (Thomson et al., 2003). In contrast, an experimental setting designed to test the effects of partial de-budding on photosynthesis, stomatal conductance and nitrogen in *Picea jezoensis* seedlings led to the conclusion that the enhanced photosynthetic rate observed in de-budded seedlings was the result of an increased root/leaf ratio that reduced the stomatal limitation of photosynthetic rate, rather than of an altered sink-source relationship or increased leaf nitrogen content (Ozaki et al., 2004).

A pair of studies aimed at determining the carbon costs of herbivory by phloem-feeding scale insects on tress found that infested trees had a greater annual photosynthesis, as determined by measuring parameters such as Vc max, the maximum rate of Rubisco-catalysed carboxylation, J max, the rate of electron transport when irradiance is saturating and/or chlorophyll fluorescence (Retuerto et al., 2004; Dungan et al., 2007). The small negative effect on tree growth and reproduction and increased photosynthetic efficiency observed were taken as an indication that damaged trees were able to compensate fully for the relatively large loss of carbon to herbivory caused by the honeydew insects. According to these workers, the amelioration of carbon loss resulting from the additional sinks for photosynthates created by scale insect feeding was achieved by increased photosynthetic rates. These results were in agreement with previous data suggesting that defoliation, as well as removal of reproductive and other vegetative sinks, may improve photosynthesis in remaining leaf tissue by increasing carboxylation efficiency and the rate of Rubisco regeneration (Holman & Oosterhuis, 1999; Thomson et al., 2003; Ozaki et al., 2004; Turnbull et al., 2007). However, they were in contradiction with data generated from the meta-analysis of a collection of reports showing that sap feeding insects have an almost universal negative effect on growth, photosynthesis, and reproduction of woody plants (Zvereva et al., 2010). The discrepancy detected was adjudicated to experimental biases introduced by the utilization of improper controls (e.g selective assignation as controls to undamaged plant sections that were avoided by herbivores or herbivore preference for hosts with higher rates of photosynthesis). Other important findings of the above meta-analysis were the following: i) sap-feeders did not change the resource allocation in plants; ii) mesophyll and phloem feeders produced stronger effects than xylem feeders, whereas generalist sap-feeders reduced plant performance to a greater extent than did specialists; iii) methodology (e.g. greenhouse vs. field settings; natural vs. imposed herbivory and short-term vs. long-term feeding) was a significant factor influencing the outcome of the experiments, and iv) sap feeding was more detrimental at higher temperatures. Thus, sap-feeders were considered to exert a more severe overall negative impact on woody plant performance than defoliators, mostly due to the latter's lower ability to compensate for sap-feeders' damage in terms of both growth and photosynthesis.

Another study in which the effect of high and low soil nutrient levels on biomass re-growth and photosynthetic up-regulation, among genotypes of the Mediterranean annual grass *Avena barbata* subjected to simulated herbivory, obtained rather unexpected results. They showed that tolerance in this species was positively correlated only with pre-defoliation photosynthetic efficiency at high nutrients, since no evidence for photosynthetic up-regulation in defoliated compared to control plants was observed regardless of nutrient treatment (Suwa & Maherali, 2008). In this context, a rather infrequent report describing compensatory responses to herbivory to the root system suggested a novel tolerance mechanism for insect herbivory. This welcome contribution to the rather unexplored area of plant root-insect interactions was designed to understand the high tolerance to root herbivory by bio-control agents shown by *Centaurea maculosa*, an invasive North American plant species. The use of ^{15}N labeling indicated that infested plants were able to sustain growth and maintain a constant shoot N status under potentially devastating conditions characterized by a drastic reduction of whole plant and root N uptake as a result of herbivory, by shifting N allocation to the shoot, away from the reach of root herbivores (Newingham et al., 2007).

Compensatory photosynthesis is also deemed to play an important role in plants that utilize carbon-based defense strategies, by increasing the availability of carbohydrates that can potentially be allocated to defense. A recent report focused on the possible effects that diverse tritrophic interactions, involving browsing herbivores and several species of resident ants, could have on foliar photosynthetic rates, measured as net photosynthesis (Pn), transpiration and water use efficiency (WUE), and concomitant availability of carbon pools for metabolism and defense in *Acacia drepanolobium*. This species is an east-African, savannah-resident tree, that is known to exhibit carbon-based investments in direct defense (e.g. erection of physical barriers and accumulation of toxic chemicals), indirect defense (e.g. housing and feeding of beneficial ants that guard the plant from herbivores) and tolerance (e.g. stimulated rates of leaf and shoot growth) (King & Caylor, 2010). Their results, which represent the first evidence that indirect defenders of plants can also benefit plants by increasing their photosynthetic rates, indicated first, that *A. drepanolobium* trees exhibited elevated photosynthetic rates in response to browsing only when occupied by strongly mutualistic ants, and second, that this photosynthetic up-regulation mitigated the costs of herbivory by increasing pools of photosynthate available for additional defense or for re-growth of lost tissue.

A unique example of positive manipulation of plant photosynthesis by insect herbivores is represented by the so-called green-island phenotype induced by leaf-miners in deciduous leaves in the autumn season, and persisting long after leaf abscission. These green-islands are characterized by photosynthetically active green patches in otherwise senescing leaves, and correspond to regions with an increased concentration in cytokinins, which are hormones involved in a variety of biological processes, many pertinent to the phenomenon in question, such as the inhibition of senescence, maintenance of chlorophyll and control of source-sink relationships for nutrient mobilization, and maintenance of enriched nutritional environments (Gan & Amasino, 1995; Balibrea Lara et al., 2004; Walters & McRoberts, 2008; Giron et al., 2007). The concentrated levels of nutrients that characterize green-islands in senescent leaves favor growth and reproduction of the leaf miners with only a limited consumption of leaf tissues. This, in turn, allows areas of uneaten tissue to be employed for thermal regulation and parasitoid avoidance (Djemaï et al., 2000; Giron et al., 2007). However, the origin of cytokinins in leaf-miner systems has not yet been determined, although several lines of evidence initially suggested that cytokinins were derived from the insect. Such a concept was questioned by recent findings suggesting that cytokinins might originate from bacterial endosymbionts known establish an intimate association with leaf mining insects. This was evidenced by the negative effects on insect fitness derived from curing leaf-miners of their symbiotic partner, which also abolished green-island formation on leaves (Kaiser et al., 2010).

A number of selected examples in which insect herbivory has been shown to have a positive influence on photosynthesis, including many already described above, are shown in Table1.

Plant species	Herbivore species	Damage type/ feeding guild	Results	Method	Reference
Alder (Alnus incana, A. glutinosa) and birch (Betula pendula)	Alder beetle (Agelastica alni)	Foliage-chewing feeder	Photosynthetic rates of grazed leaves increased following herbivory in alder; by contrast birch exhibited a decline in net photosynthesis. Differences related to the beetle's feeding behavior that often cut midribs only in birch.	Gas exchange (GE)	Oleksyn et al., 1998
Willow tree (Salix viminalis)	Aphids (Tuberolachnus salignus; Pterocomma salicis)	Sap feeder (stem-feeding)	Photosynthetic rate and leaf nitrogen content were significantly raised by T. salignus feeding.	GE	Collins et al., 2001
Cucumber (Cucumis sativum)	Brown garden snail (Helix aspersa)	Foliage-chewing feeder	Higher compensation in terms of vegetative biomass and fruit production was correlated with an increase in photosynthetic efficiency and capacity, and with less dissipation of light energy.	Chlorophyll fluorescence (ChlF)	Thompson et al.,2003
European holly trees (Ilex aquifolium)	Scale insects (Coccus sp.)	Phloem feeder	Insect infestation increased photosynthetic efficiency; effect enhanced by high temperature and light. Insects altered the photosynthesis of leaves not directly affected by the insects.	ChlF	Retuerto et al., 2004
Ezo spruce Picea jezoensis	Manual de-budding (in nature Choristoneura jezoensis; aphids Adelges japonicus)	Bud feeder	Partial de-budding enhanced photosynthetic rates in 1-year-old needles but not in current-year needles. Greater photosynthetic rate was accompanied by increased stomatal conductance.	GE	Ozaki et al., 2004
Cotton (Gossypium hirsutum)	Cotton aphid (Aphis gossypii); thrips (Thrips tabaci and Frankliniella schultzei)	Phloem feeder	Photosynthesis, respiration rates or non-structural carbohydrates on leaves were not affected by short-term aphid feeding. No increase in net photosynthesis during thrips infestation or recovery phases revealed the lack of compensation in affected leaves.	GE	Lei & Wilson 2004; Gomez et al., 2006

Plant species	Herbivore species	Damage type/ feeding guild	Results	Method	Reference
Soybean (*Glycine max*)	Japanese beetles (*Popillia japonica*); corn earworm (*Helicoverpa zea*)	Foliage-chewing feeder	Herbivory increased transpiration without affecting carbon assimilation rates or photosynthetic efficiency. Reductions in net photosynthesis and stomatal conductance occurred only when midvein was disrupted.	GE, ChlF and thermal imaging (TI)	Aldea et al., 2005
Beech trees *Nothofagus solandri*	Scale insects (*Ultracoelostoma assimil*)	Phloem feeder	Infested trees had a greater annual photosynthesis measured as Vc max, J max and chlorophyll content. Consequently, annual canopy photosynthesis was 4% greater for infested trees.	GE	Dungan et al., 2007
Wheat (*Triticum aestivum*)	Armyworm (*Spodoptera frugiperda*)	Foliage-chewing feeder	Photosynthesis, intercellular CO_2 and transpiration of injured leaves were not significantly affected; however, stomatal conductance values were higher. Spatial pattern of defoliation differentially affected photosynthesis; leaves defoliated at the basal portion had lower rates.	GE and ChlF	Macedo et al., 2007
Acacia drepanolobium	Resident ants (*Crematogaster mimosa; C. nigriceps; C. sjostedti; Tetraponera penzigi*)		Trees exhibited elevated photosynthetic rates in response to browsing only when occupied by strongly mutualistic ants (*Crematogaster mimosa; C. nigriceps*). Photosynthetic up-regulation mitigated the costs of herbivory by increasing pools for additional defense or for re-growth of lost tissue.	GE	King and Caylor, 2010

Plant species	Herbivore species	Damage type/ feeding guild	Results	Method	Reference
Nicotiana attenuata	Mirid bugs (Tupiocoris notatus)	Foliage-chewing feeder	Elevated CO_2 assimilation rate was sufficient to compensate for loss of photosynthetic active tissue. Stomatal conductance and intercellular CO_2 were not affected. Mirid salivary secretions treatment also increased photosynthetic activity.	GE and fluorescence imaging (FI)	Halitschke et al., 2011

Table 1. Some examples of positive or neutral effects on photosynthesis after herbivory damage.

4.3 Negative effects of insect herbivory on photosynthesis

In the absence of compensatory mechanisms, insect herbivory causing removal and/or injury of plant tissues most frequently leads to a direct suppression of photosynthetic activity. A seminal report describing the outcome of an extensive examination of the pertinent literature available at the time, indicated that over 50% of all plant-insect interactions, predominantly involving leaf-mining, stem-boring, galling or sucking leaf injury, resulted in a loss of photosynthetic capacity, frequently manifested as decreased photosynthetic rate (Pn) (Welter, 1989). The reduction in chlorophyll content in response to insect damage, frequently reported in plants attacked by phloem feeding insects, has been also reported to result in a decrease in photosynthesis (Kaakeh et al., 1992; Cabrera et al., 1994), with even small reductions leading to a drastic reduction in the photosynthetic rate (Nagaraj et al., 2002). Interestingly, changes in leaf pigment composition caused by insect herbivory were found to have potential application for remote sensing pest detection in Australia. Thus, the reduction in leaf chlorophyll content occurring concomitantly with an increase in photoprotective pigments, known to be a sensitive indicator of plant stress caused by root feeding phylloxera grapevine pests, was exploited for the development of a phylloxera-specific remote detection system (Blanchfield et al., 2006). Additional factors contributing to decreased photosynthesis include changes in the nutrient status of leaves caused by competition between plant sinks and additional sinks created by insect herbivores, mostly sap-feeders or gall-formers (see above), decreased stomatal conductance, which is coupled to reduced WUE and altered water transport, stomatal aperture and/or sucrose transport and loading. Most of these conditions are also known to influence indirect suppression of photosynthesis, as described below.

However, as it has been mentioned already, a plant's response to herbivory is often variable and usually depends on the combined contribution of several factors including the type of tissue injured and the extent tissue damage. An illustrative example for this effect is given by reported data showing that the removal of leaf tissue from soybean by herbivores such as Japanese beetles (Popillia japonica), corn earworm caterpillars (Helicoverpa zea) (Aldea et al., 2005), cabbage loopers (Trichoplusia ni), and green clover-worms (Plathypena scabra) (Hammond & Pedigo, 1981; Ostlie & Pedigo, 1984) caused an increase in water loss from damaged tissue, but had a minimal effect on net photosynthesis. Conversely, chewing damage by skeletonizing Mexican bean beetles (*Epilachna varivestis*) caused substantial losses of photosynthesis in the remaining leaf tissue (Peterson et al., 1998). It was hypothesized that the scraping and crushing of interveinal leaf tissue caused by feeding adults and larvae of Mexican bean beetles may have exacerbated localized water stress, ultimately causing tissue desiccation and photosynthesis repression. The timing of damage is also considered to be an important factor. In this regard, early season damage has been usually found to cause more pronounced changes in plants than late season damage, which is in accordance with the assumption that vigorously growing foliage has a greater capacity to respond to various stimuli, including damage. Also, seedlings are generally more susceptible to photosynthetic damage because of a shortage of reserves due to their smaller size or to limitations in nutrient acquisition (Nykänen & Koricheva, 2004; Hódar et al., 2008). A report recording the response of potted fruitless grapevines (*Vitis labrusca* var. Niagara) to early and late season mechanical and insect defoliation was in accordance with this concept by showing that growth, single leaf photosynthesis, and whole-vine photosynthesis were more tolerant to foliar injury late in the season than early in the season (Mercader & Isaacs, 2003). Similar results were obtained from a series of experiments performed to examine a possible trade-off between photosynthesis with defense or reproduction in the common milkweed *Asclepias syriaca*, which is a plant that accumulates toxic cardenolides in a constitutive or inducible manner and is also susceptible to insect damage during its relatively long reproduction period (Delaney et al., 2009). The results of this study showed that leaf Pn impairment after partial leaf defoliation had a seasonal pattern which correlated with *A. syriaca* reproductive phenology but not with cardenolide accumulation. In this regard, the small or absent Pn impairment occurring in leaves of pre-flowering or maturing seed pod plants, contrasted with the moderate to severe leaf Pn impairment detected in leaves of flowering and early seed pod formation plants. Such a behavior led the authors to suggest that a physiological 'cost of reproduction' might be an additional susceptibility factor leading to Pn impairment after herbivory injury on a leaf. Another important aspect to consider is that photosynthesis will be usually more affected when plants are attacked by generalist herbivores, against which they show a higher susceptibility. This is believed to be derived from the lack of a previous and selective co-evolutionary process leading to adaptation (Parker et al., 2006; however, see above). The plant's capacity to tolerate injury, its phenotypic plasticity and the type of environment with which the plant is interacting may be important factors too (Alward & Joern, 1993; Trumble et al., 1993; Delaney & Macedo, 2001; García & Ehrlén, 2002; Zvereva et al., 2010).

The nitrogen status of the plant is also considered to influence the way photosynthesis is affected by herbivory in plants. This is because of the strong positive correlation that is usually observed between photosynthesis rate and nitrogen concentration in plants, predominantly sequestered in the Rubisco enzyme (Field & Mooney, 1986; Evans, 1989; see above). It is not surprising then, that one of the mechanisms offered to explain why the localized decrease in N content negatively affects photosynthesis, a circumstance that has been frequently reported in damaged leaves of woody plants subjected to insect herbivory, is precisely that N deficiency directly affects CO_2 assimilation rates by lowering Rubisco levels (Reich et al., 1999; Mediavilla et al., 2001).

Another little studied aspect of plant-insect interations is the effect that insect oviposition might have on photosynthesis. Most of the available data suggest, however, that the effect is predominantly negative. A recent study reported that net photosynthetic rate, J max, and Vc max of pine needles laden with eggs of an herbivorous sawfly were lower than in egg-free control plants that were not attacked. The negative effect was deemed to have occurred as the result of an egg deposition process that involved wounding of the plant tissue by the sawflies' ovipositor prior to the laying of eggs into its ovipositional wound (Schröder et al., 2005). In a more recent report, the oviposition and wounding effects were separated by employing two pentatomid insects (*Murgantia histrionica* and *Nezara viridula*) having different feeding habits but known not to cut or otherwise physically damage the host substrate during the oviposition procedure. In this process, the eggs are laid in clusters on the leaf surface and adhere to it by a sticky oviduct secretion (Bin et al., 1993; Colazza et al., 2004). Nevertheless, a surprisingly large inhibition of photosynthesis was detected in leaves of *Brassica oleracea*, one of the plant models employed together with common bean, in response to oviposition by *M. histrionica*, even when oviposition was not associated with feeding activity. High resolution chlorophyll fluorescence imaging revealed that the damage to photochemistry caused by feeding and oviposition was restricted to the affected areas, whereas an increase in photochemical yield detected temporarily in the neighboring intact areas of the attacked leaves, indicated the onset of a compensatory response. To date, the way(s) in which insect oviposition affects photosynthesis, occurring either with or without ovipositional plant wounding, remain(s) unknown. However, reduced diffusion rates of CO_2 in the mesophyll cells was suggested as a possible mechanism leading to photosynthesis inhibition during oviposition in the absence of plant wounding (Velikova et al., 2010).

4.4 Indirect effects of insect herbivory on photosynthesis

Leaf area removal not only affects photosynthesis in the damaged tissue but may have a *hidden* or indirect effect in tissues not directly damaged by the herbivore which undergo an additional reduction in photosynthetic capacity and alterations in transpiration (Welter, 1989; Zangerl et al., 2002; Aldea et al., 2006; Berger et al., 2007; Bilgin et al., 2008; Nabity et al., 2009). The discovery that herbivory-induced alterations to photosynthesis and transpiration propagate into remaining undamaged leaf tissue was greatly favored by the development of imaging techniques tools. These proved capable of performing spatially-resolved measurements of the component processes of photosynthesis across leaf surfaces in order to provide direct estimates of the magnitude of local and systemic damage in a quantitative, multi-layered or complementary and non-invasive way. The ability to measure chlorophyll fluorescence by imaging techniques was a pivotal development, considering that this is by far the most important indicator of photosynthetic stress and damage to the photosynthetic apparatus. Thus, chlorophyll fluorescence provides a precise measure of the quantum yield of photosystem II in light-adapted leaves (ΦPSII), which is, in turn, related to the rate of carbon fixation (Genty et al., 1989), and may be used to calculate the photosynthetic electron transport rate driving photosynthesis and photorespiration (Di Marco et al., 1990). Chlorophyll fluorescence data can also be used to assess damage to the photochemical aspect of photosynthesis by measurements of the quantum yield in dark-adapted leaves (by determining the ratio between variable and maximal fluorescence, Fv/Fm) and by the amount of photochemical energy lost as heat (by measuring the non-photochemical quenching of fluorescence, NPQ) (Genty & Harbinson, 1996). Fluorescence imaging, which more often than not correlates with photosynthetic capacity measured by gas exchange, further improved the already superior suitability of this technique to assess damage to the photosynthetic apparatus, by providing a topographical panorama of damage in the leaf, including systemic damage produced in tissue sections not affected directly by the stressor (Chaerle et al., 2007). Moreover, this tool can be combined with thermal imaging, a powerful technique for mapping changes in temperature caused by variations in latent heat flux across leaf surfaces, which can be converted into maps of variable stomatal conductance (Omasa & Takayama, 2003; Jones, 2004; Bajons et al., 2005; Grant et al., 2006). Water limitations in leaves can result, for example, from the disruption in water transport caused by herbivore-damage of water-conducting xylem elements or by midrib vein cutting insects (Tang et al., 2006; Delaney & Higley, 2006; see below). Additional spatial patterning of other components of the photosynthetic machinery, including chlorophyll content and activation of the xanthophyll cycle can be mapped with a technique not frequently employed called hyper-spectral imaging (Nabity et al., 2009). The combined use of these techniques to measure changes in the same leaf in a given experiment, although technically challenging, has provided a deeper insight into the mechanisms by which herbivory indirectly reduces photosynthesis in the remaining undamaged leaf tissue, particularly if combined with physiological data, as reported in several related studies (Zangerl et al., 2002; West et al., 2005; Leinonen & Jones, 2004; Aldea et al., 2006; Tang et al., 2006, 2009).

The systemic suppression of photosynthesis in leaf tissues not directly damaged during insect herbivory has been found to extend to an area that greatly exceeds the actual leaf area removed or damaged by the herbivore. For example, the removal of only 5% of the area of an individual wild parsnip leaf by cabbage looper (*T. ni*) caterpillars reduced photosynthesis by 20 %, determined as ΦPSII and the rate of CO_2 uptake, in the remaining foliage (Zangerl et al., 2002). The indirect effect was observed to extend to a relatively considerable distance from the cut edges and was still detected for at least 3 days after the caterpillars were removed. Moreover, the size of the indirect effects was positively correlated with defense-related synthesis of auto-toxic furanocoumarins, suggesting that costs of chemical defense may be one factor that accounts for the deleterious indirect effects of herbivory on plants. Similarly, the decline in photosynthesis in the remaining leaf tissue of damaged oak saplings was equal to the decrease in photosynthesis associated with the actual removal of leaf tissue (Aldea et al., 2006). Chewing damage by cabbage looper larvae also caused substantial reductions in ΦPSII in Arabidopsis' leaves at some distance from the tissue removed (Tang et al., 2006). Interestingly, this study demonstrated that the degree of photosynthetic impairment caused by herbivory depended on the nature of the damage inflicted. Therefore, damage caused by first instar larvae, which typically make small holes and avoid veins, led to photosynthetic depression in the remaining leaf tissue near the holes, whereas fourth instars, that make larger perforations while feeding, had little effect on photosynthesis (Tang et al., 2006). The workers concluded that both water stress, induced by the increased rate of water loss near the damaged tissues, and the reduced stomatal conductance produced in the tissues localized at some distance from the injuries, contributed to the inhibition of photosynthesis in the remaining leaf tissues, although subsequent data suggested that induction of defense responses in areas near the holes may have also contributed to the observed decrease in photosynthesis (Tang et al., 2009). The above studies provided evidence suggesting that defense induced auto-toxicity or defense-induced down-regulation of photosynthesis contributed to the indirect repression of photosynthesis. Another contributing factor to indirect suppression of photosynthesis is vasculature tissue severance, which generally leads to a disruption in fluid or nutrient transport, and altered sink demand. The collected data indicates that the probable mechanisms responsible for reducing photosynthesis in remaining leaf tissues are multifaceted, ranging from disruptions in fluid or nutrient transport to self-inflicted reductions in metabolic processes. However, the magnitude of their contribution to indirect impairment of photosynthesis will vary depending in large part on the type of feeding damage and the mode of defense deployed by the plant under attack (Nabity et al., 2009). In this respect, chewing damage and fungal and gall infections were found to differentially affect the component processes of photosynthesis of nearby leaf tissue in several hard-wood tree species, with fungal infections and galls causing large depressions (>25%) of photosynthetic effciency (as ΦPSII) over extended areas of the leaf around the visible damage, whereas chewing damage resulted in minor (≈7%) depressions of ΦPSII that were restricted to a 1 mm perimeter around the perforations. Although similar in their effect on electron transport through PSII, the indirect effects of fungal and gall infections on photosynthesis were found to operate through different mechanisms. A reduction in stomatal conductance with an associated decline in intercellular CO_2 concentration may have contributed to the depression of ΦPSII around fungal spots but not in gall surrounding areas. On the other hand, the mild and localized suppressions of the photosynthetic efficiency in tissue surrounding chewing damage was attributed to the desiccation of tissue along the edges of damage, similarly to other reports (Aldea et al., 2005; Tang et al., 2006). One sobering conclusion reached by this study was that the indirect, negative, effects of photosynthesis caused by biotic stress on tress were exacerbated by elevated CO_2, exposing yet another damaging element of the ongoing global trend towards higher CO_2 levels in the earth's atmosphere.

Defoliation injury which severs venation indiscriminately (e.g by fouth instar, but not first instar, *T. ni* larvae; see above) can damage xylem and/or phloem, leading to altered water transport, stomatal aperture, and sucrose transport and loading. All these changes can strongly contribute to reduce photosynthesis in the remaining leaf tissue. Severing veins and inter-veinal tissue also alters the hydraulic construction of leaves as the result of an exponential reduction in resistance occurring with increasing damage (Nardini & Salleo, 2005). Long- or short-term leaf desiccation can also occur in the absence of alternative pathways for water transport. If insect feeding is subtle enough to avoid outright cell rupture (e.g. by phloem-feeders), modulation of nutrients sequestered by feeding will alter plant osmotica or sink/source relationships (Girousse et al., 2005; Dorchin et al., 2006). Feeding may physically obstruct fluid flow with insect mouthparts (stylets) or cell fragments and alter photosynthesis and water balance in remaining leaf tissue (Reddall et al., 2004; Delaney & Higley, 2006). A particular mechanism of plant vasculature disruption is midrib vein cutting, a little-known type of specialized herbivory suggested to have evolved as a strategy to avoid trapping leaf latex or toxic cardenolide defenses in plant species mostly restricted to the Asclepiadaceae (milkweed), Apocynaceae (dogbane), Polygonaceae and Fabaceae families. This type of damage was found to impair several leaf gas exchange parameters, but only downstream from the injury location. Photosynthesis impairment caused by midrib herbivory was more severe than manually imposed and actual insect defoliation, was relatively long-lasting and became most severe as the injury location came closer to the petiole (Delaney & Higley, 2006). As mentioned above, a form of defoliation in soybean plants known as skeletonization, is characterized by the removal of patches of tissue, reduced photosynthesis in remaining tissue on damaged leaves and on adjacent undamaged leaflets (Peterson et al., 1998). A related study reported results that were in agreement with the high water losses associated with skeletonizing damage, by showing that the cut edges of soybean leaves damaged by Japanese beetles known to follow this mode of herbivory, suffered a very substantial dehydration (Aldea et al., 2005). However, their data showed that although damage to the inter-veinal tissue increased transpiration by 150 % for up to 4 days post-injury, it had no detectable effect on CO_2 exchange, and even induced a short-lived increase in photosynthetic efficiency in undamaged tissue of damaged leaves. Such a contradictory outcome was deemed to have happened as a result of a transient decoupling of photosynthetic electron transport from carbon assimilation caused by insect damage (Aldea et al., 2005). Regarding the above, it is not surprising to know that plants can increase WUE as a strategy to ameliorate the negative effects of herbivory, as was recently found in apple tress infested by leaf-mining moths (Pincebourde et al., 2006). Thus, WUE was found to be about 200% higher in the mined apple leaf tissues in comparison to intact leaf portions, prompting the proposal that minimizing water losses reduces the negative impact on photosynthesis derived from herbivore attacks, by avoiding severe reductions in the CO_2 assimilated to water loss ratio.

Autotoxicity by resident plant metabolites having potential biocidal properties that can directly affect the host plant may represent an important fitness cost. This adds to the investment in energy and C and N sources already employed for their synthesis, which could have otherwise been used for growth and reproduction (Zangerl & Bazzaz, 1993). Autotoxicity has been recorded in cases where secondary compounds having biocidal properties that severely affect the photosynthetic machinery of the plant are either released from specialized storage tissues that confine them (e.g. glands, trichomes or oil tubes) or accumulate as a consequence of leaf damage. An early study investigating autotoxicity in defense-related metabolites, linked nicotine toxicity to the reduction in photosynthesis in a number Solanaceous plants (Baldwin & Callahan, 1993). Some time later, the suppression of ΦPSII in regions of the leaf near the tissue removed by caterpillars was related to an increased production of toxic furanocoumarins (Zangerl et al., 2002; see above). A subsequent study tested the autotoxicity of several essential oil components, including several monoterpenes and sesquiterpenes and myristicin, an essential oil component derived from the phenylpropanoid pathway, in three plant species known to produce them (i.e. *Pastinaca sativa*, *Petroselinum crispum*, and *Citrus jambhiri*) (Gog et al., 2005). The toxic effects, which were assessed by quantifying reductions in photosynthetic capacity as measured by chlorophyll fluorescence imaging, were examined both by exogenous applications of pure compounds and by the release, by slicing, of endogenous essential oils known to contain these compounds, among others. Monoterpenes, but not the caryophyllene and farnesene sesquiterpenes or myristicin, produced a rapid and spatially extensive decline in photosynthetic capacity that was detected within a time frame of 200 s. On the other hand, the release of endogenous essential oils significantly reduced photosynthetic activity in all three plant species examined, an effect that was more pronounced in *P. sativa* and *P. crispum*. The auto-toxic effect of monoterpenes was assumed to be related to the loss of cell and organelle integrity associated with their known capacity to disrupt membranes (Harrewijn et al., 2001; Maffei et al., 2001). Conversely, coumarins and furanocoumarins have been long known to negatively affect photosynthesis in several plant species. Photo-phosphorylation uncoupling, energy transfer inhibition and/or Hill reaction inhibition have been identified as the probable mechanisms responsible for their suppression of photosynthesis in higher plants (Macías et al., 1999; Veiga et al., 2007).

A number of selected examples in which insect herbivory has been shown to directly or indirectly influence photosynthesis in a negative way, including many already described above, are shown in Table2.

Plant species	Herbivore species	Damage type / feeding guild	Results	Method	Reference(s)
Goldenrod (*Solidago altissima*)	Aphid (*Uroleucon caligatum*); beetle (*Trirhabda* sp.); spittlebug (*Philaenus spumarius*)	Phloem, foliage and xylem-sap feeders	Photosynthetic rates per unit area of damaged leaves were reduced by spittlebug feeding, but not by beetle or aphid feeding. Spittlebug feeding did not cause stomatal closure, but impaired C fixation within the leaf.	Gas exchange (GE)	Meyer & Whitlow, 1992
Barley (*Hordeum vulgare*)	Aphid (*Schizophis graminum*)	Sap feeder	Chlorophyll content and photosynthesis decreased 75 and 45% respectively after infestation.	GE	Cabrera et al., 1994
Cotton (*Gossypium hirsutum*)	Aphid (*Aphis gossypii*)	Phloem feeder	Photosynthetic depression and transpiration increase were quantitatively related to initial aphid infestation densities and to the length of feeding	GE	Shannag et al., 1998
Soybean (*Glycine max*); dry bean (*Phaseolus vulgaris*)	Mexican bean beetles (*Epilachna varivestis*)	Foliage-scraping chewing feeder (skeletoniz er)	Adults and larvae reduced photosynthetic rates of the remaining tissue of the injured leaflet. A significant linear relationship between photosynthetic rate and percentage injury was observed. Light reactions of photosynthesis were not affected.	GE	Peterson et al., 1998
Cotton (*Gossypium hirsutum*)	Silverleaf whitefly (*Bemisia argentifolii*)	Phloem feeder	Photosynthetic rate was decreased 50%; associated with reductions in chlorophyll fluorescence and fluorescence yield. No changes were found in stomatal conductance, intercellular CO_2 concentration, and leaf chlorophyll content.	GE and chlorophyll content (ChlC)	Lin et al., 1999
Rice (*Oryza sativa*)	Planthopper (*Nilaparvata lugens*)	Phloem feeder	Suppressed photosynthetic rate after infestation, especially at lower leaf positions. Chlorophyll content and total plant dry weight were also reduced.	Carbon isotope ratios (CIR)	Watanabe & Kitagawa 2000

Plant species	Herbivore species	Damage type /feeding guild	Results	Method	Reference(s)
Nicotiana attenuata, N. longiflora	Horn worm (*Manduca sexta*); *Tupiocoris notatus*; aphid (*Myzus nicotianae*)	Foliage-chewing, single cell- and phloem feeders	Up-regulation of defense-related genes and down-regulation of primary metabolism and photosynthesis-related genes. CO_2 assimilation and photosystem II efficiency reduced by 16% and 8% respectively in the remaining tissue of damaged leaves.	GE, fluorescen ce imaging (FI) and Microarray analyses (MA)	Hermsmeier et al., 2001; Izaguirre et al., 2003; Voelckel & Baldwin 2004a,b; Voelckel et al., 2004; Halitschke et al., 2003, 2011
Wild parsnip (*Pastinaca sativa*)	Cabbage looper (*T. ni*)	Foliage-feeder	Decreased efficiency of photosystem II that extended beyond the area directly damaged.	GE and chlorophyll fluore-scence (ChlF)	Zangerl et al., 2002
Tobacco (*Nicotiana tabacum*); soybean (*Glycine max*)	Tobacco budworm (*Heliothis virescens*); oblique-banded leaf roller (*Choristoneura rosaceana*)	Foliage-chewing feeder	Insect locomotion and herbivory across leaf surfaces reduced photosynthesis and increased production of ROS and signaling molecule 4-aminobutyrate.	ChlF	Bown et al., 2002
Soybean (*Glycine max*)	Soybean aphid (*Aphis glycines*); two-spotted spider mite (*Tetranychus urticae*)	Phloem feeder	Reduction up to 50% in photosynthetic rates was not the consequence of stomatal limitation and photoelectron transport was not impaired. Spider mites decreased photosynthesis, stomatal conductance, transpiration and clorophyll content.	ChlF, GE and CIR	Macedo et al., 2003a; Haile & Higley, 2003
Sorghum (*Sorghum bicolor*)	Greenbugs (*Schizaphis graminae*)	Phloem feeder	Photosynthesis-related genes were suppressed strongly by MeJA, and to a lesser extent by SA and aphids.	MA	Zhu-Salzman et al., 2004

Plant species	Herbivore species	Damage type / feeding guild	Results	Method	Reference(s)
Cotton (*Gossypium hirsutum*)	Two-spotted spider mite (*Tetranychus urticae*)	Mesophyll feeders or foliage-chewing feeder	Reduced light-saturated photosynthesis occurred only with high infestation levels. No significant reductions in photosynthetic rates were detected at all initial infestation levels. Photosynthesis declined with crop age and was faster in mite-infested leaves. A minor enhancement of photosynthesis was observed in bottom leaves due to greater light penetration in canopies severely defoliated by mite damage.	GE and photo-synthetic photon flux density	Reddall et al., 2004, 2007
Apple trees (*Malus domestica*)	European red mites (*Panonychus ulmi*); Spotted tentiform (*Phyllonorrycter blancardella*)	Mesophyll feeders	Feeding activity reduces leaf net CO_2 exchange rates; even when green patches maintain levels close to those in intact leaves. Stomatal conductance and hence transpiration rates were highly affected.	GE	Pincebourde et al., 2006
Conifer (*Picea sitchensis*)	Spruce budworms (*Choristoneura occidentalis*) white pine weevils (*Pissodes strobi*)	Floem feeders, stem boring or foliage-chewing feeder	Photosynthesis gene expression (e.g. photosystem I and II, chlorophyll a-b binding proteins and ferredoxin) was down-regulated by budworm feeding.	MA	Ralph et al., 2006
Hardwood saplings (*Carya tomentosa, C. glabra, Quercus alba, Q. velutina Ulmus ulata, Acer rubrum*)	Polyphemus (*Antheraea polyphemus*); redhumped caterpillar (*Schizura concinna*); wasps (*Caryomyia, Eriophyes, Cecidomyia* and *Cynipid* spp.)	Foliage-feeder or gall inductor	Decreased efficiency of photosystem II extended further form visible damage. Gall damage had the greatest depression and chewing rarely affected efficiency of photosystem II surrounding tissue over small distances.	ChlF and thermal imagining (TI)	Aldea et al., 2006

Plant species	Herbivore species	Damage type /feeding guild	Results	Method	Reference(s)
Arabidopsis thaliana	Cabbage looper (T. ni)	Foliage-chewing feeder	Decreased efficiency of photosystem II determined by the mode of feeding by different larvae instars and water stress associated with herbivore damage. Corresponding induction of defense gene expression (cynnamate-4-hydroxylase) with a photosynthesis reduction, but photosynthetic damage spread further into surrounding wounded areas.	GE, ChlF and gene expression in transgenic plants harboring a C4H:GUS fusion	Tang et al., 2006, 2009
Common milkweed (Asclepias syriaca)	Milkweed tussock (Euchaetes egle); monarch butterfly larvae (Danaus plexippus); salt marsh tiger moth (Estigmene acrea)	Foliage-chewing feeder	Partial tissue consumption by insect herbivores caused photosynthetic impairment on remaining tissue. Reduction in photosynthetic rates lasted >5 days. Neighboring uninjured leaves had a small degree (10%) of compensatory photosynthesis. Complete photosynthetic recovery observed at one day post-injury.	GE	Delaney et al., 2008
Tomato (Solanum lycopersicum)	Whitefly (Bemisia tabaci)	Phloem feeder	General repression of photosynthetic genes in an apparent infestation-stage-dependent mode.	SSH	Délano-Frier & Estrada-Hernández 2009; Estrada-Hernández et al., 2009
Wheat (Triticum aestivum)	Russian wheat aphid (Diuraphis noxia); bird cherry-oat aphid (Rhopalosiphum padi)	Phloem feeder	Both aphids negatively affected net photosynthesis; D. noxia had a greater impact than R. padi. Reduction was not related to the light reaction via pigment losses.	GE and ChlC	Macedo et al., 2009
Savoy cabbage (Brassica oleracea); Common bean Phaseolus vulgaris	harlequin cabbage bug (Murgantia histrionica); Nezara viridula	Sap feeder	Photosynthesis decreased rapidly and substantially by feeding and oviposition in the attacked areas. Stomatal conductance did not decrease with photosynthesis. Oviposition did not induce photoinhibitory damage.	ChlF and GE	Velikova et al., 2010

Table 2. Some examples of direct or indirect negative effects on photosynthesis after herbivory damage.

5. Conclusion

This chapter explored the highly complex mechanisms employed by plants to adapt to the ever-changing conditions of an environment that is becoming progressively more unpredictable as the consequences of global warming become painfully apparent. Much progress has been made in the identification of many important players in this vital regulatory process, but many areas remain obscure and will require active research to be elucidated. The manifold relationships between plants and insect herbivores from the photosynthetic perspective were also examined. From this perusal it is clearly evident that almost all contact with insects, even when no damage is involved, as in some oviposition processes or in tritrophic interactions that indirectly benefit the plant, will have an impact on photosynthesis. This highlights the importance of the photosynthetic process in plant-insect interactions, which can be variously manipulated to either favor or impair the plant, or the insect or both.

6. References

Aldea, M., Hamilton, J.G., Resti, J.P., Zangerl, A.R., Berenbaum, M.R. & DeLucia, E.H. (2005). Indirect effects of insect herbivory on leaf gas exchange in soybean, *Plant Cell & Environment* 28: 402-411.

Aldea, M., Hamilton, J.G., Resti, J.P., Zangerl, A.R., Berenbaum, M.R., Frank, T.D. & DeLucia, E.H. (2006). Comparison of photosynthetic damage from arthropod herbivory and pathogen infection in understory hardwood samplings, *Oecologia* 149: 221-232.

Allen, J.F. & Forsberg, J. (2001). Molecular recognition in thylakoid structure and function, *Trends in Plant Science* 6: 317-326.

Allen, J.F. & Pfannschmidt, T. (2000). Balancing the two photosystems: photosynthetic electron transfer governs transcription of reaction centre genes in chloroplasts, *Philosophical Transactions of the Royal Society of London Series B* 355: 1351-1357.

Alward, R.D. & Joern, A. (1993) Plasticity and overcompensation in grass responses to herbivory, *Oecologia* 95: 358-364.

Andersson, B. & Aro, E.M. (2001). Photodamage and D1 protein turnover in Photosystem II, in Aro, E.M. & Andersson, B. (eds.), *Regulation of photosynthesis*, The Netherlands, Kluwer Academic Publishers, pp. 377-393.

Anten, N.P.R., Martínez-Ramos, M. & Ackerley, D.D. (2003). Defoliation and growth in an understorey palm: quantifying the contributions of compensatory responses, *Ecology* 84: 2905-2918.

Apel, K. & Hirt, H. (2004). Reactive oxygen species: metabolism, oxidative stress, and signal transduction, *Annual Review of Plant Biology* 55: 373-399.

Asada, K. (2006). Production and scavenging of reactive oxygen species in chloroplasts and their functions, *Plant Physiology* 141: 391-396.

Baena-Gonzales, E. & Aro, E.M. (2002). Biogenesis, assembly and turnover of photosystem II units, *Philosophical Transactions of the Royal Society of London Series B* 357: 1451-1460.

Baier, M. & Dietz, K.J. (2005). Chloroplasts as source and target of cellular redox regulation: a discussion on chloroplast redox signals in the context of plant physiology, *Journal of Experimental Botany* 56: 1449-1462.

Bailey, S., Thompson, E., Nixon, P.J., Horton, P., Mullineaux, C.W., Robinson, C. & Mann, N.M. (2002) A critical role for the Var2 FtsH homologue of *Arabidopsis thaliana* in the Photosystem II repair cycle *in vivo*, *Journal of Biological Chemistry* 277: 2006-2011.

Bajons, P., Klinger, G. & Schlosser, V. (2005). Determination of stomatal conductance by means of thermal infrared thermography, *Infrared Physics and Technology* 46: 429-439.

Baldwin, I.T. & Callahan, P. (1993). Autotoxicity and chemical defense: nicotine accumulation and carbon gain in solanaceaous plants, Oecologia 94: 534-541.

Baldwin, I.T. (2001). An ecologically motivated analysis of plant-herbivore interactions in native tobacco, *Plant Physiology* 127: 1449-1458.

Balibrea Lara, M.E., Gonzalez Garcia, M.C., Fatima, T., Ehness, R., Lee, T.K., Proels, R., Tanner, W. & Roitsch, T. (2004). Extracellular invertase is an essential component of cytokinin-mediated delay of senescence, *The Plant Cell* 16: 1276-1287.

Ballare, C.L. (2009). Illuminated behaviour: phytochrome as a key regulator of light foraging and plant anti-herbivore defence, *Plant Cell & Environment* 32: 713-725.

Belhaj, K., Lin, B. & Mauch, F. (2009). The chloroplast protein RPH1 plays a role in the immune response of Arabidopsis to *Phytophthora brassicae*, *The Plant Journal* 58: 287-298.

Bellafiore, S., Bameche, F., Peltier, G. & Rochaix, J.D. (2005). State transitions and light adaptation require chloroplast thylakoid protein kinase STN7, *Nature* 433: 892-895.

Bennett, J. (1977). Phosphorylation of chloroplast membrane polypeptides, *Nature* 269: 344-346.

Bennett, J. (1979). Chloroplast phosphoproteins. Phosphorylation of polypeptides of the light-harvesting chlorophyll protein complex, *European Journal of Biochemistry* 99: 133-137.

Berger, S., Benediktyova, Z., Matous, K., Bonfig, K., Mueller, M.J., Nedbal, L. & Roitsch, T. (2007). Visualization of dynamics of plant-pathogen interaction by novel combination of chlorophyll fluorescence imaging and statistical analysis: differential effects of virulent and avirulent strains of *P. syringae* and of oxylipins on *A. thaliana*, *Journal of Experimental Botany* 58: 797-806.

Biehl, A., Richly, E., Noutsos, C., Salamini, F. & Leister, D. (2005). Analysis of 101 nuclear transcriptomes reveals 23 distinct regulons and their relationship to metabolism, chromosomal gene distribution and co-ordination of nuclear and plastid gene expression, *Gene* 344: 33-41.

Bienert, G.P., Møller, A.L.B., Kristiansen, K.A., Schulz, A., Møller, I.M., Schjoerring, J.K. & Jahn, T.P. (2007). Specific aquaporins facilitate the diffusion of hydrogen peroxide across membranes, *Journal of Biological Chemistry* 282: 1183-1192.

Bilgin, D.D., Aldea, M., O'Neill, B.F., Benitez, M., Li, M., Clough, S.J. & DeLucia, E.H. (2008). Elevated ozone alters soybean-virus interaction, *Molecular Plant-Microbe Interactions* 21: 1297-1308.

Bilgin, D.D., Zavala, J.A., Zhu, J., Clough, S.J., Ort, D.R. & DeLucia, E.H. (2010). Biotic stress globally down-regulates photosynthesis genes, *Plant Cell & Environment* 33: 1597-1613.

Bin, F., Vinson, S.B., Strand, M.R., Colazza, S. & Jones, W.A. (1993). Source of an egg kairomone for *Trissolcus basalis*, a parasitoid of *Nezara viridula*, *Physiological Entomology* 18: 7-15.

Blanchfield, A. L., Robinson, S. A.M Renzullo, L. J. & Powell, K. S. (2006). Phylloxera infested grapevines have reduced chlorophyll and increased photoprotective pigment content-can leaf pigment composition aid pest detection? http://ro.uow.edu.au/scipapers/40

Bonardi, V., Pesaresi, P., Becker, T., Schleiff, E., Wagner, R., Pfannschmidt, T., Jahns, P. & Leister, D. (2005). Photosystem II core phosphorylation and photosynthetic acclimation require two different protein kinases, *Nature* 437: 1179-1182.

Bonaventure, G., Gfeller, A., Proebsting, W.M., Hoerstensteiner, S., Chetelat, A., Martinoia, E., Farmer, E.E. (2007). A gain of function allele of TPC1 activates oxylipin biogenesis after leaf wounding in Arabidopsis, *The Plant Journal* 49: 889-898.

Booij-James, I.S., Edelman, M. & Mattoo, A.K. (2009). Nitric oxide donor-mediated inhibition of phosphorylation shows that light-mediated degradation of photosystem II D1 protein and phosphorylation are not tightly linked, *Planta* 229: 1347-1352.

Bown, A.W., Hall, D.E. & MacGregor, K.B. (2002). Insect footsteps on leaves stimulate the accumulation of 4-aminobutyrate and can be visualized through increased chlorophyll fluorescence and superoxide production, *Plant Physiology* 129: 1430-1434.

Bräutigam, K., Dietzel, L., Kleine, T., Ströher, E., Wormuth, D., Dietz, K.J., Radke, D., Wirtz, M., Hell, R., Dörmann, P., Nunes-Nesi, A., Schauer, N., Fernie, A.R., Oliver, S.N., Geigenberger, P., Leister, D. & Pfannschmidt, T. (2009). Dynamic plastid redox signals integrate gene expression and metabolism to induce distinct metabolic states in photosynthetic acclimation in Arabidopsis, *The Plant Cell* 21: 2715-2732.

Breyton, C. (2000). Conformational changes in the cytochrome b6f complex induced by inhibitor binding, *Journal of Biological Chemistry* 275: 13195-13201.

Bukhov, N.G., Heber, U., Wiese, C. & Shuvalov, V.A. (2001). Energy dissipation in photosynthesis: does the quenching of chlorophyll fluorescence originate from antenna complexes of photosystem II or from the reaction center? *Planta* 212: 749-758.

Cabrera, H.M., Argandona, V.H. & Corcuera, L.J. (1994). Metabolic changes in barley seedlings at different aphid infestation levels, *Phytochemistry* 35: 317-319.

op den Camp, R.G.L., Przybyla, D., Ochsenbein, C., Laloi, C., Kim, C.H. & Danon A, Wagner, D., Hideg, E., Göbel, C., Feussner, I., Nater, M. & Apel, K. (2003). Rapid induction of distinct stress responses after the release of singlet oxygen in Arabidopsis, *The Plant Cell* 15: 2320-2332.

Carlberg, I., Hansson, M., Kieselbach, T., Schröder, W.P., Andersson, B. & Vener, A.V. (2003). A novel plant protein undergoing light-induced phosphorylation and release from the photosynthetic thylakoid membranes, *Proceedings of the National Academy of Sciences of the USA* 100: 757-762.

Chaerle, L., Leinonen, I., Jones, H.G. & Van Der Straeten, D. (2007). Monitoring and screening plant populations with combined thermal and chlorophyll fluorescence imaging, *Journal of Experimental Botany* 58: 773-784.

Chaves, M.M., Flexas, J. & Pinheiro, C. (2009). Photosynthesis under drought and salt stress: regulation mechanisms from whole plant to cell, *Annals of Botany* 103: 551-560.

Cipollini, D. (2005). Consequences of the overproduction of methyl jasmonate on seed production, tolerance to defoliation and competitive effect and response of *Arabidopsis thaliana*, *New Phytologist* 173: 146-153.

Friend or Foe? Exploring the Factors that Determine the Difference Between Positive and Negative
Effects on Photosynthesis in Response to Insect Herbivory

203

Colazza, S., Fucarino, A., Peri, E., Salerno, G., Conti, E. & Bin, F. (2004). Insect oviposition induces volatiles emission in herbaceous plant that attracts egg parasitoids, *Journal of Experimental Biology* 207: 47-53.

Collins, C.M., Rosado, R.G. & Leather S.R. (2001). The impact of the aphids *Tuberolachnus salignus* and *Pterocomma salicis* on willow trees, *Annals of Applied Biology* 138: 133–140.

Creelman, R.A. & Mullet, J.E. (1997). Jasmonic acid distribution and action in plants: Regulation during development and response to biotic and abiotic stress, *Proceedings of the National Academy of Sciences of the USA* 92: 4114-4119.

Cyr, H. & Pace, M.L. (1993). Magnitude and patterns of herbivory in aquatic and terrestrial systems, *Nature* 361: 148-150.

Danon, A., Miersch, O., Felix, G., den Camp, R.G.L.O. & ApeL, K. (2005). Concurrent activation of cell death-regulating signaling pathways by singlet oxygen in *Arabidopsis thaliana*, *The Plant Journal* 41: 68-80.

Délano-Frier, J.P. & Estrada-Hernández, M.G. (2009). *Bemisia tabaci*-infested tomato plants show phase-specific pattern of photosynthetic gene expression indicative of disrupted electron flow leading to photo-oxidation and plant death, *Plant Signaling & Behavior* 4: 1-4.

Delaney, K.J. & Macedo, T.B. (2001). The impact of herbivory on plants: yield, fitness, and population dynamics, in Peterson, R.K.D. & Higley, L.G. (eds.), *Biotic stress and yield loss*, CRC, Boca Raton, FL. pp. 135-160.

Delaney, K.J. & Higley, L.G. (2006). An insect countermeasure impacts plant physiology: midrib vein cutting, defoliation and leaf photosynthesis, *Plant, Cell & Environment* 29: 1245-1258.

Delaney, K.J., Haile, F.J., Peterson, R.K.D. & Highley, L.G. (2009). Seasonal patterns of leaf photosynthesis after insect herbivory on common milkweed, *Asclepias syriaca*: reflection of a physiological cost of reproduction, not defense?, *The American Midland Naturalist* 162: 224-238.

Depége, N., Bellafiore, S. & Rochaix, J.D. (2003). Role of chloroplast protein kinase Stt7 in LHCII phosphorylation and state transition in *Chlamydomonas*, *Science* 299: 1572-1575.

De Vos, M., Van Zaanen, W., Koornneef, A., Korzelius, J.P., Dicke, M., Van Loon, L.C., Pieterse, C.M.J. (2005). Herbivore-induced resistance against microbial pathogens in Arabidopsis, *Plant Physiology* 142: 353-363.

Dietz, K-J., Vogel, M.O. & Viehhauser, A. (2010). AP2/EREBP transcription factors are part of gene regulatory networks and integrate metabolic, hormonal and environmental signals in stress acclimation and retrograde signaling, *Protoplasma* 245: 3-14.

Dietzel, L., Bräutigam, K. & Pfannschmidt, T. (2008). Photosynthetic acclimation: state transitions and adjustment of photosystem stoichiometry – functional relationships between short-term and long-term light quality acclimation in plants, *FEBS Journal* 275: 1080-1088.

Di Marco, G., Manes, F.S., Tricoli, D. & Vitale, E. (1990). Fluorescence parameters measured concurrently with net photosynthesis to investigate chloroplastic CO_2 concentration in leaves of *Quercus ilex* L., *Journal of Plant Physiology* 136: 538-543.

Djemaï, I., Meyhöfer, R. & Casas, J. (2000). Geometrical games between a host and a parasitoid, *The American Naturalist* 156: 257-265.

Dorchin, N., Cramer, M.D. & Hoffmann, J.H. (2006). Photosynthesis and sink activity of wasp-induced galls in *Acacia pycnantha, Ecology* 87: 1781-1791.

Dungan, R.J., Turnbull, M.H. & Kelly, D. (2007). The carbon costs for host trees of a phloem-feeding herbivore, *Journal of Ecology* 95: 603-613.

Dynowski, M., Schaaf, G., Loque, D., Moran, O. & Ludewig, U. (2008). Plant plasma membrane water channels conduct the signalling molecule H_2O_2, *Biochemical Journal* 414: 53-61.

Edelman, M. & Mattoo, A.K. (2008). D1-protein dynamics in photosystem II: the lingering enigma, *Photosynthesis Research* 98: 609-620.

Ehness, R., Ecker, M., Godt, D.E. & Roitsch, T. (1997). Glucose and stress independently regulate source and sink metabolism and defense mechanisms via signal transduction pathways involving protein phosphorylation, *The Plant Cell* 9: 1825-1841.

Ensminger, I., Busch, F. & Hüner, N.P.A. (2006). Photostasis and cold acclimation: sensing low temperature through photosynthesis, *Physiologia Plantarum* 126: 28-44.

Estrada-Hernández, M.G., Valenzuela-Soto, J.H., Ibarra-Laclette, E. & Délano-Frier, J.P. (2009). Differential gene expression in whitefly *Bemisia tabaci*-infested tomato (*Solanum lycopersicum*) plants at progressing developmental stages of the insect's life cycle, *Physiologia Plantarum* 137: 44-60.

Evans, J.R. (1989.) Photosynthesis and nitrogen relationship in leaves of C_3 plants, *Oecologia* 78: 9-19.

Fey, V., Wagner, R., Brautigam, K. & Pfannschmidt, T. (2005). Photosynthetic redox control of nuclear gene expression, *Journal of Experimental Botany* 56: 1491-1498.

Field, C.B. & Mooney, H.A. (1986). The photosynthesis nitrogen relationship in wild plants in Givinish, T. (ed), *On the economy of plant form and function,* Cambridge University Press, Cambridge, pp. 25-55.

Finazzi, G., Zito, F., Barbagallo, R.P. & Wollman, F.A. (2001). Contrasted effects of inhibitors of cytochrome b(6)f complex on state transitions in *Chlamydomonas reinhardtii*- the role of Q(o) site occupancy in LHCII kinase activation, *Journal of Biological Chemistry* 276: 9770-9774.

Fischer, B.B., Krieger-Liszkay, A., Hideg, E., Ŝnyrychová, I., Wiesendanger, M. & Eggen, R.I.L. (2007). Role of singlet oxygen in chloroplast to nucleus retrograde signaling in *Chlamydomonas reinhardtii, FEBS Letters* 581: 5555-5560.

Fleischmann, M.M., Ravanel, S., Delosme, R., Olive, J., Zito, F., Wollman, F.A. & Rochaix, J.D. (1999). Isolation and characterization of photoautotrophic mutants of *Chlamydomonas reinhardtii* deficient in state transition, *Journal of Biological Chemistry* 274: 30987-30994.

Flors, C. & Nonell, S. (2006). Light and singlet oxygen in plant defense against pathogens: phototoxic phenalone phytoalexins, *Accounts of Chemical Reserach* 39: 293-300.

Flügge, U.I., Fischer, K., Gross, A., Sebald, W., Lottspeich, F. & Eckerskorn, C. (1989). The triose phosphate-3-phosphoglyceratephosphate translocator from spinach chloroplasts: nucleotide sequence of a full-length cDNA clone and import of the *in vitro* synthesized precursor protein into chloroplasts, *EMBO Journal* 8: 39-46.

Foyer, C.H. & Noctor, G. (2009). Redox regulation in photosynthetic organisms: signaling, acclimation, and practical implications, *Antioxidants & Redox Signaling* 11: 862-889.

Friend or Foe? Exploring the Factors that Determine the Difference Between Positive and Negative
Effects on Photosynthesis in Response to Insect Herbivory
205

Frenkel, M., Bellafiore, S., Rochaix, J.D. & Jansson, S. (2007). Hierarchy amongst photosynthetic acclimation responses for plant fitness, *Physiologia Plantarum* 129: 455-459.

Fuhrmann, E., Gathmann, S., Rupprecht, E., Golecki, J. & Schneider, D. (2009). Thylakoid membrane reduction affects the photosystem stoichiometry in the cyanobacterium *Synechocystis* sp. PCC 68031, *Plant Physiology* 149: 735-744.

Gadjev, I., Vanderauwera, S., Gechev, T.S., Laloi, C., Minkov, I.N., Shulaev, V., Apel, K., Inzé, D., Mittler, R. & Van Breusegem, F. (2006). Transcriptomic footprints disclose specificity of reactive oxygen species signaling in Arabidopsis, *Plant Physiology* 141: 436-445.

Gan, S. & Amasino, R.M. (1995). Inhibition of leaf senescence by auto-regulated production of cytokinin, *Science* 270: 1986-1988.

García, M.B. & Ehrlén, J. (2002). Reproductive effort and herbivory timing in a perennial herb: fitness components at the individual and population levels, *American Journal of Botany* 89: 1295-1302.

Genty, B., Briantais, J.M. & Baker, N.R. (1989). The relationship between the quantum yield of photosynthetic electron transport and quenching of chlorophyll fluorescence, *Biochimica et Biophysica Acta* 990: 87-92.

Genty, B. & Harbinson, J. (1996). Regulation of light utilization for photosynthetic electron transport, in Baker, N.R. (ed.), *Photosynthesis & the Environment*. Kluwer Academic, Dordrecht. pp. 69–99.

Giri, A.P., Wunsche, H., Mitra, S., Zavala, J.A., Muck, A., Svatos, A. & Baldwin, I.T. (2006). Molecular interactions between the specialist herbivore *Manduca sexta* (Lepidoptera, Sphingidae) and its natural host *Nicotiana attenuata*. VII. Changes in the plant's proteome, *Plant Physiology* 142: 1621-1641.

Giron, D., Kaiser, W., Imbault, N. & Casas, J. (2007). Cytokinin-mediated leaf manipulation by a leaf-miner caterpillar, *Biology Letters* 3: 340-343.

Girousse, C., Moulia, B., Silk, W. & Bonnemain, J.L. (2005). Aphid infestation causes different changes in carbon and nitrogen allocation in alfalfa stems as well as different inhibitions of longitudinal and radial expansion, *Plant Physiology* 137: 1474-1484.

Gog, L., Berenbaum, M.R., DeLucia, E.H. & Zangerl, A.R. (2005). Autotoxic effects of essential oils on photosynthesis in parsley, parsnip, and rough lemon, *Chemoecology* 15: 115-119.

Gomez, S.K., Oosterhuis, D.M., Hendrix, D.L., Johnson D.R. & Steinkraus, D.C. (2006). Diurnal pattern of aphid feeding and its effect on cotton leaf physiology, *Environmental and Experimental Botany* 55: 77-86.

Grant, O.M., Caves, M.M. & Jones, H.G. (2006). Optimizing thermal imaging as a technique for detecting stomatal closure induced by drought stress under greenhouse conditions, *Physiologia Plantarum* 127: 507-518.

Green, E.S., Zangerl, A.R. & Berenbaum, M.R. (2001). Effects of phytic acid and xanthotoxin on growth and detoxification in caterpillars, *Journal of Chemical Ecology* 27: 1763-1773.

Hahlbrock, K., Bednarek, P., Ciolkowski, I., Hamberger, B., Heise, A., Liedgens, H., Logemann, E., Nürnberger, T., Schmelzer, E., Somssich, I.E. & Tan, J. (2003). Non-self recognition, transcriptional reprogramming, and secondary metabolite accumulation during plant/pathogen interactions, *Proceedings of the National Academy of Sciences of the USA* 100: 14569-14576.

Haile, F.J., Higley, L.G., Ni, X. & Quisenberry, S.S. (1999). Physiological and growth tolerance in wheat to Russian wheat aphid (Homoptera: Aphididae) injury, *Environmental Entomology* 28: 787-794.

Haile, F.J. & Higley, L.G. (2003). Changes in soybean gas-exchange after moisture stress and spider mite injury, *Environmental Entomology* 32: 433-440.

Haldrup, A., Jensen, P.E., Lunde, C. & Scheller, H.V. (2001). Balance of power: a view of the mechanism of photosynthetic state transitions, *Trends in Plant Science* 6: 301-305.

Halitschke, R., Gase, K., Hui, D., Schmidt, D.D. & Baldwin, I.T. (2003). Molecular interactions between the specialist herbivore *Manduca sexta* (Lepidoptera, Sphingidae) and its natural host *Nicotiana attenuata*. VI. Microarray analysis reveals that most herbivore-specific transcriptional changes are mediated by fatty acid–amino acid conjugates, *Plant Physiology* 131: 1894-1902.

Halitschke, R., Hamilton, J.G. & Kessler, A. (2011). Herbivore-specific elicitation of photosynthesis by mirid bug salivary secretions in the wild tobacco *Nicotiana attenuata*, *New Phytologist* 191: 528-535.

Hamel, P., Olive, J., Pierre, Y., Wollman, F.A. & de Vitry, C. (2000). A new subunit of cytochrome b6f complex undergoes reversible phosphorylation upon state transition, *Journal of Biological Chemistry* 275: 17072-17079.

Hammond, R.B. & Pedigo, L.P. (1981). Effects of artificial and insect defoliation on water loss from excised soybean leaves, *Journal of the Kansas Entomological Society* 54: 331-336.

Harrewijn, P., van Oosten, A.M. & Piron, P.M. (2001). Natural Terpenoids as Messengers: A Multidisciplinary Study of Their Production, Biological Functions and Practical Applications. Kluwer, Dordrecht.

Haussuhl, K., Andersson, B. & Adamska, I. (2001). A chloroplast DegP2 protease performs the primary cleavage of the photodamaged D1 protein in plant photosystem II, *EMBO Journal* 20: 713-722.

Heiber, I., Stroher, E., Raatz, B., Busse, I., Kahmann, U., Bevan, M.W., Dietz, K.J. & Baier M. (2007). The redox imbalanced mutants of Arabidopsis differentiate signaling pathways for redox regulation of chloroplast antioxidant enzymes, *Plant Physiology* 143: 1774-1788.

Heidel, A.J. & Baldwin, I.T. (2004). Microarray analysis of salicylic acid- and jasmonic acid-signaling in responses of *Nicotiana attenuata* to attack by insects from multiple feeding guilds, *Plant Cell & Environment* 27: 1362-1373.

Heineke, D., Riens, B. Grosse, H., Hoferichter, P., Peter, U., Flügge, U-I. & Heldt, H.W. (1991). Redox transfer across the inner chloroplast envelope membrane, *Plant Physiology* 95: 1131-1137.

Heng-Moss, T., Macedo, T., Franzen, L., Baxendale, F., Higley, L. & Sarath, G. (2006). Physiological responses of resistant and susceptible buffalo grasses to *Blissus occiduus* (Hemiptera: Blissidae) feeding, *Journal of Economic Entomology* 99: 222-228.

Henkes, G.J., Thorpe, M.R., Minchin, P.E.H., Schurr, U. & Röse, U.S.R. (2008). Jasmonic acid treatment to part of the root system is consistent with simulated leaf herbivory, diverting recently assimilated carbon towards untreated roots within an hour, *Plant Cell & Environment* 31: 1229-1236.

Friend or Foe? Exploring the Factors that Determine the Difference Between Positive and Negative
Effects on Photosynthesis in Response to Insect Herbivory
207

Hermsmeier, D., Schittko, U. & Baldwin, I.T. (2001). Molecular interactions between the specialist herbivore *Manduca sexta* (Lepidoptera: Sphingidae) and its natural host *Nicotiana attenuata*. I. Large-scale changes in the accumulation of growth- and defense-related plant mRNAs, *Plant Physiology* 125: 683-700.

Hihara, Y., Kamei, A., Kanehisa, M., Kaplan, A. & Ikeuchi M. (2001). DNA microarray analysis of cyanobacterial gene expression during acclimation to high light, *The Plant Cell* 13: 793-806.

Hideg, E., Kalai, T., Kós, P.B., Asada, K. & Hideg, K. (2006). Singlet oxygen in plants-its significance and possible detection with double (fluorescent and spin) indicator reagents, *Photochemistry & Photobiology* 82: 1211-1218.

Hjalten, J., Danell, K. & Ericson, L. (1993). Effects of simulated herbivory and intra-specific competition on the compensatory ability of birches, *Ecology* 74: 1136-1142.

Hochwender, C.G., Marquis, R.J. & Stowe, K.A. (2000). The potential for and constraints on the evolution of compensatory ability in *Asclepias syriaca*, *Oecologia* 122: 361-370.

Hódar, J.A., Zamora, R., Castro, J., Gómez, J.M. & García, D. (2008). Biomass allocation and growth responses of Scots pine saplings to simulated herbivory depend on plant age and light availability, *Plant Ecology* 197: 229-238.

Holman, E.M. & Oosterhuis, D.M. (1999). Cotton photosynthesis and carbon partitioning in response to floral bud loss due to insect damage, *Crop Science* 39: 1347-1351.

Holt, N.E., Fleming, G.R. & Niyogi, K.K. (2004). Toward an understandingof the mechanism of non-photochemical quenching in green plants, *Biochemistry* 43: 8281-8289.

Holt, N.E., Zigmantas, D., Valkunas, L., Li, X-P., Niyogi, K.K. & Fleming, G.R. (2005). Carotenoid cation formation and the regulation of photosynthetic light harvesting, *Science* 307: 433-436.

Horton, P., Ruban, A.V. & Young, A.J. (1999). Regulation of the structure and function of the light harvesting complexes of photosystem II by the xanthophyll cycle, in Frank, H.A.,Young, A.J., Britton, G. & Cogdell, R.J. (eds) *Advances in Photosynthesis and Respiration: the Photochemistry of Carotenoids* Vol. 8. Kluwer Academic Publishers, The Netherlands, pp 271-291.

Horton, P., Johnson, M.P., Pérez-Bueno, M., Kiss, A.Z. & Ruban, A.V. (2008). Does the structure and macro-organisation of photosystem II in higher plant grana membranes regulate light harvesting states?, *The FEBS Journal* 275: 1069-1079.

Howe, G.A. & Jander, G. (2008). Plant immunity to insect herbivores, *Annual Review of Plant Biology* 59: 41-66.

Hui, D.Q., Iqbal, J., Lehmann, K., Gase, K., Saluz, H.P. & Baldwin, I.T. (2003). Molecular interactions between the specialist herbivore *Manduca sexta* (Lepidoptera: Sphingidae) and its natural host *Nicotiana attenuata*. V. Microarray analysis and further characterization of large-scale changes in herbivore-induced mRNAs, *Plant Physiology* 131: 1877-1893.

Hüner, N.P.A., Öquist, G. & Sarhan, F. (1998). Energy balance and acclimation to light and cold, *Trends in Plant Science* 3: 224-230.

Ishiga, Y., Uppalapati, S.R., Ishiga, T., Elavarthi, S., Martin, B. & Bender, C.L. (2009). The phytotoxin coronatine induces light-dependent reactive oxygen species in tomato seedlings, *New Phytologist* 181:147-160.

Ivanov, A.G., Sane, P.V., Krol, M., Gray, G.R., Balseris, A., Savitch, L.V., Öquist, G. & Huner, N.P.A. (2006). Acclimation to temperature and irradiance modulates PS II charge recombination, *FEBS Letters*, 580: 2797-2802.

Ivanov, A., Sane, P.V., Hurry, V., Öquist G. & Huner, N.P.A. (2008). Photosystem II reaction centre quenching: mechanisms and physiological role, *Photosynthesis Research* 98: 565-574.

Izaguirre, M.M., Scopel, A.L., Baldwin, I.T. & Ballaré, C.L. (2003). Convergent responses to stress, solar ultraviolet-B radiation and *Manduca sexta* herbivory elicit overlapping transcriptional responses in field-grown plants of *Nicotiana longiflora*, *Plant Physiology* 132: 1755-1767.

Johnson, M.P., Pérez-Bueno, M.L., Zia, A., Horton, P. & Ruban, A.V. (2009). The zeaxanthin-independent and zeaxanthin-dependent qE components of nonphotochemical quenching involve common conformational changes within the Photosystem II antenna in Arabidopsis, *Plant Physiology* 149: 1061-1075.

Jones, H.G. (2004). Application of thermal imaging and infrared sensing in plant physiology and ecophysiology, *Advances in Botanical Research incorporating Advances in Plant Pathology* 41: 107-163.

Kaakeh, W., Pfeiffer, D.G. & Marini, R.P. (1992). Combined effects of *Spirea* aphid (Homoptera, Aphididae) and nitrogen fertilization on net photosynthesis, total chlorophyll content, and greenness of apple leaves, *Journal of Economical Entomology* 85: 939-946.

Kaiser, W., Huguet, H., Casas, J., Commin, C. & Giron, D. (2010). Plant green-island phenotype induced by leaf miners is mediated by bacterial symbionts, *Philosophical Transactions of the Royal Society of London Series B* 277: 2311-2319.

Kanervo, E., Spetea C., Nishiyama, Y., Murata, N., Andersson, B. & Aro, E.M. (2003). Dissecting a cyanobacterial proteolytic system: efficiency in inducing degradation of the D1 protein of photosystem II in cyanobacteria and plants, *Biochimica et Biophysica Acta* 1607: 131-140.

Kato, Y., Miura, E., Ido, K., Ifuku, K. & Sakamoto, W. (2009). The variegated mutants lacking chloroplastic FtsHs are defective in D1 degradation and accumulate reactive oxygen species, *Plant Physiology* 151: 1790-1801.

Kempema, L.A., Cui, X., Holzer, F.M. & Walling, L.L. (2007). Arabidopsis transcriptome changes in response to phloem-feeding silverleaf whitefly nymphs: similarities and distinctions in responses to aphids, *Plant Physiology* 143: 849-865.

King, E.G. & Caylor, K.K. (2010). Herbivores and mutualistic ants interact to modify tree photosynthesis, *New Phytologist* 187: 17-21.

Kleine, T., Voigt, C. & Leister, D. (2009). Plastid signalling to the nucleus: messengers still lost in the mists?, *Trends in Genetics* 25: 185-192.

Koussevitzky, S., Nott, A., Mockler, T.C., Hong, F., Sachetto-Martins, G., Surpin, M., Lim, J., Mittler, R. & Chory, J. (2007). Signals from chloroplasts converge to regulate nuclear gene expression, *Science* 316: 715-719.

Kovtun, Y., Chiu, W.L., Tena, G. & Sheen J. (2000). Functional analysis of oxidative stress-activated mitogen-activated protein kinase cascade in plants, *Proceedings of the National Academy of Sciences of the USA* 97: 2940-2945.

Krause, G.H. & Weis, E. (1991). Chlorophyll fluorescence and photosynthesis: the basics, *Annual Review of Plant Physiology & Plant Molecular Biology* 42: 313-349.

Kruse, O., Nixon, P.J., Schmid, G.H. & Mullineaux, C.W. (1999). Isolation of state transition mutants of *Chlamydomonas reinhardtii* by fluorescence video imaging, *Photosynthesis Research* 61: 43-51.

Külheim, C., Agren, J. & Jansson, S. (2002). Rapid regulation of light harvesting and plant fitness in the field, *Science* 297: 91-93.

Laloi C., Przybyla, D. & Apel, K. (2006). A genetic approach towards elucidating the biological activity of different reactive oxygen species in *Arabidopsis thaliana*, *Journal of Experimental Botany* 57: 1719-1724.

Laloi, C., Stachowiak, M., Pers-Kamczyc, E., Warzych, E., Murgia, I. & Apel, K. (2007). Cross-talk between singlet oxygen- and hydrogen.peroxide-dependent signaling of stress responses in *Arabidopsis thaliana*, *Proceedings of the National Academy of Sciences of the USA* 104: 672-677.

Larkin, R.M., Alonso, J.M., Ecker, J.R. & Chory, J. (2003). GUN4, a regulator of chlorophyll synthesis and intracellular signaling, *Science* 299: 902-906.

Lee, K.P., Kim, C., Landgraf, F. & Apel, K. (2007). EXECUTER1- and EXECUTER2-dependent transfer of stress-related signals from the plastid to the nucleus of *Arabidopsis thaliana*, *Proceedings of the National Academy of Sciences of the USA* 104: 10270-10275.

Lei, T.T. & Wilson, L.J. (2004). Recovery of leaf area through accelerated shoot ontogeny in thrips-damaged cotton seedlings, *Annals of Botany* 94: 179-186.

Leimu, R. & Koricheva, J. (2006). A meta-analysis of tradeoffs between plant tolerance and resistance to herbivores: combining the evidence from ecological and agricultural studies, *Oikos*, 112: 1-9.

Leinonen, I. & Jones, H.G. (2004). Combining thermal and visible imagery for estimating canopy temperature and identifying plant stress, *Journal of Experimental Botany* 55: 1423-1431.

Lemeille, S., Willig, A., Depege-Fargeix, N., Delessert, C., Bassi, R. & Rochaix, J.D. (2009). Analysis of the chloroplast kinase STT7 during state transitions, *PLoS Biology* 7: e45.

Lemeille, S., Turkina, M., Vener, A.V. & Rochaix, J.D. (2010). Stt7- dependent phosphorylation during state transitions in the green alga *Chlamydomonas reinhardtii*, *Molecular & Cellular Proteomics* 9.6: 1281-1295.

Lim, P.O., Kim, H.J., Nam & H.G. (2007). Leaf senescence, *Annual Review of Plant Biology* 58: 115-136.

Lin, T.B., Schwartz, A. & Saranga, Y. (1999). Photosynthesis and productivity of cotton under silverleaf whitefly stress, *Crop Science* 39: 174-184.

Lindahl, M., Spetea, C., Hundal, T., Oppenheim, A.B., Adam, Z. & Andersson, B. (2000). The thylakoid FtsH protease plays a role in the light-induced turnover of the photosystem II D1 protein, *The Plant Cell* 12: 419-431.

Logemann, E., Parniske, M. & Hahlbrock, K. (1995). Modes of expression and common structural features of the complete phenylalanine ammonia-lyase gene family in parsley, *Proceedings of the National Academy of Sciences of the USA* 92: 5905-5909.

Lorenzo, O., Chico, J.M., Sanchez-Serrano, J.J. & Solano, R. (2004). *JASMONATE INSENSITIVE1* encodes a MYC transcription factor essential to discriminate between different jasmonate-regulated defense responses in Arabidopsis, *The Plant Cell* 16: 1938-1950.

Lunde, C., Jensen, P.E., Haldrup, A., Knoetzel, J. & Scheller, H.V. (2000). The PSI-H subunit of photosystem I is essential for state transitions in plant photosynthesis, *Nature* 408: 613-615.

Mabry, C.M. & Wayne, P.W. (1997). Defoliation of the annual herb *Abutilon theophrasti* - mechanisms underlying reproductive compensation, *Oecologia* 111: 225-232.

Macedo, T.B., Bastos, C.S., Higley, L.G., Ostlie, K.R. & Madhavan, S. (2003a). Photosynthetic responses of soybean to soybean aphid (Homoptera: Aphididae) injury, *Journal of Economic Entomology* 96: 188-193.

Macedo, T.B., Higley, L.G., Ni, X. & Quisenberry, S. (2003b). Light activation of Russian wheat aphid-elicited physiological responses in susceptible wheat, *Journal of Economic Entomology* 96: 194-201.

Macedo, T.B., Peterson, R.K.D., Weaver D.K. & Morrill, W.L. (2005) Wheat stem sawfly, *Cephus cinctus* Norton, impact on wheat primary metabolism: an ecophysiological approach, *Environmental Entomology* 34: 719-726.

Macedo, T.B., Peterson, R.K.D., Dausz, C.L. & Weaver, D.K. (2007). Photosynthetic responses of wheat, *Triticum aestivum* L., to defoliation patterns on individual leaves, *Environmental Entomology* 36: 602-608.

Macedo, T.B., Peterson, R.K.D., Weaver, D.K. & Ni, X.Z. (2009). Impact of *Diuraphis noxia* and *Rhopalosiphum padi* (Hemiptera: Aphididae) on primary physiology of four near-isogenic wheat lines, *Journal of Economic Entomology* 102: 412–421.

Macías, M.L., Rojas, I.S., Mata, R. & Lotina-Hennsen, B. (1999). Effect of selected coumarins on spinach chloroplast photosynthesis, *Journal of Agricultural and Food Chemistry* 47: 2137-2140.

Maffei, M., Camusso, W. & Sacco, S. (2001). Effect of *Mentha* x *piperita* essential oil and monoterpenes on cucumber root membrane potential, *Phytochemistry* 58: 703-707.

Maslenkova, L. & Zeinalov, Y. (1999). Thermo-luminescence and oxygen evolution in JA-treated barley (Hordeum vulgare L.), *Bulgarian Journal of Plant Physiology* 25: 58-64.

Mattoo, A.K., Hoffman-Falk, H., Marder, J.B. & Edelman, M. (1984). Regulation of protein metabolism: coupling of photosynthetic electron transport to *in vivo* degradation of the rapidly metabolized 32-kilodalton protein of the chloroplast membranes, *Proceedings of the National Academy of Sciences of the USA* 81: 1380-1384.

Mattoo, A.K., Marder, J.B. & Edelman, M. (1989). Dynamics of the photosystem II reaction center, *Cell* 56: 241-246.

Matsubara, S. & Chow, W-S. (2004). Populations of photo-inactivated photosystem II reaction centers characterized by chlorophyll a fluorescence lifetime *in vivo*, *Proceedings of the National Academy of Sciences of the USA* 101: 18234-18239.

Mattson, W.J. & Addy, N.D. (1975). Phytophagous insects as regulators of forest primary production, *Science* 190: 515-522.

Mediavilla, S., Escudero, A. & Heilmeier, H. (2001). Internal leaf anatomy and photosynthetic resource-use efficiency: interspecific and intraspecific comparisons, *Tree Physiology* 21: 251-259.

Melis, A. (1999). Photosystem II damage and repair cycle in chloroplasts: what modulates the rate of photodamage *in vivo*?, *Trends in Plant Science* 4: 130-135.

Mercader, R.J. & Isaacs, R. (2003). Phenology-dependent effects of foliar injury and herbivory on the growth and photosynthetic capacity of nonbearing *Vitis labrusca* (Linnaeus) var. Niagara, *American Journal of Enology and Viticulture* 54: 252-260.

Meskauskiene, R., Nater, M., Goslings, D., Kessler, F., den Camp, R. & Apel, K. (2001). FLU: a negative regulator of chlorophyll biosynthesis in *Arabidopsis thaliana*, *Proceedings of the National Academy of Sciences of the USA* 98: 12826-12831.

Meyer, G.A. & Whitlow, T.H. (1992). Effects of leaf and sap feeding insects on photosynthetic rates of goldenrod, *Oecologia* 92: 480-489.

Mittler, R., Vanderauwera, S., Gollery, M. & Van Breusegem, F. (2004). Reactive oxygen gene network of plants, *Trends in Plant Science* 9: 490-498.

Mitra, S. & Baldwin, I.T. (2008). Independently silencing two photosynthetic proteins in *Nicotiana attenuata* has different effects on herbivore resistance, *Plant Physiology* 148: 1128-1138.

Mochizuki, N., Brusslan, J.A., Larkin, R., Nagatani, A. & Chory, J. (2001). Arabidopsis *genomes uncoupled 5* (*GUN5*) mutant reveals the involvement of Mg-chelatase H subunit in plastid to- nucleus signal transduction, *Proceedings of the National Academy of Sciences of the USA* 98: 2053-2058.

Mochizuki, N., Tanaka, R., Tanaka, A., Masuda, T. & Nagatani, A. (2008). The steady-state level of Mg-protoporphyrin IX is not a determinant of plastid-to-nucleus signaling in Arabidopsis, *Proceedings of the National Academy of Sciences of the USA* 105: 15184-15189.

Montesano, M., Scheller, H.V., Wettstein, R. & Palva, E.T. (2004). Down-regulation of photosystem I by *Erwinia carotovora* derived elicitors correlates with H_2O_2 accumulation in chloroplasts of potato, *Molecular Plant Pathology* 5: 115-123.

Moreno, J.E., Tao, Y., Chory, J. & Ballare, C.L. (2009). Ecological modulation of plant defense via phytochrome control of jasmonate sensitivity, *Proceedings of the National Academy of Sciences of the USA* 106: 4935-4940.

Moulin, M., McCormac, A.C., Terry, M.J. & Smith, A.G. (2008). Tetrapyrrole profiling in Arabidopsis seedlings reveals that retrograde plastid nuclear signaling is not due to Mg protoporphyrin IX accumulation, *Proceedings of the National Academy of Sciences of the USA* 105: 15178-15183.

Mubarakshina, M.M., Ivanov, B.I., Naydov, I.A., Hillier, W., Badger, M.R. & Krieger-Liszkay, A. (2010). Production and diffusion of chloroplastic H_2O_2 and its implication to signaling, *Journal of Experimental Botany* 61: 3577-3587.

Mühlenbock, P., Szechyńska-Hebda, M., Płaszczyca, M., Baudo, M., Mateo A., Mullineaux, P.M., Parker, J.E., Karpińska, B., & Karpińskie, S. (2008). Chloroplast signaling and *LESION SIMULATING DISEASE1* regulate crosstalk between light acclimation and immunity in *Arabidopsis*, *The Plant Cell* 20: 2339-2356.

Mullineaux, P.M. & Rausch, T. (2005). Glutathione, photosynthesis and the redox regulation of stress-responsive gene expression, *Photosynthesis Research* 86: 459-474.

Muramatsu, M., Sonoike, K. & Hihara, Y. (2009). Mechanism of down-regulation of photosystem I content under high-light conditions in the cyanobacterium *Synechocystis* sp. PCC 6803, *Microbiology* 155: 989-996.

Murgia, I., Tarantino, D., Vannini, C., Bracale, M., Carravieri, S. & Soave, C. (2004). *Arabidopsis thaliana* plants overexpressing thylakoidal ascorbate peroxidase show increased resistance to Paraquat-induced photooxidative stress and to nitric oxide-induced death, *The Plant Journal* 38: 940-953.

Nabity, P.D., Zavala, J.A. & DeLucia, E.H. (2009). Indirect suppression of photosynthesis on individual leaves by arthropod herbivory, *Annals of Botany* 103: 655-663.

Nagaraj, N., Reese, J.C., Kirkham, M.B., Kofoid, K., Campbell, L.R. & Loughin, T.M. (2002). Relationship between chlorophyll loss and photosynthetic rate in greenbug (Homoptera: Aphididae) damaged *Sorghum*, *Journal of the Kansas Entomological Society* 75: 101-109.

Nanba, O. & Satoh, K. (1987). Isolation of a photosystem II reaction center consisting of D-1 and D-2 polypeptides and cytochrome b-559, *Proceedings of the National Academy of Sciences of the USA* 84: 109-112.

Nardini, A. & Salleo, S. (2005). Water stress-induced modifications of leaf hydraulic architecture in sunflower: co-ordination with gas exchange, *Journal of Experimental Botany* 56: 3093-3101.

Nelson, N. & Yocum, C.F. (2006). Structure and function of Photosystems I and II, *Annual Review of Plant Biology* 57: 521-565.

Newingham, B.A., Callaway, R.M. & Bassirirad, H. (2007). Allocating nitrogen away from a herbivore: a novel compensatory response to root herbivory, *Oecologia* 153: 913-920.

Neill, S.J., Desikan, R., Clarke, A., Hurst, R.D. & Hancock, J.T. (2002). Hydrogen peroxide and nitric oxide as signalling molecules in plants, *Journal of Experimental Botany* 53: 1237-1247.

Niyogi, K.K., Li, X-P., Rosenberg, V. & Jung, H-S. (2005). Is PsbS the site of non-photochemical quenching in photosynthesis?, *Journal of Experimental Botany* 56: 375-382.

Nykänen, H. & Koricheva, J. (2004). Damage-induced changes in woody plants and their effects on insect herbivore performance: a meta-analysis, *Oikos* 104: 247-268.

Ochsenbein C., Przybyla D., Danon A., Landgraf F., Göbel, C., Imboden, A., Feussner, I. & Apel, K. (2006). The role of *EDS1* (*enhanced disease susceptibility*) during singlet oxygen-mediated stress responses of Arabidopsis, *The Plant Journal* 47: 445-456.

Oelmuller, R. & Mohr, H. (1986). Photooxidative destruction of chloroplasts and its consequences for expression of nuclear genes, *Planta* 167: 106-113.

Oerke, E-C. & Dehne, H-W. (1997). Global crop production and the efficacy of crop protection-current situation and future trends, *European Journal of Plant Pathology* 103: 203-215.

Oleksyn, J., Karolewki, P., Giertych, M. J., Zytkowiak, R., Reich P.B. & Tjoelker G. (1998). Primary and secondary host plants differ in leaf-level photosynthetic response to herbivory: evidence from *Alnus and Betula* grazed by the alder beetle, *Agelastica alni*, *New Phytologist* 140: 239-249

Omasa, K. & Takayama, K. (2003). Simultaneous measurement of stomatal conductance, non-photochemical quenching, and photochemical yield of photosystem II in intact leaves by thermal and chlorophyll fluorescence imaging, *Plant & Cell Physiology* 44: 1290-1300.

Öquist, G. & Huner, N.P.A. (2003). Photosynthesis of overwintering evergreen plants, *Annual Review of Plant Biology* 54: 329-355.

Ostlie, K.R. & Pedigo, L.P. (1984). Water loss from soybeans after simulated and actual insect defoliation, *Environmental Entomology* 13: 1675-1680.

Oswald, O., Martin, T., Dominy, P.J. & Graham, I.A. (2001). Plastid redox state and sugars: Interactive regulators of nuclear-encoded photosynthetic gene expression, *Proceedings of the National Academy of Sciences of the USA* 98: 2047-2052.

Overmyer, K., Brosché, M. & Kangasjärvi, J. (2003). Reactive oxygen species and hormonal control of cell death, *Trends in Plant Science* 8: 335-342.

Ozaki, K., Saito, H. & Yamamuro, K. (2004). Compensatory photosynthesis as a response to partial debudding in ezo spruce, *Picea jezoensis*, seedlings, *Ecological Research* 19: 225-231.

Ozaki, H., Ikeuchi, M., Ogawa, T., Fukuzawa, H. & Sonoike, K. (2007). Large scale analysis of chlorophyll fluorescence kinetics in *Synechocystis* sp. PCC 6803: identification of the factors involved in the modulation of photosystem stoichiometry, *Plant Cell & Physiology* 48: 451-458.

Parker, J.D., Burkepile, D.E. & Hay, M.E. (2006). Opposing effects of native and exotic herbivores on plant invasions, *Science* 311: 1459-1461.

Paul, M.J. & Foyer, C.H. (2001). Sink regulation of photosynthesis, *Journal of Experimental Botany* 52: 1383-1400.

Pesaresi, P., Hertle, A., Pribil, M., Kleine, T., Wagner, R., Strissel, H., Ihnatowicz, A., Bonardi, V., Scharfenberg, M., Schneider, A., Pfannschmidt, T. & Leister, D. (2009). Arabidopsis STN7 kinase provides a link between short- and long-term photosynthetic acclimation, *The Plant Cell* 21: 2402-2423.

Pesaresi, P., Hertle, A., Pribil, M., Schneider, A., Kleine, T. & Leister, D. (2010). Optimizing photosynthesis under fluctuating light: the role of the Arabidopsis STN7 kinase, *Plant Signaling & Behavior* 5: 21-25.

Peterhansel, C., Horst, I., Niessen, M., Blume, C., Kebeish, R., Kürkcüoglu, S. & Kreuzaler, F. (2010). Photorespiration, *The Arabidopsis Book:* e130.

Peterson, R.K.D., Higley, L.G., Haile, F.J. & Barrigossi, J.A.F. (1998). Mexican bean beetle (Coleoptera: Coccinelidae) injury affects photosynthesis of *Glycine max* and *Phaseolus vulgaris*, *Environmental Entomology* 27: 373-381.

Pfannschmidt, T. (2003). Chloroplast redox signals: how photosynthesis controls its own genes, *Trends in Plant Science* 8: 33-41.

Pfannschmidt, T., Bräutigam, K., Wagner, R., Dietzel, L., Schröter, Y., Steiner, S. & A. Nykytenko, A. (2009). Potential regulation of gene expression in photosynthetic cells by redox and energy state: approaches towards better understanding, *Annals of Botany* 103: 599-607.

Pincebourde, S., Frak, E., Sinoquet, H., Regnard, J.L &, Casas, J. (2006). Herbivory mitigation through increased water-use efficiency in a leaf-mining moth-apple tree relationship, *Plant Cell & Environment* 29: 2238-2247.

Portis, A.R. (1995). The regulation of Rubisco by Rubisco activase, *Journal of Experimental Botany* 46: 1285-1291.

Prasil, O., Adir, N. & Ohad, I. (1992). Dynamics of Photosystem II: mechanisms of photoinhibition and recovery process, in Barber, J. (ed.). *The photosystems: structure, function and molecular biology*, vol. 11. Amsterdam, Elsevier Science Publishers, pp. 295-348.

Pribil, M., Pesaresi, P., Hertle, A., Barbato, R. & Leister, D. (2010). Role of plastid protein phosphatase TAP38 in LHCII dephosphorylation and thylakoid electron flow, *PLoS Biology* 8:e 1000288.

Puthiyaveetil, S., Kavanagh, T.A., Cain, P., Sullivan, J.A., Newell, C.A., Gray, J.C., Robinson, C., van der Giezen, M., Rogers, M.B. & Allen, J.F. (2008). The ancestral symbiont sensor kinase CSK links photosynthesis with gene expression in chloroplasts, *Proceedings of the National Academy of Sciences of the USA* 105: 10061-10066.

Puthiyaveetil, S., Ibrahim, I.M. & Allen, J.F. (2011). Oxidation-reduction signalling components and regulatory pathways of state transitions and photosystem stoichiometry adjustment in chloroplasts, *Plant Cell & Environment* doi: 10.1111/j.1365-3040.2011.02349.x

Qubbaj, T., Reineke, A. & Zebitz C.P.W. (2005). Molecular interactions between rosy apple aphids, *Dysaphis plantaginea*, and resistant and susceptible cultivars of its primary host *Malus domestica*, *Entomologia Experimentalis et Applicata* 115: 145-152.

Ralph, S.G., Yueh, H., Friedmann, M., Aeschliman, D., Zeznik, J.A., Nelson, C.C., Butterfield, Y.S.N., Kirkpatrick, R., Liu, J. & Jones, S.J.M., Marra, M.A., Douglas, C.J., Ritland, K. & Bohlmann, J.(2006). Conifer defence against insects: microarray gene expression profiling of Sitka spruce (*Picea sitchensis*) induced by mechanical wounding or feeding by spruce budworms (*Choristoneura occidentalis*) or white pine weevils (*Pissodes strobi*) reveals large-scale changes of the host transcriptome, *Plant, Cell & Environment* 29: 1545-1570.

Rausch, T., Gromes, R., Liedschulte, V., Muller, I., Bogs, J., Galovic, V. & Wachter, A. (2007). Novel insight into the regulation of GSH biosynthesis in higher plants, *Plant Biology* 9: 565-572.

Rausher, M.D., Iwao, K., Simms, E.L., Ohsaki, N. & Hall, D. (1993). Induced resistance in *Ipomea purpurea*, *Ecology* 74: 20-29.

Reddall, A., Sadras, V.O., Wilson, L.J. & Gregg, P.C. (2004). Physiological responses of cotton to two-spotted spider mite damage, *Crop Science* 44: 835-846.

Reddall, A.A., Wilson, L.J., Gregg, P.C. & Sadras, V.O. (2007). Photosynthetic response of cotton to spider mite damage: interaction with light and compensatory mechanisms, *Crop Science* 47: 2047-2057.

Reich, P.B., Ellsworth, D.S. & Walters, M.B. (1999). Generality of leaf trait relationships: a test across six biomes, *Ecology* 80: 1955-1969.

Retuerto, R., Fernandez-Lema, B., Rodriguez-Roiloa & Obeso, J.R. (2004). Increased photosynthetic performance in holly trees infested by scale insects, *Functional Ecology*, 18: 664-669.

Reymond, P., Bodenhausen, N., Van Poecke, R.M.P., Krishnamurthy, V., Dicke, M. & Farmer, E.E. (2004). A conserved transcript pattern in response to a specialist and a generalist herbivore, *The Plant Cell* 16: 3132-3147.

Rintamaki, E., Martinsuo, P., Pursiheimo, S. & Aro, E.M. (2000). Cooperative regulation of light-harvesting complex II phosphorylation via the plastoquinol and ferredoxin-thioredoxin system in chloroplasts, *Proceedings of the National Academy of Sciences of the USA* 97: 11644-11649.

Roloff, A., Parthier, B. & Wasternack, C. (1994). Relationship between degradation of ribulose 1,5-bisphosphate carboxylase/oxygenase and synthesis of an abundant protein of 23 kDa of barley leaves induced by jasmonates, *Journal of Plant Physiology* 143: 39-46.

Saibo, N.J.M., Lourenço, T. & Oliveira, M.M. (2009). Transcription factors and regulation of photosynthetic and related metabolism under environmental stresses, *Annals of Botany* 103: 609-623.

Friend or Foe? Exploring the Factors that Determine the Difference Between Positive and Negative
Effects on Photosynthesis in Response to Insect Herbivory
215

Shannag, H.K., Thorvilson, H. & El-Shatnawi, M.K. (1998).Changes in photosynthetic and transpiration rates of cotton leaves infested with the cotton aphid, *Aphis gossypii*: unrestricted infestation, *Annals of Applied Biology* 132: 13-18.

Shapiguzov, A., Ingelsson, B., Samol, I., Andres, C., Kessler, F., Rochaix, J.D., Vener, A.V. & Goldschmidt-Clermont, M. (2010). The PPH1 phophatase is specifically involved in LHCII dephosphorylation and state transitions in Arabidopsis, *Proceedings of the National Academy of Sciences of the USA* 107: 4782-4787.

Schröder, R., Forstreuter, M. & Hilker, M. (2005). A plant notices insect egg deposition and changes its rate of photosynthesis, *Plant Physiology* 138: 470-477.

Schwachtje, J., Minchin, P.E.H., Jahnke, S., van Dongen, J.T., Schittko, U. & Baldwin, I.T. (2006). SNF1-related kinases allow plants to tolerate herbivory by allocating carbon to roots, *Proceedings of the National Academy of Sciences of the USA* 103: 12935-12940.

Silva, P., Thompson, E., Bailey, S., Kruse, O., Mullineaux, C.W., Robinson, C., Mann, N.H. & Nixon, P.J. (2003). FtsH is involved in the early stages of repair of Photosystem II in *Synechocystis* sp PCC 6803, *The Plant Cell* 15: 2152-2164.

Singh, M., Yamamoto, Y., Satoh, K., Aro, E-V., Kanervo, E. (2005). Post-illumination-related loss of photochemical efficiency of Photosystem II and degradation of the D1 protein are temperature-dependent, *Journal of Plant Physiology* 162: 1246-1253.

Silla, F. & Escudero, A. (2003). Uptake, demand and internal cycling of nitrogen in saplings of Mediterranean *Quercus* species, *Oecologia* 136: 28-36.

Snyders, S. & Kohorn, B.D. (1999). TAKs, thylakoid membrane protein kinases associated with energy transduction, *Journal of Biological Chemistry* 274: 9137-9140.

Snyders, S. & Kohorn, B.D. (2001). Disruption of thylakoid-associated kinase 1 leads to alteration of light harvesting in Arabidopsis, *Journal of Biological Chemistry* 276: 32169-32176.

Solanke, A.U. & Sharma, A.K. (2008). Signal transduction during cold stress in plants. *Physiological & Molecular Biology of Plants* 14: 69-79.

Spoel, S.H. & Dong, X. (2008). Making sense of hormone crosstalk during plant immune responses, *Cell Host & Microbe* 3: 348-351.

Staswick, P.E., Tiryaki, I. & Rowe, M.L. (2002). Jasmonate response locus *JAR1* and several related Arabidopsis genes encode enzymes of the firefly luciferase superfamily that show activity on jasmonic, salicylic, and indole-3-acetic acids in an assay for adenylation, *The Plant Cell* 14: 1405-1415.

Stowe, K.A., Marquis, R.J., Hochwender, C.J. & Simms, E.L. (2000). The evolutionary ecology of tolerance to consumer damage, *Annual Review of Ecology and Systematics* 31: 565-595.

Strand, A., Asami, T., Jose Alonso, J., Ecker, J.R. & Chory, J. (2003). Chloroplast to nucleus communication triggered by accumulation of Mg-protoporphyrinIX, *Nature* 421: 79-83.

Strauss, S.Y. & Agrawal, A.A. (1999). The ecology and evolution of plant tolerance to herbivory, *Trends in Ecology & Evolution* 14: 179-185.

Streatfield, S.J., Weber, A., Kinsman, E.A., Häusler, R.E., Li, J., Post-Beittenmiller, D., Kaiser, W.M., Pyke, K.A., Flügge, U-I. & Chory, J. (1999). The phosphoenolpyruvate/phosphate translocator is required for phenolic metabolism, palisade cell development, and plastid-dependent nuclear gene expression, *The Plant Cell* 11: 1609-1622.

Sun, S., Yu, J.P., Chen, F., Zhao, T.J., Fang, X.H., Li, Y.Q. & Sui, S.F. (2008). TINY, a dehydration-responsive element (DRE)-binding protein-like transcription factor connecting the DRE- and ethylene-responsive element-mediated signaling pathways in Arabidopsis, *Journal of Biological Chemistry* 283: 6261-6271.

Suwa, T. & Maherali, H. (2008). Influence of nutrient availability on the mechanisms of tolerance to herbivory in an annual grass, *Avena barbata* (Poaceae), *American Journal of Botany* 95: 434-440.

Szyszka, B., Ivanov, A.G. & Hüner, N.P.A. (2007). Psychrophily is associated with differential energy partitioning, photosystem stoichiometry and polypeptide phosphorylation in *Chlamydomonas raudensis*, *Biochimica et Biophysica Acta* 1767: 789-800.

Tang, J.Y., Zielinski, R.E., Zangerl, A.R., Crofts, A.R., Berenbaum, M.R. & DeLucia, E.H. (2006). The differential effects of herbivory by first and fourth instars of *Trichoplusia ni* (Lepidoptera: Noctuidae) on photosynthesis in *Arabidopsis thaliana*, *Journal of Experimental Botany* 57: 527-536.

Tang, J., Zielinskia, R., Aldea, M. & DeLucia, E. (2009). Spatial association of photosynthesis and chemical defense in *Arabidopsis thaliana* following herbivory by *Trichoplusia ni*, *Physiologia Plantarum* 137: 115-124.

Taniguchi, M., Taniguchi, Y., Kawasaki, M., Takeda, S., Kato, T., Sato, S., Tabata, S., Miyake, H. & Sugiyama, T. (2002). Identifying and characterizing plastidic 2-oxoglutarate/malate and dicarboxylate transporters in *Arabidopsis thaliana*, *Plant & Cell Physiology* 43: 706-717.

Thivierge, K., Prado, A., Driscoll, B.T., Bonneil, E., Thibault, P. & Bede, J.C. (2010). Caterpillar- and salivary-specific modification of plant proteins, *Journal of Proteome Research*, 9: 5887-5895.

Thomson, V.P., Cunningham, S.A., Ball, M.C. & Nicorta, A.B. (2003). Compensation for herbivory by *Cucumis sativus* through increased photosynthetic capacity and efficiency, *Oecologia*, 134: 167-175.

Tiffin, P. (2000). Mechanisms of tolerance to herbivore damage: what do we know?, *Evolutionary Ecology* 14: 523-526.

Tikkanen, M. Piippo, M., Suorsa, M., Sirpio, S., Mulo, P., Vainonen, J. Vener, A.V., Allahverdiyeva Y., & Aro, E.M. (2006). State transitions revisited-a buffering system for dynamic low light acclimation of Arabidopsis, *Plant Molecular Biology* 62: 779-793.

Tikkanen, M., Nurmi, M., Kangasjarvi, S. & Aro, E.M. (2008). Core protein phosphorylation facilitates the repair of photodamaged photosystem II at high light, *Biochimica et Biophysica Acta* 1777: 1432-1437.

Tognetti, V.B., Mühlenbock, P. & Van Breusegem, F. (2011). Stress homeostasis-the redox and auxin perspective, *Plant Cell & Environment* doi: 10.1111/j.1365-3040.2011.02324.x

Triantaphylidés, C. & Havaux, M. (2009). Singlet oxygen in plants: production, detoxification and signaling, *Trends in Plant Science* 14: 219-228.

Trumble, J.T., Kolodny-Hirsch, D.M. & Ting, I.P. (1993). Plant compensation for arthropod herbivory, *Annual Review of Entomology* 38: 93-119.

Turnbull, T.L., Adams, M.A. & Warren, C.R. (2007). Increased photosynthesis following partial defoliation of field-grown *Eucalyptus globulus* is not caused by increased leaf nitrogen, *Tree Physiology* 27: 1481-1492.

Vainonen, J.P., Hansson, M. & Vener, A.V. (2005). STN8 protein kinase in *Arabidopsis thaliana* is specific in phosphorylation of photosystem II core proteins, *Journal of Biological Chemistry* 280: 33679-33686.

Veiga, T.A.M., González-Vázquez, R., Oiano Neto, J., Silva, M.F.G.F., King-Díaz, B. & Lotina-Hennsen, B. (2007). Siderin from *Toona ciliata* (Meliaceae) as photosystem II inhibitor on spinach thylakoids, *Archives of Biochemistry and Biophysics* 465: 38-43.

Velikova, V., Salerno, G., Frati, F., Peri, E., Conti, E., Colazza, S. & Loreto, F. (2010) Influence of feeding and oviposition by phytophagous pentatomids on photosynthesis of herbaceous plants, *Journal of Chemical Ecology* 36: 629–641.

Voelckel, C. & Baldwin, I.T. (2004a) Generalist and specialist lepidopteran larvae elicit different transcriptional responses in *Nicotiana attenuata*, which correlate with larval FAC profiles, *Ecology Letters* 7: 770-775.

Voelckel, C. & Baldwin, I.T. (2004b). Herbivore-induced plant vaccination. Part II. Array-studies reveal the transience of herbivore-specific transcriptional imprints and a distinct imprint from stress combinations, *The Plant Journal* 38: 650–663.

Voelckel, C., Weisser, W.W. & Baldwin, I.T. (2004). An analysis of plant-aphid interactions by different microarray hybridization strategies, *Molecular Ecology* 13: 3187-3195.

Vranova, E., Inze, D. & Van Breusegem, F. (2002). Signal transduction during oxidative stress, *Journal of Experimental Botany* 53: 1227-1236.

Wachter, A., Wolf, S., Steininger, H., Bogs, J. & Rausch, T. (2005). Differential targeting of GSH1 and GSH2 is achieved by multiple transcription initiation: implications for the compartmentation of glutathione biosynthesis in the Brassicaceae, *The Plant Journal* 41: 15-30.

Wagner, D., Przybyla, D., den Camp, R., Kim, C., Landgraf, F., Lee, K.P., Würsch, M. Laloi, C., Nater, M., Hideg, E. & Apel, K. (2004). The genetic basis of singlet oxygen-induced stress responses of *Arabidopsis thaliana*, *Science* 306: 1183-1185.

Walling, L.L. (2008). Avoiding effective defenses: strategies employed by phloem-feeding insects, *Plant Physiology* 146: 859-866.

Walters, D.R., McRoberts, N. & Fitt, B.D. (2008). Are green-islands red herrings? Significance of green-islands in plant interactions with pathogens and pests, *Biological Reviews of the Cambridge Philosophical Society* 83: 79-102.

Wasternack, C. & Parthier, B. (1997). Jasmonate-signalled plant gene expression, *Trends in Plant Science* 2: 302-307.

Watanabe, T. & Kitagawa, H. (2000). Photosynthesis and translocation of assimilates in rice plants following phloem feeding by the planthopper *Nilaparvata lugens* (Homoptera: Delphacidae), *Journal of Economic Entomology* 93: 1192-1198.

Welter, S.C. (1989). Arthropod impact on plant gas exchange, in: Bernays, E.A, (ed.), *Insect-plant interactions*. Boca Raton, FL: CRC Press, pp. 135-151.

West, J.D., Peak, D., Peterson, J.Q. & Mott, K.A. (2005). Dynamics of stomatal patches for a single surface of *Xanthium strumarium* L. leaves observed with fluorescence and thermal images, *Plant, Cell & Environment* 28: 633-641.

Wollman, F.A. (2001). State transitions reveal the dynamics and flexibility of the photosynthetic apparatus, *EMBO Journal* 20: 3623-3630.

Yamaguchi-Shinozaki, K. & Shinozaki, K. (2006). Responses and tolerance to dehydration and cold stresses, *Annual Review of Plant Biology* 57: 781-803.

Yamazaki, J-Y., Suzuki, T., Maruta, E. & Kamimura, Y. (2005). The stoichiometry and antenna size of the two photosystems in marine green algae, *Bryopsis maxima* and *Ulva pertusa*, in relation to the light environment of their natural habitat, *Journal of Experimental Botany* 56: 1517-1523.

Yan, Y., Stolz, S., Chetelat, A., Reymond, P., Pagni, M., Dubugnon, L. & Farmer, E.E. (2007). A downstream mediator in the growth repression limb of the jasmonate pathway, *The Plant Cell* 19: 2470-2483.

Yokthongwattana, K. & Melis, A. (2006). Photoinhibition and recovery in oxygenic photosynthesis: mechanism of a photosystem II damage and repair cycle, in Demmig-Adams, B., Adams, W.W. III & Mattoo, A.K. (eds) *Photoprotection, photoinhibition, gene regulation, and environment*. Springer, The Netherlands, pp 175-191.

Yuan, H.Y., Chen, X.P., Zhu, L.L. & He, G.C. (2005). Identification of genes responsive to brown planthopper *Nilaparvata lugens* Stal (Homoptera, Delphacidae) feeding in rice, *Planta* 221: 105-112.

Zangerl, A.R. & Bazzaz, F.A. (1993). Theory and pattern in plant defense allocation, in Fritz, S.R. & Simms, E.L. (eds) *Plant Resistance to Herbivores and Pathogens*, University of Chicago Press, Chicago pp. 363-391.

Zangerl, A.R., Hamilton, J.G., Miller, T.J., Crofts, A.R., Oxborough, K., Berenbaum, M.R. & DeLucia, E.H. (2002). Impact of folivory on photosynthesis is greater than the sum of its holes, *Proceedings of the National Academy of Sciences of the USA* 99: 1088-1091.

Zavala, J.A. & Baldwin, I.T. (2006). Jasmonic acid signaling and herbivore resistance traits constrain regrowth after herbivore attack in *Nicotiana attenuata*, *Plant Cell & Environment* 29: 1751-1760.

Zhang, Z., Huang, L., Shulmeister, V.M., Chi, Y.I., Kim, K.K., Hung, L.W., Crofts, A.R., Berry, E.A. & Kim, S.H. (1998). Electron transfer by domain movement in cytochrome bc1, *Nature* 392: 677-684.

Zhang, S. & Scheller, H.V. (2004). Light-harvesting complex II binds to several small subunits of photosystem I, *Journal of Biological Chemistry* 279: 3180-3187.

Zhu-Salzman, K., Salzman, R.A., Ahn, J.E. & Koiwa H. (2004) Transcriptional regulation of sorghum defense determinants against a phloem-feeding aphid, *Plant Physiology* 134: 420-431.

Zou, J., Rodriguez-Zas, S., Aldea, M., Li, M., Zhu, J., Gonzalez, D.O., Vodkin, L.O., DeLucia, E. & Clough, S.J. (2005). Expression profiling soybean response to *Pseudomonas syringae* reveals new defense-related genes and rapid HR-specific down regulation of photosynthesis, *Molecular Plant-Microbe Interactions* 18: 1161-1174.

Zvereva, E.L., Lanta, V. & Kozlov, M.V. (2010). Effects of sap-feeding insect herbivores on growth and reproduction of woody plants: a meta-analysis of experimental studies, *Oecologia* 163: 949-960.

10

Instinctive Plant Tolerance Towards Abiotic Stresses in Arid Regions

Mohamed Mohamed Ibrahim
Botany and Microbiology Department, Faculty of Science, Alexandria University
Egypt

1. Introduction

Arid environments are extremely diverse in terms of their land forms, soils, fauna, flora, water balances, and human activities. Because of this diversity, no practical definition of arid environments can be derived. However, the one binding element to all arid regions is aridity. Aridity is usually expressed as a function of rainfall and temperature. A useful "representation" of aridity is the following climatic aridity index: p/ETP, where P = precipitation; ETP = potential evapotranspiration, calculated by method of Penman, taking into account atmospheric humidity, solar radiation, and wind. Three arid zones can be delineated by this index: namely, hyper-arid, arid and semi-arid. Of the total land area of the world, the hyper-arid zone covers 4.2 percent, the arid zone 14.6 percent, and the semiarid zone 12.2 percent. Therefore, almost one-third of the total area of the world is arid land.

Arid climate, is a climate that does not meet the criteria to be classified as a polar climate, and in which precipitation is too low to sustain any vegetation at all, or at most a very scanty scrub. An area that features this climate usually (but not always) experiences less than 250 mm (10 inches) per year of precipitation and in some years may experience no precipitation at all. In some instances an area may experience more than 250 mm of precipitation annually, but is still considered a desert climate because the region loses more water via evapotranspiration than falls as precipitation. Although different classification schemes and maps differ in their details, there is a general agreement about the fact that large areas of the Earth are arid. These include the hot deserts located broadly in sub-tropical regions, where the accumulation of water is largely prevented by either low precipitations, or high evaporation, or both. Abiotic disorders are associated with non-living causal factors such as weather, soils, chemicals, mechanical injuries, cultural practices and, in some cases, a genetic predisposition within the plant itself. Abiotic disorders may be caused by a single extreme environmental event such as one night of severe cold following a warm spell or by a complex of interrelated factors or events. A biotic plant problems are sometimes termed "physiological disorders" that reflects the fact that the injury or symptom, such as reduced growth, is ultimately due to the cumulative effects of the causal factors on the physiological processes necessary for plant growth and development (Schutzki & Cregg, 2007).

Abiotic stresses, such as drought, salinity, extreme temperatures, chemical toxicity and oxidative stress are serious threats to agriculture and the natural status of the environment. Increased salinization of arable land is expected to have devastating global effects, resulting in 30% land loss within the next 25 years, and up to 50% by the year 2050. Therefore, breeding for drought and salinity stress tolerance in crop plants (for food supply) and in forest trees (a central component of the global ecosystem) should be given high research priority in plant biotechnology programs(Wang et al., 2003).

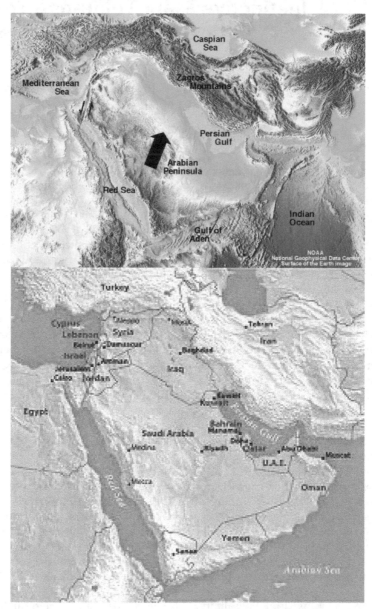

Fig. 1. The Arabian Peninsula (land-surface image formatted and labelled by Bruce Rails back); National Geophysical Data Center (NOAA).

Desert plants generally follow two main strategies i.e., they tolerate the stresses through phonologic and physiological adjustments referred to as tolerance or avoidance mechanisms contribute to the ability of a plant to survive stress but it also depends on the frequency and severity of the stress periods. Xeromorphic characteristics of desert plants have developed as the result of adaptation to drought, temperature divergence, salinity, poor nutrition, strong wind, sand movement and high light intensity (Fahn,1964,1990; Fahn and Cutler 1992; Huang et al.,1997). Plants in many habitats have various physiological mechanisms for responding to environmental changes, and the ability to tolerate environmental disturbances often contributes to their success in communities (Gutterman, 2001). In addition to genetic adaptation, the survival of a certain species is often determined by its ability to acclimate to environmental changes (Gutterman, 2002). Acclimation is known to be a widespread phenomenon in nature, and long-term responses can be observed in the course of a season.

Fig. 2. Xeromorphic characteristics of desert plants in Arid environment (Ibrahim & El-Gaely, 2011)

2. Convergent abiotic stress

More than one abiotic stress including drought, dust, salinity, heavy metals and UV can occur at one time. For example, high temperature and high photon irradiance often accompany low water supply, which can in turn be exacerbated by subsoil mineral toxicities that constrain root growth. Furthermore, one abiotic stress can decrease a plant's ability to resist a second stress. For example, low water supply can make a plant more susceptible to damage from high irradiance due to the plant's reduced ability to reoxidize NADPH and thus maintain an ability to dissipate energy delivered to the photosynthetic light-harvesting reaction centers (Mark & Bacic, 2005). If a single abiotic stress is to be identified as the most common in limiting the growth of crops worldwide, it most probably be low water supply (Boyer, 1982; Araus et al., 2008). The Arabian peninsula is one of the five major regions where dust originates (Idso, 1976). The Sahara and dry lands around the Arabian peninsula are the main source of airborne dust, with some contributions from Iran, Pakistan and India into the Arabian Sea, and China's storms deposit dust in the Pacific. Dust affects photosynthesis and transpiration physically when it accumulates on leaf surfaces. Covering and plugging stomata, shading and removing cuticular wax were reported as physical effects of dust (Luis et al., 2008).

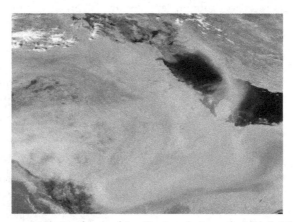

Fig. 3. Dust plumes swept across the Arabian Peninsula in early March 2009. The Moderate Resolution Imaging Spectroradiometer (MODIS) on NASA's Aqua satellite captured this image on March 11, 2009.

In arid environments, decreased water use efficiency because of dust deposition, could therefore contribute substantially to drought stress. The physical effects of dust accumulating on leaf surfaces, on leaf physiology, such as photosynthesis, transpiration, stomatal conductance and leaf temperature of cucumber and kidney bean plants were investigated by Hirano et al., 1995. It was found that dust decreased stomatal conductance in the light, and increased it in the dark by plugging the stomata, when the stomata were open during dusting. When dust of smaller particles was applied, the effect was greater (Hirano et al. 1995). However, the effect was negligible when the stomata were closed during dusting. The dust decreased the photosynthetic rate by shading the leaf surface. The dust of smaller particles had a greater shading effect. Moreover, it was found that the additional absorption of incident radiation by the dust increased the leaf temperature, and consequently changed the photosynthetic rate in accordance with its response curve to leaf temperature. The increase in leaf temperature also increased the transpiration rate (Hirano et al., 1995). Dust may allow the penetration of phytotoxic gaseous pollutants into plant leaves. Visible injury symptoms may occur and generally there is decreased productivity.

Correia et al., 2004 studied the deposition of dust on the foliar surface of the evergreen *Olea europaea* and a semi-deciduous (*Cistus laurifolius*). They found that the affect mainly on the reflectance, it increased with increasing deposition levels, causing a complementary decrease in light absorbance by the leaves of both species. As a consequence, the energy balance of the leaves and net photosynthesis may be altered, thus reducing the productivity of the affected vegetation. However, this effect seems to be more pronounced in *C. laurifolius* compared to *O. europaea*. This could mean that some species maybe more susceptible to dust pollution. In this sense, one could expect an alteration on the specific composition of the vegetation of the affected areas in response to dust pollution (Correia et al., 2004).

On photosynthesis, however, almost all the previous studies only guessed the physical effects in their discussions. Dust deposition has been found to affect photosynthesis, stomatal functioning and productivity (Luis et al., 2008, Ibrahim & El-Gaely, 2011). Chlorophyll fluorescence, an indication of the fate of excitation energy in the photosynthetic apparatus, has been used as an early, in vivo, indication of many types of plant stress (Maxwell & Johnson, 2000, Ibrahim and Bafeel, 2008). Photoinhibition is evident through the reduction in the quantum yield of photosystem 2 (PSII) and a decrease in variable chlorophyll (Chl) a fluorescence (Demmig-Adams and Adams, 1993; Ibrahim & El-Gaely, 2011).

Fig. 4. Influence of dust deposition and its physical effect on blocking stomata in some plants (A-F) in arid environment (Ibrahim & El-Gaely, 2011).

The decrease of efficiency of PSII photochemistry under stress may reflect not only the inhibition of PSII function, but also an increase in the dissipation of thermal energy (Demmig-Adams& Adams 1993), the latter is often considered as a photo-protective mechanism.

Fig. 5. Visual symptoms of some abiotic stress(including drought, dust accumulation and heavy metal pollution) on some desert plants in arid environment (Ibrahim & El-Gaely, 2011).

3. Spontaneous relationship between abiotic stress and oxidative stress

The reactive oxygen species (ROS) that arise from normal metabolic processes are kept under tight control by various antioxidant mechanisms. ROS are important signal molecules that regulate many physiological processes, including environmental stress responses. Under steady state conditions, the ROS molecules are scavenged by various antioxidative defense mechanisms (Foyer & Noctor, 2005). The equilibrium between the production and the scavenging of ROS may be perturbed by various biotic and abiotic stress factors such as salinity, UV radiation, drought, heavy metals, temperature extremes, nutrient deficiency, air pollution, herbicides and pathogen attacks. The ability to utilize oxygen has provided plants with the benefit of metabolizing fats, proteins and carbohydrates for energy; however, it does not come without cost.

Oxygen is a highly reactive atom that is capable of becoming part of the potentially damaging molecules commonly called "free radicals" which appear to be a major contributor to aging and damage the cell. Fortunately, free radical formation is controlled naturally by various beneficial compounds known as antioxidants that protect cellular membranes and organelles from the damaging effects of active species. Antioxidants are the first line of defense against free radical damage, and are critical for maintaining optimum health and well being of the plant cells. The need for antioxidant becomes even more critical with increased exposure to free radicals. Each organelle has potential targets for oxidative stress as well as mechanisms for eliminating the noxious oxyradicals. Therefore, plants are equipped with complex antioxidant systems composed of low molecular weight antioxidants non enzymatic compounds, like lipid soluble and membrane-associated tocopherol; ascorbate and glutathione (Foyer 1993), (Foyer & Noctor, 2005) as well as protective antioxidant enzymes such as superoxide dismutase (SOD, EC 1.15.1.1), catalase (CAT, EC 1.11.1.6), peroxidases (APX, EC 1.11.1.11) and glutathione reductase (GR, EC 1.6.4.2). Other components of this system, monodehydroascorbate radical reductase, and glutathione reductase serve to maintain the antioxidants in their reduced functional state (Schwanz et al.,1996) Whether this is the case or not, the antioxidant defenses appear to provide crucial protection against oxidative damage in cellular membranes and organelles in plants grown under unfavorable conditions (Smirnoff 1993 and Kocsy et al.,2000).

Ibrahim & Sameera, 2011 showed that the activity of peroxidise (POD) and CAT of *Lepidium sativum* treated with lead mainly displayed biphasic responses due to increased Pb2+ level. SOD activity under elevated lead stress was steadily stimulated with increasing metal ions level in medium up to 600 ppm. The results showed that, under high metal stress, POD and CAT activities were inhibited, while SOD activity was stimulated, indicating that those enzymes are located at different cellular sites, which had different resistance to heavy metals. Thus, the deterioration of cellular system functions by high metal stress might result in inhibition of enzyme activity (Fig. 5)

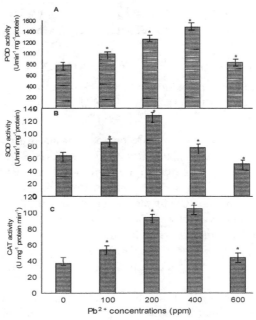

Fig. 6. Antioxidant enzyme activities POD (A), SOD (B) and CAT(C) of *Lepidium sativum* leaves subjected to various concentrations of Pb^{2+}. Each value represents the mean ±SE of five replicates. Significant differences (P<0.05) between treatments according to LSD test are shown by an asterisk (Ibrahim & Bafeel, 2011).

Oxygen free radicals or activated oxygen has been implicated in diverse environmental stresses in plants and animals and appears to be a common participation in most, if not all, degenerative conditions in eukaryotic cells. The peroxidation of lipid, the cross-linking and inactivation of proteins and mutations in DNA are typical consequences of free radicals, but because the reactions occur quickly and often are components of complex chain reactions, we usually can only detect their "footprints". Accumulation of ROS as a result of various environmental stresses is a major cause of loss of crop productivity worldwide (Mittler, 2002), (Apel and Hirt, 2004) , (Khan & Singh, 2008), (Mahajan & Tuteja, 2005) , (Tuteja , 2007; 2010).

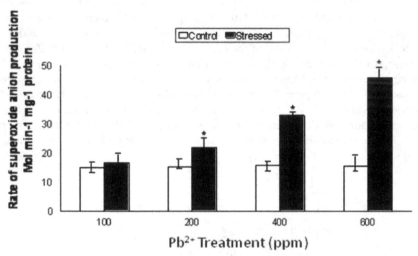

Fig. 7. Changes in the rate of superoxide production rate in roots of untreated and plants of *Lepidium sativum* subjected to various concentrations of Pb^{2+} for 10 days (Ibrahim & Bafeel, 2009).

Oxidative stress is a condition in which ROS or free radicals, are generated extra- or intracellular, which can exert their toxic effects to the cells. These species may affect cell membrane properties and cause oxidative damage to nucleic acids, lipids and proteins that may make them non functional. It is well documented that various abiotic stresses lead to the overproduction of ROS in plants which are highly reactive and toxic and ultimately results in oxidative stress. In an environment of molecular oxygen (O_2), all living cells are confronted with the reactivity and toxicity of active and partially reduced forms of oxygen: singlet oxygen ($^1O_2$), superoxide anion ($O_2.^-$), hydroxyl radical (HO.), and hydrogen peroxide (H_2O_2), which can lead to the complete destruction of cells (Mittler et al., 2004).

These reactive oxygen species (ROS) can show acute production under conditions such as ultraviolet light, environmental stress, or anthropic action through xenobiotics such as herbicides. However, their production is also directly and constantly linked with fundamental metabolic activities in different cell compartments, especially peroxisomes, mitochondria, and chloroplasts. In plants, the links between ROS production and photosynthetic metabolism are particularly important (Rossel et al., 2002).

4. Examples of oxidative stress indices

4.1 Lipid peroxidation

It has been recognized that during lipid peroxidation (LPO), products are formed from polyunsaturated precursors that include small hydrocarbon fragments such as ketones, malondialdehyde (MDA), etc and compounds related to them (Garg & Manchanda, 2009) . Some of these compounds react with thiobarbituric acid (TBA) to form colored products called thiobarbituric acid reactive substances (TBARS) (Heath & Packer, 1968). LPO, in both cellular and organelle membranes, takes place when above-threshold ROS levels are reached, thereby not only directly affecting normal cellular functioning, but also aggravating the oxidative stress through production of lipid-derived radicals (Montillet et al., 2005).

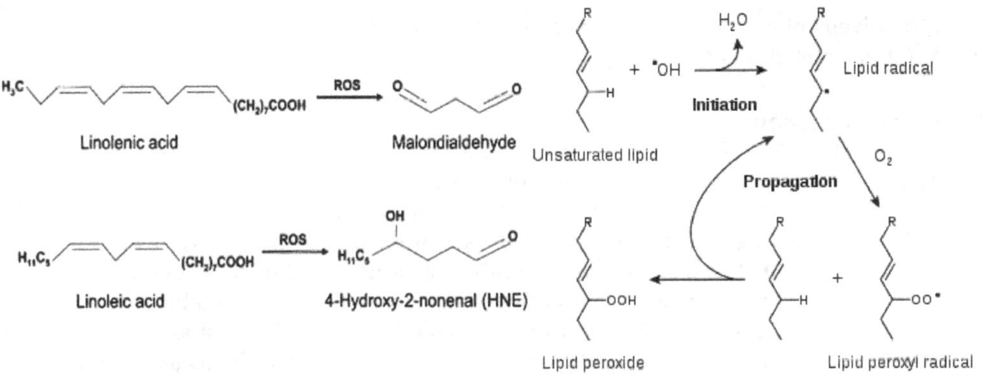

Fig. 8. Production of lipid-derived radicals via lipid peroxidation

4.2 Hydrogen peroxide

Hydrogen peroxide (H_2O_2) plays a dual role in plants: at low concentrations, it acts as a signal molecule involved in acclimatory signaling triggering tolerance to various biotic and abiotic stresses and, at high concentrations, it leads to programmed cell death (PCD) (Quan et al., 2008). H_2O_2 has also been shown to act as a key regulator in a broad range of physiological processes, such as senescence (Peng et al., 2005), photorespiration and photosynthesis (Noctor & Foyer, 1998), stomatal movement (Bright et al., 2006), cell cycle (Mittler et al., 2004) and growth and development (Foreman et al., 2003).

Also, H_2O_2 is starting to be accepted as a second messenger for signals generated by means of ROS because of its relatively long life and high permeability across membranes (Quan et al., 2008). In an interesting study the response of pre-treated citrus roots with H_2O_2 (10 mM for 8 h) or sodium nitroprusside (SNP; 100 mM for 48 h) was investigated to know the antioxidant defense responses in citrus leaves grown in the absence or presence of 150 mM NaCl for 16d (Tanoua et al., 2009). It was noted that H_2O_2 and SNP increased the activities of leaf antioxidant enzymes such as, superoxide dismutase (SOD), catalase (CAT), ascorbate peroxidase (APX) and glutathione reductase (GR) along with the induction of related-isoform(s) under non-NaCl-stress conditions.

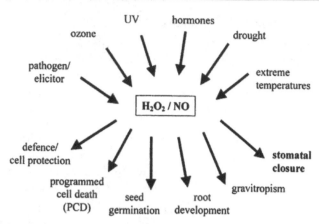

Fig. 9. Involvement of H_2O_2 and NO in cellular responses to various stresses and stimuli(Desikan et al., 2004).

4.3 Protein oxidation

Protein oxidation is defined as covalent modification of a protein induced by ROS or byproducts of oxidative stress. Most types of protein oxidations are essentially irreversible, whereas, a few involving sulfur-containing amino acids are reversible (Ghezzi &Bonetto, 2003). Protein carbonylation is widely used marker of protein oxidation (Moller et al., 2007) and (Job et al., 2005). The oxidation of a number of protein amino acids particularly Arg, His, Lys, Pro, Thr and Trp give free carbonyl groups which may inhibit or alter their activities and increase susceptibility towards proteolytic attack (Moller et al., 2007). Protein carbonylation may occur due to direct oxidation of amino acid side chains (e.g. proline and arginine to γ-glutamyl semialdehyde, lysine to amino adipic semialdehyde, and threonine to aminoketobutyrate) (Shringarpure& Davies, 2002).

5. Physiological, biochemical and molecular responses of plant to abiotic stresses

5.1 Photosynthetic responses toward oxidative stress

In higher plants, photosynthesis takes place in chloroplasts, which contain a highly organized thylakoid membrane system that harbours all components of the light-capturing photosynthetic apparatus and provides all structural properties for optimal light harvesting. Oxygen generated in the chloroplasts during photosynthesis can accept electrons passing through the photosystems, thus forming O_2^-. Through a variety of reactions, O_2^- leads to the formation of H_2O_2, OH and other ROS. The ROS comprising O_2^-, H_2O_2, $^1O_2$, HO_2^-, OH, ROOH, ROO, and RO are highly reactive and toxic and causes damage to proteins, lipids, carbohydrates, DNA which ultimately results in cell death (Bryan , 1996; Downs et al., 1999).

In chloroplast activated oxygen species can be generated by direct transfer of excitation energy from chlorophyll to produce singlet oxygen, or by univalent oxygen reduction at PSI, in the Mehler reaction (Asada et al.,1998). The latter process results in the formation of the superoxide anion radical ($O_2\cdot^-$), singlet oxygen ($^1O_2$) and eventually H_2O_2 and the highly toxic hydroxyl radical ($\cdot$OH). It is well known that Cu^{2+} catalyze the formation of OH. from the non-enzymatic chemical reaction between superoxide and H_2O_2.

Thylakoids are considered to be one of the major sites of superoxide production because of the simultaneous presence in chloroplasts of a high oxygen level and an electron transport system. Most of the superoxide is produced by photosystem I via the univalent reduction of oxygen through the ferredoxin / ferredoxin NADP+ oxidoreductase system (Mehler reaction). The use of DCMU, the known inhibitor of photosynthetic electron transport, and the use of the new spin trap DEPMPO have demonstrated that photosystem II also contributes to superoxide production (Navari-Izzo et al.,1998).

The modifications of the chloroplast in response to various environmental stresses have been widely studied in different laboratories and, thus the literature in the area is vast. The stress is sensed at the levels of pigment composition, structural organization, primary photochemistry and the CO_2 fixation(Biswal et al., 2003; Biswal, 2005).

Spatial and temporal complexity of photosynthesis makes photostasis prone to stress. The sequence of photosynthesis is known to cover a wide time-span and begins with photophysical and photochemical events, i.e. light absorption, excitation energy transfer and charge separation in the timescale of femtoseconds (10^{-15} s) to nanoseconds (10^{-9} s). This is followed by electron transport in the microseconds (10^{-6} s) to milliseconds (10^{-3} s) range, and finally by enzyme mediated reactions in the milliseconds to seconds range. Relatively slow reactions are rate-limiting and thus, incompatible with the fast reactions. Further, the fast primary photochemical reactions are relatively stress-resistant compared to temperature-dependent, slow, enzyme-mediated reactions associated with the electron transport system and carbon dioxide fixation in the Calvin–Benson cycle (Krause & Jahns, 2004). This results in the development of excitation pressure at the source. Since plants are photoautotrophs, light at any intensity in combination with other environmental stresses can bring a change in photostasis in terms of accumulation of excess unutilized quanta because of weakened sink demand induced by stress. In addition, high light always accumulates excess energy at the 'source'. NPQ of excess quanta at the source is one of the major processes for restoration of the balance and maintenance of photostasis(Biswal et al., 2011).

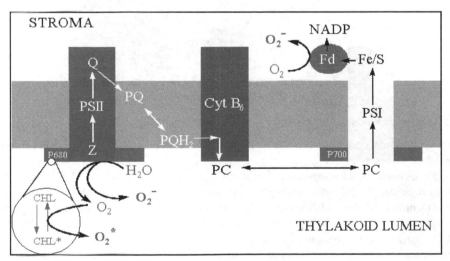

Fig. 10. Electron transport system in the thylakoid membrane showing three possible sites of activated oxygen production (Elstner, 1991; Bryan, 1996).

5.2 Plant responses toward temperature divergence

The climatic pattern in the arid zones is frequently characterized by a relatively "cool" dry season, followed by a relatively "hot" dry season, and ultimately by a "moderate" rainy season. In general, there are significant diurnal temperature fluctuations within these seasons. Quite often, during the "cool" dry season, daytime temperatures peak between 35 and 45 centigrade and fall to 10 to 15 centigrade at night. Daytime temperatures can approach 45 centigrade during the "hot" dry season and drop to 15 centigrade during the night. During the rainy season, temperatures can range from 35 centigrade in the daytime to 20 centigrade at night. In many situations, these diurnal temperature fluctuations restrict the growth of plant species.

Fig. 11. Different kinds of desert plants in arid environment (Ibrahim 2011).

Arid region plants are adapted to cope with temperature divergence between the prolonged annual hot and dry period in summer and the cooled winter. Plants evolved different survival mechanism including activation of antioxidant system, up-regulation of early light-induced proteins (ELIPs), and xanthophyll-cycle-dependent heat energy dissipation, among others (Demmig-Adams and Adams, 1993; Verhoeven et al., 2005). Increases in temperature raise the rate of many physiological processes such as photosynthesis in plants, to an upper limit. Extreme temperatures can be harmful when beyond the physiological limits of a plant. Decreasing photosynthesis seems to be the major cause of the chill induced reduction in the growth of plant in temperate climates (Baker et al., 1994). Several indicators support this assumption: periods of low temperature were accompanied by a lower chlorophyll content

(Leipner et al., 1999; Fryer et al., 1998), an increased pool size of xanthophyll cycle pigments, reduced photosynthetic capacity (Baker et al., 1994; Fryer et al., 1998).

Leaf antioxidant systems can prevent or alleviate the damage caused by reactive oxygen species (ROS) under stress conditions, and include enzymes such as superoxide dismutase (SOD), catalase (CAT), peroxidase (POD), ascorbate peroxidase (APX), and metabolites including ascorbate acid (AsA) and glutathione (GSH) (Asada, 1999; Xu et al., 2008). Phenolics are ubiquitous secondary metabolites in plants including large group of biologically active components, from simple phenol molecules to polymeric structures with molecular mass above 30 kDa (Dreosti, 2000, Ibrahim et al., 2011).

Artemisia monosperma showed the lowest activities for Guaiacol peroxidase(GuPx) and polyphenol oxidase (PPO) at 38°C and at 47°C in comparison with activities on plants collected at 9 and 15 °C (Table 1). Moreover, the relationship between GuPx and PPO activities and soluble phenolics concentration in A. monosperma plants appear to indicate that 47°C and 9°C caused heat and cold stress, by subjecting the plants to a super-optimal and suboptimal temperatures respectively(Ibrahim et al., 2011).

The metabolism of phenolic compounds includes the action of oxidative enzymes such as GuPx and PPO, which catalyze the oxidation of phenols to quinones (Thypyapong et al., 1995; Vaughn and Duke, 1984). Some studies have reported that these enzyme activities increase in response to different types of stress, both biotic and abiotic (Ruiz et al., 1998, 1999). More specifically, both enzymes have been related to the appearance of physiological injuries caused by thermal stress (Grace et al., 1998).

Phenylalanine ammonia-lyase (PAL) is considered to be the principal enzyme of the phenylpropanoid pathway (Kacperska, 1993) catalyzing the transformation, by deamination, of L-Phenylalanine into trans-cinnamic acid, which is the prime intermediary in the biosynthesis of phenolics (Levine et al., 1994). This enzyme increases in activity in response to thermal stress and is considered by most authors to be one of the main lines of cell acclimation against stress in plants (Leyva et al., 1995). Phenols are oxidized by peroxidase (POD) and primarily by polyphenol oxidase (PPO), this latter enzyme catalyzing the oxidation of the o-diphenols to o-diquinones, as well as hydroxylation of monophenols(Thypyapong et al., 1995). These activities of enzymes increase in response to different types of stress, both biotic and abiotic (Ruiz et al., 1998, 1999). More specifically, both enzymes have been related to the appearance of physiological injuries caused in plants by different stress (Grace et al., 1998; Ruiz et al., 1998; Ibrahim et al., 2011).

Sampling date	Temp.	U mg protein min^{-1}			PPO μmol caffeic acid mg^{-1} protein min^{-1}	POD μmol guaiacol mg^{-1} protein min^{-1}
		SOD	APX	CAT		
15 June	38°C	14.7±1.12	4.32±0.66	3.27±0.06	11.4±0.83	12.7±0.99
15 Aug	47°C	16.6±1.22	5.31±0.71	4.87±0.07	22.3±2.30	16.9±1.23
15 Dec	9°C	22.9±2.08	20.8±2.14	18.74±1.32	38.8±3.86	37.3±2.98
15 Feb	15°C	19.0±2.03	17.5±1.65	13.58±0.98	28.2±2.19	25.3±2.07

Table 1. Variation of antioxidant enzymes activities(superoxide dismutase, SOD; ascorbate peroxidase, APX; catalase, CAT; phenol peroxidase, PPO and guaiacol peroxidase, POD in Artemisia monosperma plant in response to temperature divergence in Riyadh (Saudi Arabia)(Ibrahim et al., 2011).

Over-expression of ROS scavenging enzymes like isoforms of SOD (Mn-SOD, Cu/Zn-SOD, Fe-SOD), CAT, APX, GR, DHAR, GST and GPX resulted in abiotic stress tolerance in various plants due to efficient ROS scavenging capacity. Pyramiding of ROS scavenging enzymes may also be used to obtain abiotic stress tolerance plants. Therefore, plants with the ability to scavenge and/or control the level of cellular ROS may be useful in future to withstand harsh environmental conditions.

5.3 Osmotic adjustment in stressed plants

Osmotic response and their adjustment was considered as a biochemical marker in plants subjected to abiotic stress such as salinity can occur by the accumulation of high concentrations of either inorganic ions or low molecular weight organic solutes. Although both of these play a crucial role in higher plants grown under saline conditions, their relative contribution varies among species, among cultivars and even between different compartments within the same plant (Greenway & Munns, 1980). The compatible osmolytes generally found in higher plants are low molecular weight sugars, organic acids, polyols, and nitrogen containing compounds such as amino acids, amides, imino acids, ectoine (1,4,5,6-tetrahydro-2-methyl-4-carboxylpyrimidine), proteins and quaternary ammonium compounds. According to Murakeozy et al.(2003), of the various organic osmotica, sugars contribute up to 50% of the total osmotic potential in glycophytes subject to saline conditions. The accumulation of soluble carbohydrates in plants has been widely reported as a response to salinity or drought, despite a significant decrease in net CO_2 assimilation rate (Carm, 1976; Popp & Smirnoff, 1995).

5.4 Role of amino acids and amides on the avoidance of abiotic stress

Amino acids have been reported to accumulate in higher plants under salinity stress (Bielski, 1983; Moller, 2001; Mahajan and Tuteja, 2005). The important amino acids include alanine, arginine, glycine, serine, leucine, and valine, together with the imino acid, proline, and the non-protein amino acids, citrulline and ornithine (Mahajan and Tuteja, 2005, Hu, 2007) .Proline, which occurs widely in higher plants, accumulates in larger amounts than other amino acids in salt stressed plants (Bielski et al., 1983; McDowell and Dangl, 2000; Navrot et al., 2007; Pastore et al., 2002; Reumann et al., 2004). Proline accumulation is one of the common characteristics in many monocotyledons under saline conditions (Dybing et al., 1978; Grant and Loake, 2000), although in barley seedlings, NaCl stress did not affect proline accumulation (Bolwell & Woftastek, 1997). However, proline accumulation occurs in response to water deficit as well as to salt. Thus, synthesis of proline is a non-specific response to low growth medium water potential (Navrot et al., 2007). Proline regulates the accumulation of useable N, is osmotically very active (Bielski et al., 1983; Moller, 2001), contributes to membrane stability (Heath, & Packer, 1968; Garg and Manchanda, 2009; Montillet et al., 2005) and mitigates the effect of NaCl on cell membrane disruption (Fam and Morrow, 2003). Even at supra-optimal levels, proline does not suppress enzyme activity (Hayashi and Nishimura, 2003; Moller et al., 2007).

6. Conclusion

According to our investigations, Ibrahim & Bafeel, 2008 concluded that dark chilling imposes metabolic limitation on photosynthesis and ROS are involved, to some degree, in the limiting photosynthetic capacity of alfalfa leaves. After recovery period the alfalfa plants showed physiological and biochemical changes that contribute to its superior dark chilling resistance and prevent the leaves from undergoing photooxidation damage and eventual death. Also our results showed that high cellular levels of H_2O_2 accumulated during the dark chilling treatment can induce the activation of a defense mechanism against chilling stress or programmed cell death. The accumulation of H_2O_2 can be induced by the increase in SOD activity. Therefore, during the recovery treatment the accumulated H_2O_2, in turn, may activate a protective mechanisms that increase the activities of several antioxidant enzymes such as APX, CAT and GR .Also induce alterations in the relative concentration of several non-enzymatic antioxidant compounds such as phenolics and tocopherols. (Bafeel & Ibrahim, 2008).

Results reported by Ibrahim & Alaraidh, 2010 demonstrated that changes in gene expression do occur in the two cultivars of Triticum aestivum in response to drought, and these differentially expressed genes, though functionally not known yet, may play important roles for cultivars to exhibit its response to drought stress before and after rehydration. Moreover, Ibrahim & Bafeel, 2009 concluded that prolonged stress induced by Pb^{2+} concentrations, can result into the activation of antioxidative enzymes and also enhance the gene expression of these antioxidant enzymes.

Although oxidative stress is potentially a lethal situation, it is also clear that plant systems exploit the interaction with oxygen. The production and destruction of active oxygen species is intimately involved with processes such as the hypersensitive responses and the regulation of photosynthetic electron flow. There are numerous sites of oxygen activation in the plant cell, which are highly controlled and tightly coupled to prevent release of intermediate products. Under stress situations, it is likely that this control or coupling breaks down and the process "dysfunctions" leaking activated oxygen. This is probably a common occurrence in plants especially when we consider that a plant has minimal mobility and control of its environment. Activated forms of oxygen are important in the biosynthesis of "complex" organic molecules, in the polymerization of cell wall constituents, in the detoxification of xenobiotic chemicals and in the defense against pathogens. Thus, the plant's dilemma is not how to eliminate the activation of oxygen, but how to control and manage the potential reactions of activated oxygen.

Genetic engineering also offer advantages in terms of the study of the physiological roles of enzymes where a classical genetic approach, such as selection of enzyme-deficient mutants, is difficult or almost impossible to carry out. In plant systems, the situation is often considerably complicated by the presence of a large number of isoenzyme forms, for example, the large GR and SOD families of isoenzymes, encoded by different genes. In the future, however, the use of antisense technology combined with selection of specific cDNA clones for isoenzymes may facilitate investigation of such enzyme-deficient mutants. Current observations suggest that increasing the level of stress tolerance by reinforcing the plant's defense system with new genes is an attainable goal.

7. References

Araus, J.L.; Slafer, G.A.; Royo, C. & Dolores Serret, M. (2008). Breeding for Yield Potential and Stress Adaptation in Cereals. Critical Reviews in Plant Science, 27, pp. 377–412.

Asada, K. (1999). The water-water cycle in chloroplasts: scavenging of active oxygens and dissipation of excess photons. Annu Rev Plant Physiol Plant Mol Biol 50, pp. 601–639.

Apel, K. & Hirt, H. (2004). Reactive oxygen species: metabolism, oxidative stress, and signal transduction, Annu. Rev. Plant Biol. 55, pp. 373–399.

Bafeel, S.O. & Ibrahim M. M. (2008).Activities of antioxidant enzymes and accumulation of α-tocopherol in *Medicago sativa* seedlings subjected to dark and chilling conditions, Int. J. Agric. Biol. 10(6), pp. 593-598.

Baker, A.J.M.; McGrath, S.P.; Sidoli, C.M.D.& Reeves, R.D. (1994). The possibility of in situ heavy metal decontamination of polluted soils using crops of metal-accumulating plants. Resources, Conservation, and Recycling 11, pp. 41-49.

Bielski, B.H.; Arudi, R.L. & Sutherland, M.W. (1983). A study of the reactivity of HO_2/O_2^- with unsaturated fatty acids, J. Biol. Chem. 258, pp. 4759-4761.

Biswal, U. C., Biswal, B. & Raval, M. K.(2003). Chloroplast Biogenesis: From Proplastid to Gerontoplast, Springer, Dordrecht, The Netherlands.

Biswal, B.(2005). Photosynthetic response of green plants to environmental stress: Inhibition of photosynthesis and adaptational mechanisms. In Handbook of Photosynthesis (ed. Pessarakli, M.) CRC Press, Florida, USA, 2nd edn, pp. 739–749.

Biswal, B.; Joshi, P. N.; Raval, M. K. & Biswal, U. C.(2011). Photosynthesis, a global sensor of environmental stress in green plants: stress signalling and adaptation. CURRENT SCIENCE, VOL. 100(12), pp. 25.

Bolwell, G.P. & Woftastek, P. (1997). Mechanism for the generation of reactive oxygen species in plant defense-broad perspective, Physiol. Mol. Plant Pathol. 51, pp. 347–349.

Boyer, J. S. (1982). Plant Productivity and Environment, Science. . 218 no. 4571 pp. 443-448.

Bright, J.; Desikan, R.; Hancock, J.T.; Weir, I.S. & Neill, S.J. (2006). ABA-induced NO generation and stomatal closure in Arabidopsis are dependent on H2O2 synthesis, The Plant J. 45, pp. 113–122.

Bryan, D.M. (1996).
http://www.plantstress.com/Articles/Oxidative%20Stress.htm#biological

Correia, O.; Brugnoli, E.; Nunes, A. & Mcguas, C. (2004). Effect of dust deposition on foliar absorbance of mediterranean species. Revista de Biologia, 22 (1/4), pp. 143-151.

Cram, W.J. (1976). Negative feedback regulation of transport in cells. The maintenance of turgor, volume and nutrient supply, in: U. Luttge, M.G. Pitman (Eds.), Encyclopaedia of Plant Physiology, New Series, vol. 2, Springer-Verlag, Berlin, pp. 284-316.

Demmig-Adams, B & Adams, W. I. (1993). Chlorophyll and Carotenoid Composition in Leaves of Euonymus kiautschovicus Acclimated to Different Degrees of Light Stress in the Field *Australian Journal of Plant Physiology* 23 (5) 649 - 659

Desikan, R.; Cheung, M.K.; Bright, J.; Henson, D.;. Hancock, J. T. & Neill, S. J. (2004). ABA, hydrogen peroxide and nitric oxide signaling in stomatal guard cells. 55(395), pp. 205-212.

Downs, C. A., Ryan, S. L. and Heckathorn, S. A. (1999b). The chloroplast small heat-shock protein: evidence for a general role in protecting photosystem II against oxidative stress and photoinhibition. J. Plant Physiol. 155: 488-496.

Dreosti, I.E. (2000). Antioxidant polyphenols in tea, cocoa and wine. Nutrition 16, pp. 7 – 8.

Dybing, E.; Nelson, J.R.; Mitchell, J.R.; Sesame, H.A. & Gillette, J.R. (1976). Oxidation of a methyldopa and other catechols by chytochromes R450-generated superoxide anion: possible mechanism of methyldopa hepatitis, Mol. Pharmacol. 12, pp. 911–920.

Elstner, E.F. (1991). Mechanisms of oxygen activation in different compartments of plant cells In: Active oxygen/oxidative stress and plant metabolism. Pell E.J. and Steffen K.L. (eds) American Soc. Plant Physiol. Rockville, M.D. pp. 13-25.

Fahn, A. (1964). Some anatomical adaptations of desert plants. Phytomorphology, 14, pp. 93–102.

Fahn,A. (1990). Plant Anatomy,3rd Edition. Pergamon, Oxford,588pp.

Fahn,A. & Cutler,D. (1992). Xerophytes. Borntraeger, Berlin 176pp

Fam, S.S. & Morrow, J.D. (2003). The isoprostanes: unique products of arachidonic acid oxidation-a review, Curr. Med. Chem. 10, pp. 1723-1740.

Foreman, J.; Demidchik, V.; Bothwell, J.H.; Mylona, P.; Miedema, H.; Torres, M.A.; Linstead, P.; Costa, S.; Brownlee, C.; Jones, J.D.; Davies, J.M. & Dolan, L. (2003). Reactive oxygen species produced by NADPH oxidase regulate plant cell growth, Nature 422, pp. 442–446.

Foyer, C. (1993). Ascorbic acid. In: R.G. Alscher and J.L. Hess (eds.), Antioxidants in higher plants. CRC Press, Boca Raton, FL., pp. 31-58.

Foyer, C.H. & Noctor, G. (2005). Redox homeostis and antioxidant signaling: a metabolic interface between stress perception and physiological responses, Plant Cell 17, pp. 1866–1875.

Fryer, M.J.; Andrews, J.R.; Oxborough, K.; Blowers, D.A. & Baker, N.R. (1998). Relationship between CO_2 assimilation, photosynthetic electron transport and active O_2 metabolism in leaves of maize in the field during periods of low temperature. Plant Physiology 116, pp. 571–580.

Garg, N. & Manchanda, G. (2009). ROS generation in plants: boon or bane?, Plant Biosys. 143, pp. 8–96.

Ghezzi, P. & Bonetto, V.(2003). Redox proteomics: identification of oxidatively modified proteins, Proteomics 3, pp. 1145-1153.

Grace, J.; Lloyd, J.; Miranda, A.C.; Miranda, H.S. & Gash, J.H.C. (1998). Fluxes of carbon dioxide and water vapour over a C4 pasture in south-western Amazonia (Brazil). Australian Journal of Plant Physiology, 25, pp. 519–530.

Grant, J.J. & Loake, G.J. (2000). Role of reactive oxygen intermediates and cognate redox signaling in disease resistance, Plant Physiol. 124, pp. 21–29.

Greenway, H. & Munns, R. (1980). Mechanism of salt tolerance in nonhalophytes, Annu. Rev. Plant Physiol. 31,pp. 149–190.

Gutterman,Y.(2001). Regeneration of Plants in Arid Ecosystems Resulting from Patch Disturbance, Geobotany 27. Kluwer Academic Publishers,Dordrecht,260pp.

Gutterman,Y.(2002). Survival strategies of annual desert plants. Adaptations of desert organism. Berlin, Heidelberg, New York, Springer. 348pp.

Hayashi, M. & Nishimura, M. (2003). Entering a new era of research on plant peroxisomes, Curr. Opin. Plant Biol. 6, pp. 577–582.

Heath, R.L. & Packer, L. (1968). Photoperoxidation in isolated chloroplasts: I. Kinetics and stoichiometry of fatty acid peroxidation, Arch. Biochem. Biophys. 125, pp. 180–198.

Hirano, T.a.; Kiyota, M.; & Aiga, I. (1995). Physical effects of dust on leaf physiology of cucumber and kidney bean plants. Environmental Pollution, 89 (3), pp. 255-261.

Hu, J.P. (2007). Toward understanding plant peroxisome proliferation, Plant Sig. Behav. 2, pp. 308-310.

Huang, Z.Y.; Wu, H. & Hu,Z.H. (1997). The structures of 30 species of psammophytes and their adaptation to the sandy desert environment in Xinjiang. Acta Phytoecologica Sinica 21, pp. 521-530.

Ibrahim, M. M. & Bafeel O.S. (2008). Photosynthetic Efficiency and Pigment Contents in alfalfa (*Medicago sativa*) Seedlings Subjected to Dark and Chilling Conditions, Int. J. Agric. Biolo. 10(3), pp. 306- 310.

Ibrahim, M. M. & Bafeel O.S. (2009). Alteration of gene expression, superoxide anion radical and lipid peroxidation induced by lead toxicity in leaves of Lepidium sativum. Journal of Animal & Plant Sciences, 4,(1), pp. 281 – 288.

Ibrahim, M.M. & Alaraidh,I. (2010). Differential Gene Expression and Physiological Adaptation of Two *Triticum aestivum* Cultivars for Drought Acclimation. CATRINA, 5 (1), pp. 15-21.

Ibrahim M.M. & Bafeel, S.O.(2011). Molecular and Physiological Aspects for *Lepidium sativum* Tolerance in Response to Lead Toxicity. Fresenius Environnemental bulletin 20, No 8.

Ibrahim, M. M. & El-Gaely, G.A. (2011). The dust storm in Riyadh- Saudi Arabia: its implications on some physiological parameters for some plants in Riyadh, Saudi Arabia. In press.

Ibrahim, M. M.; Boukhari, N. & El-Gaely, G.A.(2011). Molecular and Physiological behavior variations of *Artemisia monosperma* to seasonal temperature divergence in the middle region, Saudi Arabia. In press.

Idso, S.B. (1976). Dust storms. Scientific American, 235 (4): 108–114.

Job, C.; Rajjou, L.; Lovigny, Y.; Belghazi, M. & Job, D. (2005). Patterns of protein oxidation in Arabidopsis seeds and during germination, Plant Physiol. 138, pp. 790–802.

Kacperska, A. (1993). Water potential alteration-A prerequisite or a triggering stimulus for the development of freezing tolerance in overwintering herbaceous plants, in: P.H Li, L. Christerson (Eds.), Advances in Plant Cold Hardiness, CRC Press, Boca Raton, 1993, pp. 73–91.

Khan, N.A. & Singh S. (2008). In: Editors, Abiotic Stress and Plant Responses, IK International, New Delhi.

Kocsy, G., Szalai, G., Vagujfalvi, A., Stehli, L., Orosz, G. and Galiba, G. (2000). Genetic study of glutathione accumulation during cold hardening in wheat. Planta 210, pp.295-301.

Krause, G. H. & Jahns, P. (2004). Non-photochemical energy dissipation determined by chlorophyll fluorescence quenching: characterization and function. In Chlorophyll a Fluorescence: A Signature of Photosynthesis (eds Papageorgiou, G. C. and Govindjee), Springer, Dordrecht, (reprinted 2010), pp. 463–495.

Leipner, J.; Fracheboud, Y.; Stamp, P. (1999). Effect of growing season on the photosynthetic apparatus and leaf antioxidative defenses in two maize genotypes of different chilling tolerance. Environmental and Experimental Botany 42,2, pp. 129-139.

Levine, A.; Tenhaken, R.; Dixon, R.& Lamb, C. (1994). H2O2 from the oxidative burst orchestrates the plant hypersensitive disease resistance response, Cell 79, pp. 583–593.

Leyva, A.; Jarrillo, J.A.; Salinas, J.; Martı´nez-Zapater, M. (1995). Low temperature induces the accumulation of phenylalanine ammonia-lyase and chalcone synthase mRNA of Arabidopsis thaliana in light-dependent manner, Plant Physiol. 108, pp. 39–46.

Luis, M.; Igreja, A.; Casimiro, A.P. & Joao, S.P. (2008). "Carbon dioxide exchange above a Mediterranean C3/C4 grassland during two climatologically contrasting years" Global change Biology, 14 (3), pp. 539-555.

Mahajan, S. & Tuteja, N. (2005)Cold, salinity and drought stresses: an overview, Arch. Biochem. Biophys. 444, pp. 139–158.

Maxwell, K.& Johnson, G. N. (2000). Chlorophyll fluorescence—A practical guide. J. Exp. Bot. 51, pp. 659–668.

McDowell, J.M. & Dangl, J.L. (2000). Signal transduction in the plant immune response, Trends Biochem. Sci. 25, pp. 79–82.

Mittler, R.(2002). Oxidative stress, antioxidants and stress tolerance, Trends Plant Sci. 7, pp. 405–410.

Mittler, R.; Vanderauwera, S.; Gollery, M. &Van Breusegem, F. (2004) Reactive oxygen gene network of plants. Trends Plant Sci. 9, pp. 490-498

Moller, I.M.(2001). Plant mitochondria and oxidative stress: electron transport, NADPH turnover, and metabolism of reactive oxygen species, Annu. Rev. Plant Physiol. Mol. Biol. 52, pp. 561–591.

Moller, I.M.; Jensen, P.E. & Hansson, A. (2007). Oxidative modifications to cellular components in plants, Annu. Rev. Plant Biol. 58, pp. 459–481.

Montillet, J.L.; Chamnongpol, S.; Rustérucci, C.; Dat, J.; van de Cotte, B.; Agnel, J.P.; Battesti, C.; Inzé, D. Van Breusegem & C. Triantaphylides, (2005). Fatty acid hydroperoxides and H2O2 in the execution of hypersensitive cell death in tobacco leaves, Plant Physiol. 138, pp. 1516–1526.

Murakeozy, E.P.; Nagy, Z.; Duhaze, C.; Bouchereau, A.& Tuba, Z. (2003). Seasonal changes in the levels of compatible osmolytes in three halophytic species of inland saline vegetation in Hungary, J. Plant Physiol. 160,pp. 395-401.

Navari-Izzo, F., Quartacci, M. F., Pinzino, C., Dalla Vecchia, F. and Sgherri, C. L. M. (1998). Thylakoid-bound and stromal antioxidative enzymes in wheat treated with excess copper. Physiol. Plant. 104, pp. 630–638.

Navrot, N.; Rouhier, N.; Gelhaye, E. & Jaquot, J.P. (2007). Reactive oxygen species generation and antioxidant systems in plant mitochondria, Physiol. Plant. 129, pp. 185–195.

Noctor, G.& Foyer, C.H. (1998). A re-evaluation of the ATP: NADPH budget during C3 photosynthesis. A contribution from nitrate assimilation and its associated respiratory activity?, J. Exp. Bot. 49, pp. 1895–1908.

Pastore, D.; Lausa, M.N.; Di Fonzo, N. & Passarella, S. (2002). Reactive oxygen species inhibit the succinate oxidation-supported generation of membrane potential in wheat mitochondria, FEBS Lett. 516, pp. 15–19.

Peng, C. L.; Ou, Z.Y.; Liu, N. & Lin, G.Z. (2005). Response to high temperature in flag leaves of super high-yielding rice Pei'ai 64S/E32 and Liangyoupeijiu, Rice Sci. 12, pp. 179–186.

Popp, M & Smirnoff, N. (1995). Polyol accumulation and metabolism during water deficit, in: N. Smirnoff (Ed.), Environment and Plant Metabolism: Flexibility and Acclimation, Bios Scientific, Oxford, pp. 199–215.

Quan, L.J.; Zhang, B.; Shi, W.-W. & Li, H.-Y. (2008). Hydrogen peroxide in plants: a versatile molecule of the reactive oxygen species network, J. Integrat. Plant Biol. 50, pp. 2–18.

Reumann, S.; Ma, C.; Lemke, S. & Babujee, L. (2004). AraPerox: a database of putative Arabidopsis proteins from plant peroxisomes, Plant Physiol. 136, pp. 2587–2608.

Rossel, J.B.; Wilson, I.W.& Pogson, B.J. (2002) Global changes in gene expression in response to high light in Arabidopsis. Plant Physiol 130, pp. 1109–1120

Ruiz, J.M.; Bretones, G.; Baghour, M.; Ragala, L.; Belakbir, A.; Romero, L. (1998). Relationship between boron and phenolic metabolism in tobacco leaves, Phytochemistry 48, pp. 269–272.

Ruiz, J.M.; Garcı́a, P.C.; Rivero, R.M. & Romero, L. (1999). Response of phenolic metabolism to the application to the carbendazim plus boron in tobacco leaves, Physiol. Plant. 106, pp. 151–157.

Schutzki, R. & Cregg, B. (2007). Abiotic plant disorders. A diagnostic guide to problem-solving. MSU Extension Bulletin E-2996. 16 pp.

Schwanz, P., Picon, C., Vivin, P., Dreyer, E., Guehi, J-M. & Polle, A. (1996). Response of antioxidative systems to drought stress in pendunculate oak and maritime pine as modulated by elevated CO_2. Plant Physiol. 110, pp. 393 - 402.

Shringarpure, R. & Davies, K.J. (2002). Protein turnover by the proteasome in aging and disease, Free Radic. Biol. Med. 32, pp. 1084–1089.

Smirnoff, N. (1993).The role of active oxygen in the response of plants to water deficit and desiccation. New Phytol. 125, pp. 27-58.

Tanoua, G., Molassiotis, A. & Diamantidis, G. (2009). Hydrogen peroxide-and nitric oxide-induced systemic antioxidant prime-like activity under NaCl-stress and stress-free conditions in citrus plants, J. Plant Physiol. 166, pp. 1904–1913.

Thypyapong, P.; Hunt, M.D. & Steffens, J.C. (1995). Systemic wound induction of potato (Solanum tuberosum) polyphenol oxidase, Phytochemistry 40, pp. 673–676.

Tuteja, N. (2007). Mechanisms of high salinity tolerance in plants, Meth. Enzymol.: Osmosens. Osmosignal. 428, pp. 419–438.

Tuteja, N. (2010). In: H. Hirt (Ed.), Cold, salt and drought stress. in: Plant Stress Biology: From Genomics towards System Biology, Wiley-Blackwell, Weinheim, Germany, pp. 137–159.

Vaughn, K.C & Duke, S.O. (1984). Function of polyphenol oxidase in higher plants, Physiol. Plant. 60, pp. 106–112.

Verhoeven, A.S.; Swanberg, A.; Thao, M.; & Whiteman, J. (2005). Seasonal changes in leaf antioxidant systems and xanthophyll cycle characteristics in Taxus x media growing in sun and shade environments. Physiol. Plant, 123, pp. 428-434.

Wang, W.; Vinocur, B. & Altman, A. (2003). Plant responses to drought, salinity and extreme temperatures: towards genetic engineering for stress tolerance. Planta, 218, pp. 1–14.

Xu, H.; Kim, Y.K.; Jin, X.J.; Lee, S.Y. & Park, S.U. (2008). Rosmarinic Acid Biosynthesis in Callus and Cell Cultures of Agastache rugosa Kuntze. J. Med. Plants Res. 2, pp. 237–241.

11

Photoacoustics: A Potent Tool for the Study of Energy Fluxes in Photosynthesis Research

Yulia Pinchasov-Grinblat and Zvy Dubinsky
The Mina & Everard Goodman Faculty of Life Sciences,
Bar-Ilan University,
Ramat-Gan,
Israel

1. Introduction

Phytoplankton cells are ideal organisms for the study of various aspects of photosynthesis, since most of their cells are devoted to components related to the harvesting of light energy and its storage as high energy compounds. They lack flowers, roots and all of the many structures and mechanism evolved in the course of the emergence of plants from the primordial oceans and conquering land. The products of photosynthesis are synthesized while carbon from assimilated CO_2 is being reduced and oxygen from photolytically split water is evolved. In most open water bodies – freshwater and marine – the energy input of the entire ecosystem depends on microscopic free-floating photosynthetic organisms- the phytoplankton. The determination of phytoplankton biomass and its photosynthesis activity is a great interest to ecologists.

The photoacoustic method allows the direct determination of the biomass of different taxa of phytoplankton and the efficiency of their photosynthesis. The latter is accomplished by relating the energy stored photochemically by photosynthesis to the total light energy absorbed by the plant material.

The method yields rapid, direct results of the efficiency of photosynthesis, compared to standard measurements based on ^{14}C fixation and oxygen evolution, or compared to indirect results from measurements of variable fluorescence.

We review the history of the application of photoacoustics to photosynthesis research. Our results show that the pulsed photoacoustic technique provides direct information on the biomass and phytoplankton photosynthesis and demonstrate its application in the study of phytoplankton ecology and physiology and in basic research of their photobiology.

The photoacoustics has a high potential for following the effects of environmental parameters such as irradiance, nutrient status and pollution on phytoplankton communities and their photosynthetic activity.

2. The history of photoacoustic effect definition

The photoacoustic effect was first investigated in the 1880 by Alexander Graham Bell. During his experiments with the "photophone", which carried an acoustic signal with a beam of sunlight that was reflected by an acoustic modulated mirror, he noticed that a rapidly interrupted beam of sunlight focused on a solid substance produces an audible sound. He observed that the resulting acoustic signal is dependent on the composition of the sample and correctly conjectured that the effect was caused by absorption of the incident light.

Recognizing that the photoacoustic effect had applications in spectroscopy, Bell developed the "spectrophone," essentially an ordinary spectroscope equipped with a hearing tube instead of an eyepiece. (Fig. 1). Samples could then be analyzed by sound when a source of light was applied.

Fig. 1. Historical setup used by Bell (Bell, 1881).

As noted by Bell, "the ear cannot of course compete with the eye for accuracy", when examining the visible spectrum. Bell published the results in a presentation to the American Association for the Advancement on Science in 1880 (Bell 1880).

In his paper, Bell described for the first time the resonant photoacoustic effect: "When the beam was thrown into a resonator, the interior of which had been smoked over a lamp, most curious alternations of sound were observed. The interrupting disk was set rotating at a high rate of speed and was then allowed to come gradually to rest. An extremely musical tone was at first heard, which gradually fell in pitch as the rate of interruption grew less." (Fig. 2) The loudness of the sound produced varied in the most interesting manner. Minor

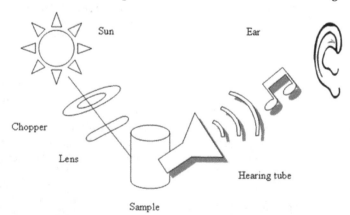

Fig. 2. Schematic setup of "photophone" used by Bell. As light source the sun (or a conventional radiation source) was employed. The acoustic signal was detected with a hearing tube and the ear (Bell, 1881).

reinforcements were constantly occurring, which became more and more marked as the true pitch of the resonator was neared.

As shown in Fig.3 the light pulse was absorbed by the sample of matter, and then converted to energy equivalents. The resulting energy will be partially radiated as heat (generation of wave), consequently a pressure wave can be detected by an acoustic sensor. The pressure waves are characteristic of the sample and are used to determine composition, concentration, and other thermophysical properties.

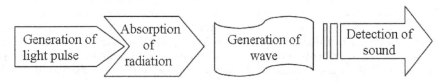

Fig. 3. The schematic of sound signal detection as used by Bell (Bell, 1881).

"When at last the frequency of interruption corresponded to the fundamental frequency of the resonator, the sound produced was so loud that it might have been heard by an audience of hundreds of people." (Bell, 1881).

Some time later Viengerov (1938, 1940) used the photoacoustic effect for the measurements of light absorption in gases and obtained quantitative estimates of concentration in gas mixtures based on signal magnitudes. He used blackbody infrared sources, for radiation input and a microphone to detect the acoustic signal. However, his results were affected by the relatively low sensitivity of his microphone as well as undesired photoacoustic effects from the glass chamber, a problem that persists in modern photoacoustic analysis.

In 1946 Gorelik suggested the use of the photoacoustic effect for the determination of energy transfer rates between vibrational and translational degrees of freedom of gas molecules. When a sample of gas in a photoacoustic cell is irradiated by photons, which it absorbs, the absorbed energy is used to excite to vibrational or vibrational-rotational energy state in the infrared, visible or ultraviolet ranges of the electromagnetic spectrum. (Rosencwajg, 1980).

Between 1950 and 1970 the photoacoustic gas analyzer employing a conventional light source gave way to the more sensitive gas chromatography technique. Similarly, the spectrophone gave way to the more versatile infrared spectrophotometer.

The development of the laser in the early 1970s had critical implications for photoacoustic spectroscopy. Lasers provided high intensity light at a tunable frequency, which allowed an increase in sound amplitude and sensitivity. In 1968 Kerr and Atwood were the first to apply a CO_2 laser to illuminate a photoacoustic cell (Kerr and Atwood, 1968). More interest in the method was generated when Kreuzer (1971) demonstrated part-per-billion (ppb) detection sensitivities of methane in nitrogen using a helium-neon laser excitation source, and later (Kreuzer 1972) sub-ppb concentrations of ammonia and other gases in mixtures, using infrared CO and CO_2 lasers.

Later, in the 1980s, Patel and Tam (Patel and Tam, 1981, Tam, 1986) have established not only the modern technological basis of the method, by using pulsed lasers as the light source and piezoelectric transducers as the photoacoustic detectors, but also provided the complete theoretical description of the photoacoustic phenomenon, based on the original concepts of Landau and Lifschitz (Landau and Lifschitz, 1959). Since then the photoacoustic method has been adapted and further developed by several groups (Rothberg 1983, Braslavksy 1986).

3. Photoacoustics and photosynthesis

In principle, photosynthesis generates three phenomena which may be detected by photoacoustics, with adequate setups. The thermal expansion of tissue, liquids and gases due to light energy converted to heath is termed as the **photothermal** signal. That is generated always, when photosynthetic tissue, or cell is exposed to a light pulse, since never is all of the light absorbed by plant tissue stored as products of the process. The unused fraction of the absorbed light energy is converted to heat, resulting in measurable pressure transient. When a leaf is illuminated by a pulse of light, the resulting photosynthetic photolysis of water causes the evolution of a burst of gaseous oxygen. That process leads to an increase in pressure, a change which is readily detected by a microphone as the **photobaric** signal. For detailed definitions and description see review by Malkin (1996). In addition to these two, absorption of light by components of the photosynthetic apparatus, such as PSI, or PSII, is accompanied by change in its spatial conformation and volume change, or **electrostriction.**

4. A leaves

Photoacoustic methods provide unique capabilities for photosynthesis research. The pulsed photoacoustic technique gives a direct measurement of the enthalpy change of photosynthetic reactions (Carpentier et al., 1984). A microphone may detect the photoacoustic waves *via* the thermal expansion in the gas phase. This method allows *in vivo* measurements of the photosynthetic thermal efficiency, or energy storage, and of the optical cross-section of the light harvesting systems (Carpentier et al., 1985, Buschmann, 1990).

Photoacoustic spectroscopy first emerged as a technique for photosynthesis research in the pioneering works of Cahen and Malkin (Malkin and Cahen 1979). Oxygen evolution by leaf tissue can be measured photoacoustically with a time resolution that is difficult to achieve by other methods (Canaani et al. 1988; Malkin 1996).

Photoacoustic measurements can achieve microsecond time resolution and allow determination of fast induction phenomena in isolated reaction centers, photosystems, thylakoid membranes intact cells and leaf tissue (Fig. 4).

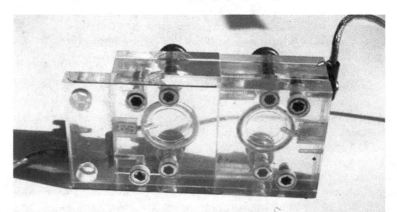

Fig. 4. Setup for photoacoustic photosynthesis measurements in air phase on leaf discs and algae collected on filters (Cha and Mauzarall 1992). Both the sample and reference twin chambers are connected by an air passage to hearing aid microphones.

These phenomena include states of the oxygen evolving complex in leaf tissue (Canaani et al. 1988) and the earliest steps of photosynthetic electron transport in photosystems (Arata and Parson 1981; Delosme et al. 1994, Edens et al. 2000).

In the work of da Silva (da Silva et al, 1995) the photoacoustic method has been demonstrated to be suitable, efficient and reliable technique to measure photosynthetic O^2 evolution in leaves.

The O^2 evolution in intact undetached leaves of dark adapted seedlings was measured during photosynthesis with the objective to detect genetic differences (da Silva et al., 1995).

Photoacoustic method can also measure state photosynthesis in intact cells and leaf tissue if the measuring pulses are given in combination with continuous background light (Kolbowski et al. 1990).

5. Phytoplankton

A simple technique based on photoacoustic measurements allowed us to determine the biomass, as well as the efficiency of photosynthesis, for different taxa of phytoplankton in situ (Dubinsky et al., 1998).

The experimental system is shown schematically in Figure 5 and 6 (Dubinsky et al. (1998), Mauzerall et al. (1998), and Pinchasov et al. (2005)).

Fig. 5. Photoacoustic phytoplankton cell (Dubinsky et al., 1998). The reddish algal culture is of *Porphyridium cruentum* and the laser pulse at 560nm.

The laser pulse is incident upon the suspension of algae, whose pigments absorb part of the laser beam. A variable fraction of the absorbed light pulse is stored in the photochemical products of photosynthesis. The remainder of the absorbed light is converted to heat, producing an acoustic wave that is intercepted by a detector (for details see Pinchasov *et al.* 2005). The signal contains a noisy background and later reflections from the walls of the vessel as well as from impedance mismatch within the detector.

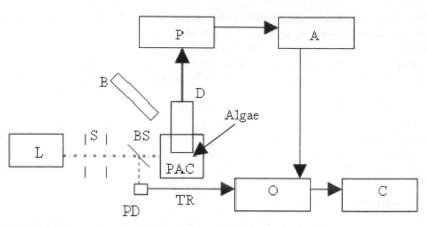

Fig. 6. Schematic representation of the experimental setup: L - Minilite Q – Switched
Nd:YAG Laser, 532 nm, S – beam-shaping slits, BS – beam splitter, PAC – photoacoustic cell
with suspension of algae (30 ml), D - stainless-steel photoacoustic detector, containing a
10-mm diameter resonating ceramic disc (BM 500, Sensor, Ontario, Canada), P – low-noise
Amptek A-250 preamplifier, A – SRS 560 – low noise amplifier, PD – photodiode,
TR - trigger signal, B – background light source, quartz-halogen illuminator
(Cole Parmer 4971), O – Tektronix TDS 430A oscilloscope, C – computer.

The use of piezoelectric films acoustically coupled to a liquid sample and a pulsed laser light
source increased the time resolution of the photoacoustic technique to the microsecond scale
(Nitsch et al., 1988; Mauzerall et al., 1995). Photoacoustic thermodynamic studies have been
carried out on isolated photosynthetic reaction centers from bacteria *Rb. sphaeroides* (Arata &
Parson, 1981), on PS I from cyanobacteria (Delosme et al., 1994), and on PS II from
Chlamydomonas reinhardtii (Delosme et al., 1994).

The resulting electric signal PA, is stored and subsequently analyzed on a computer. Thus,
the light energy storage efficiency f is determined following Eq. 1.

$$f = (PA_{light} - PA_{dark})/PA_{light} \qquad (1)$$

PA_{dark} is the photoacoustic signal generated by the weak laser pulse in the dark and PA_{light} is
the signal produced under the same pulse obtained under saturating (~3000 μmole photons
$m^{-2} s^{-1}$) continuous white light from a quartz-halogen illuminator (Pinchasov 2006).

We illustrate the application of the method by determining the effects of photoacclimation,
nutrient limitation and lead poisoning on phytoplankton cultures from different taxa.

6. The effect of nutrient limitation on photosynthesis

We were able to follow the effects of the key environmental parameter, nutrient status, on
the photosynthetic activity of phytoplankton. The nutrients examined were nitrogen,
phosphorus and iron (Pinchasov at al., 2005).

The algae for these cultures were harvested by centrifugation from the nutrient-replete
media in which they were grown, and resuspended in media from which N or P was
omitted. Cultures were followed over two weeks and compared for their photosynthetic
energy storage efficiency.

As seen in Fig. 7, all three algal species showed a sharp decrease in efficiency; by ~50±5% in the P limited, and ~60±5% in the N limited cultures, as compared to the nutrient replete controls (=100%). Fig. 8 shows the light energy storage efficiency under different ambient irradiance levels, resulting in an energy-storage curve for *Isochrysis galbana*.

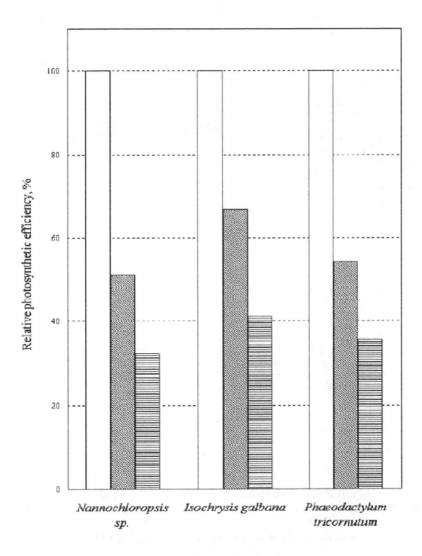

Fig. 7. The effect of nutrient limitation on relative photosynthetic efficiency. For each species photosynthetic energy efficiency of the nutrient replete control was taken as 100%. Controls (clear columns) were grown in nutrient replete media, whereas in the -P (gray) and -N (horizontal hatch) cultures, phosphorus and nitrogen were omitted from the medium. (Pinchasov et al., 2005).

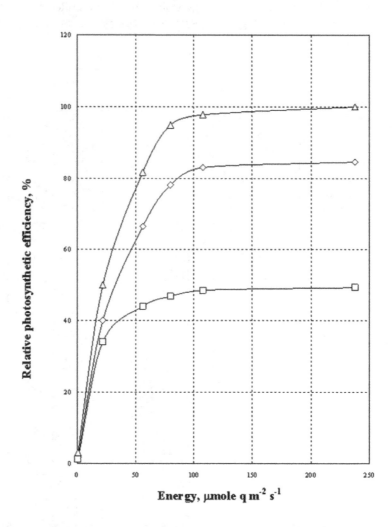

Fig. 8. The effect of nutrient limitation on the photosynthesis – irradiance relationship of *Isochrysis galbana*. Nutrient replete control (-Δ-), phosphorus limited (-◊-),and nitrogen limited (-□-).
The maximal storage in the nutrient replete control was taken as 100% (Pinchasov et al., 2006).

For the iron limitation experiments the algae were cultured in iron-replete media, under the same conditions as in the nitrogen and phosphorus depletion experiments. The photoacoustic experiments were conducted after two weeks in these media. As the iron is progressively depleted, the ability of the three species to store energy decreases Fig. 9.

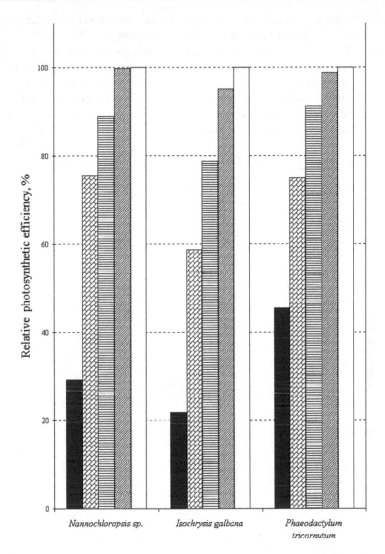

Fig. 9. The effect of iron concentration on photosynthetic efficiency. Controls (clear columns) were grown in iron replete media containing 0.6 mg L⁻¹. The iron concentration in the iron-limited cultures was (hatched columns, from left to right), 0 mg L⁻¹, 0.03 mg L⁻¹, 0.09 mg L⁻¹, and 0.18 mg L⁻¹, respectively (Pinchasov et al., 2005).

In our experiments, the exposure of the Cyanobacteria *S. leopoliensis* to different concentrations of lead resulted in some major changes in chlorophyll concentration and photosynthesis Fig. 10 (Pinchasov et al., 2006).

Figure 11 shows the changes in photosynthetic efficiency following lead application. The reduction of photosynthesis reached ~50% and ~80% with 25 ppm and 200 ppm correspondingly. It is important to emphasize that these results are similar in trend to the decrease in chlorophyll concentration. Most of the decrease seen after the first 24 hours already took place in the first 40 min, and probably even earlier.

With increasing lead concentration and duration of exposure, inhibition of photosynthesis increases. Since the photoacoustic method yields photosynthetic energy storage efficiency results that are independent of chlorophyll concentration, it means that the observed decrease in efficiency is not due to the death of a fraction of the population, but rather due to the impairment of photosynthetic function in all cells, possibly due to inactivation of increasing fractions of the photosynthetic units.

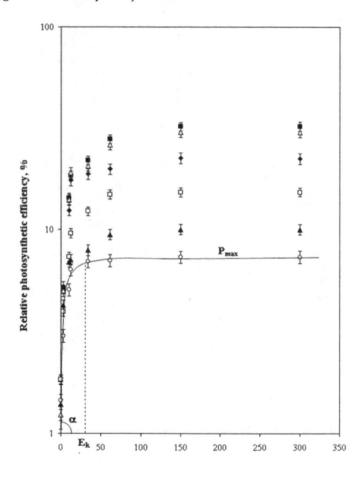

Fig. 10. Relative photosynthetic efficiency of *Synecococcus leopoliensis* (Cyanobacteria) versus irradiance after 7 days of exposure to lead. (■) MFM medium and 0 ppm, (Δ) phosphorus free medium (MFM-P) (●) MFM-P and 25 ppm, (□) MFM-P and 50 ppm, (□) MFM-P and 100 ppm, (○) MFM-P and 200 ppm (Pinchasov et al., 2006).

With increasing lead concentration and duration of exposure, inhibition of photosynthesis increases. Since the photoacoustic method yields photosynthetic energy storage efficiency results that are independent of chlorophyll concentration, it means that the observed decrease in efficiency is not due to the death of a fraction of the population, but rather due to the impairment of photosynthetic function in all cells, possibly due to inactivation of increasing fractions of the photosynthetic units.

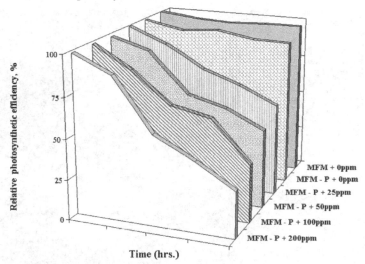

Fig. 11. The effect of lead on relative photosynthetic efficiency of *Synechococcus leopoliensis* (Pinchasov et al., 2006).

Fig. 12. An experimental photoacoustic setup, where the pulses previously obtained by a laser were produced by red light emitting diodes. The cuvette with the algal culture faces the LED array, whereas the microphone, visible on the left, is horizontal, placed at the rear window of the cuvette.

Recently Gorbunov et al., (submitted) were able to conduct photoacoustic measurements on *Chlamydomonas reihardtii* and to determine the allocation of energy to either photosystem by using PSI or PSII deficient mutants. In these experiments the brief exciting pulses hitherto produced by lasers were generated by red light emitting diodes (Fig. 12), and the saturating, continuous light was provided by blue LEDs (Fig. 13).

Fig. 13. The same setup as in fig 12. The blue LEDs provide the saturating, continuous light.

Recently Chengyi Yan et al. (submitted) were able to conduct photoacoustic measurements on *Chlamydomonas reihardtii* and to determine the allocation of energy to either photosystem by using PSI or PSII deficient mutants. In these experiments the brief exciting pulses hitherto produced by lasers were generated by red light emitting diodes (Fig. 12), and the saturating, continuous light was provided by blue LEDs (Fig. 13).

In these experiments the authors also estimated the contribution of electrostriction to the photoacoustic signal by comparing results at room temperature with ones measured at 4 oC, the temperature at which the photobaric signal is eliminated, and electrostiction is singled out.

7. Conclusions

1. Photoacoustics can be used to reliably estimate the concentration of photosynthetic pigments in phytoplankton cultures or assemblages.
2. The efficiency of energy storage by phytoplankton photosynthesis can be estimated directly, easily, rapidly and reliably by photoacoustics.

3. The effects of any environmental stressor, such as temperature, nutrient limitation, high/dim light and pollutants on the photosynthetic capacity of phytoplankton can be detected rapidly by photoacoustics.
4. Future work is likely to seek the replacement of lasers by LED sources, allowing the development of portable systems suited for field work, including submersible profilers.

8. References

Arata, H.; Parson, W.W. 1981, Enthalpy and volume changes accompanying electron transfer from P-870 to quinones in Rhodopseudomonas sphaeroides reaction centers. *Biochim. Biophys. Acta* Vol. 636, pp. 70-81.

Bell, A. G. 1881. Upon the production of sound by radiant energy. *Phil. Mag.(Fifth Series)* Vol. 11, pp.510-528.

Bell, A.G. On the production and reproduction of sound by light: the photophone. 1880 *Am. J. Sci.* Vol. 20, pp. 305-324.

Braslavsky, S. E., 1986. Photoacoustic and photothermal methods applied to the study of radiationless deactivation processes in biological systems and in substances of biological interest. *Photochem. Photobiol.*, Vol. 43, pp. 667-675.

Buschmann, C., 1990. Photoacoustic spectroscopy and its application in plant science. *Botanica Acta,* Vol. 103, pp. 9-14.

Canaani, O., Malkin, S., Mauzerall, D. 1988. Pulsed photoacoustic detection of flash-induced oxygen evolution from intact leaves and its oscillations. *Proc. Natl. Acad. Sci* U.S.A. Vol. 85, pp. 4725-4729.

Carpentier R, Larue B and Leblanc R. M., 1984. Photoacoustic spectroscopy of Anacystis nidulans III. Detection of photosynthetic activities. *Arch. Biochem. Biophys,*.Vol. 228, pp. 534-543.

Carpentier R., Nakatani H. Y., and Leblanc R. M., 1985 Photoacoustic detection of energy conversion in a photosystem II submembrane preparation from spinach. *Biochim Biophys Acta,* Vol. 808, pp.470-473.

Cha Y. and Mauzerall, D. C. 1992. Energy Storage of Linear and Cyclic Electron Flows in Photosynthesis. *Plant Physiol,*. Vol. 100, pp. 1869-1877.

Cha, Y., Mauzerall, D.C., 1992. Energy storage in linear and cyclic electron flows in photosynthesis. *Plant Physiol.* Vol. 100, pp. 1869-1877.

Chengyi Y., Schofield, O., Dubinsky, Z., Mauzerall, D., Falkowski, P. G., and M. Gorbunov. Photosynthetic energy storage efficiency in Chlamydomonas reinhardtii, based on microsecond photoacoustics. Submitted to *Photosynthesis Research*.

da Silva W.J., Prioli L.M.; Magalhaes A.C.N., Pereira A.C., Vargas H., Mansanares A.M., Cella N., Miranda L.C.M., Alvarado-Gil J., 1995. Photosynthetic O_2 evolution in maize inbreds and their hybrids can be differentiated by open photoacoustic cell technique. *Plant Science*, Vol. 104, pp. 177-181.

Delosme, R.; Beal, D.; Joliot, P., 1994. Photoacoustic detection of flash-induced charge separation in photosynthetic systems. Spectral dependence of the quantum yield. *Biochim. Biophys. Acta,* Vol. 1185, pp.56-64.

Dubinsky, Z., Falkowski, P.G., Post, A.F., van Hes, U.M., 1987. A system for measuring phytoplankton photosynthesis in a defined light field with an oxygen electrode. *J. Plankton Res.* Vol. 9, pp. 607-612.

Dubinsky, Z., J. Feitelson & D. C. Mauzerall, 1998. Listening to phytoplankton: Measuring biomass and photosynthesis by photoacoustics. *Journal of Phycology* Vol. 34, pp. 888-892.

Edens, G.J., Gunner, M.R.; Xu, Q., Mauzerall, D., 2000 The enthalpy and entropy of reaction for formation of P+QA- from excited reaction centers of *Rhodobacter sphaeroides*. *J. Am. Chem. Soc.* Vol. 122, pp. 1479-1485.

Gorelik, S., 1946, Dokl. *Akad. Nauk. SSSR* Vol. 54, p. 779

Kerr, E.L., Atwood, J.G., 1968. The laser illuminated absorptivity spectrophone: a method for measurement of weak absorptivity in gases at laser wavelengths. *Appl. Opt.*, Vol. 7, pp. 915-921.

Kolbowski, J., H., Reising, U., Schreiber, 1990. Computer-controlled pulse modulation system for analysis of photoacoustic signals in the time domain. *Photosynth. Res.*, Vol. 25, pp. 309-316.

Kreuzer, L.B. 1971 Ultralow gas concentration infrared absorption spectroscopy. *J. Appl. Phys.*, Vol. 42, pp. 2934-2943.

Kreuzer, L.B., Kenyon, N.P., Patel, C.K.N., 1972. Air pollution: sensitive detection of ten pollutant gases by carbon monoxide and carbon dioxide lasers. *Science*, Vol. 177, pp. 347-349.

Landau, L. D.; Lifschitz, E. M. *Fluid Mechanics*, Course of Theoretical Physics, Vol. 6; Pergamon Press: London, 1959.

Malkin, S., 1996. The photoacoustic method in photosynthesis-monitoring and analysis of phenomena which lead to pressure changes following light excitation. In Amesz, J. & A. Hoff (eds), *Biophysical Techniques in Photosynthesis*. Kluwer Academic Publishers, pp. 191-206.

Malkin, S., Cahen, D., 1979. Photoacoustic spectroscopy and radiant energy conversion: theory of the effect with special emphasis on photosynthesis. *Photochem. Photobiol.*, Vol. 29, pp. 803-813.

Mauzerall, D. C., J. Feitelson & Z. Dubinsky, 1998. Discriminating between phytoplankton taxa by photoacoustics. Israel Journal of Chemistry Vol. 38, pp.257-260.

Mauzerall, D.; Feitelson, J.; Prince, R., 1995. Wide band time-resolved photoacoustic study of electron transfer reactions: Difference between measured enthalpies and redox free energies. *J. Phys. Chem.* Vol.99, pp. 1090-1093.

Nitsch, C.; Braslavsky, S.E.; Schatz, G.H., 1988 Laser-induced optoacoustic calorimetry of primary processes in isolated photosystem I and photosystem II particles. *Biochim. Biophys. Acta,* Vol. 932, pp. 201-212.

Patel, C. K. N., and A. C. Tam, 1981. Opto-acoustic spectroscopy of condensed matter. *Rev. Mod. Phys.*, Vol. 53, pp. 517-550.

Pinchasov Y., Berner T., Dubinsky Z., 2006. The effect of lead on photosynthesis, as determined by photoacoustics in *Synechococcus leopoliensis* (Cyanobacteria). *Water Air and Soil pollution*, Vol. 175, pp. 117-125.

Pinchasov, Y., D. Kotliarevsky, Z. Dubinsky, D. C. Mauzerall & J. Feitelson, 2005. Photoacoustics as a diagnostic tool for probing the physiological status of phytoplankton. *Israel Journal of Plant Sciences*, Vol. 53, pp. 1-10.

Rosencwaig, A. 1980 *Photoacoustics and Photoacoustic Spectroscopy*, Wiley and Sons, New York

Rosencwaig, A., 1980 Photoacoustics ans Photoacoustic Sprectroscopy. Wiley & Sons, New York.

Rothberg, L. J.; Simon, J. D.; Bernstein, M.; Peters, K. S. *J. Am. Chem. Soc.* 1983, Vol. 105, pp. 3464-3468.

Tam A. C., 1986. Applications of photoacoustic sensing techniques. *Rev. Mod. Physics*, Vol. 58, pp. 381-431.

Viengerov, M.,L. 1938, Eine Methode der gasanalyse, beruhend auf der optisch-akustischen tyndall-röntgenerscheinung. *Dokl. Akad. Nauk SSSR* Vol. 19, p. 687

Viengerov, M.,L.1940 *Izv. Akad. Nauk* SSSR, Vol. 4, p. 94 (In Russian).

12

Photosynthesis in Extreme Environments

Angeles Aguilera[1], Virginia Souza-Egipsy[2] and Ricardo Amils[1,3]
[1]Centro de Astrobiología (INTA-CSIC)
[2]Centro de Investigaciones Agrarias (CSIC)
[3]Centro de Biología Molecular (UAM-CSIC)
Spain

1. Introduction

Our ongoing exploration of Earth has led to continued discoveries of life in environments that have been previously considered uninhabitable. For example, we find thriving communities in the boiling hot springs of Yellowstone, the frozen deserts of Antarctica, the concentrated sulfuric acid in acid-mine drainages, and the ionizing radiation fields in nuclear reactors (González-Toril et al., 2003; Lebedinsky et al., 2007; Pointing et al., 2009). We find some microbes that grow only in brine and require saturated salts to live, and we find others that grow in the deepest parts of the oceans and require 500 to 1000 bars of hydrostatic pressure (Horikoshi, 1998; Ma et al., 2010). Life has evolved strategies that allow it to survive even beyond the daunting physical and chemical limits to which it has adapted to grow. To survive, organisms can assume forms that enable them to withstand freezing, complete desiccation, starvation, high levels of radiation exposure, and other physical or chemical challenges. Furthermore, they can survive exposure to such conditions for weeks, months, years, or even centuries. We need to identify the limits for growth and survival and to understand the molecular mechanisms that define these limits.

Biochemical studies will also reveal inherent features of biomolecules and biopolymers that define the physico-chemical limits of life under extreme conditions. Broadening our knowledge both of the range of environments on Earth that are inhabitable by microbes and of their adaptation to these habitats will be critical for understanding how life might have established itself and survived.

The diversity of life on Earth today is a result of the dynamic interplay between genetic opportunity, metabolic capability, and environmental change. For most of their existence, Earth's habitable environments have been dominated by microorganisms and subjected to their metabolism and evolution. As a consequence of geological, climatologic, and microbial processes acting across geological time scales, the physical-chemical environments on Earth have been changing, thereby determining the path of evolution of subsequent life. For example, the release of molecular oxygen by cyanobacteria as a by product of photosynthesis as well as the colonization of Earth's surface by metazoan life contributed to fundamental, global environmental changes. The altered environments, in turn, posed novel evolutionary opportunities to the organisms present, which ultimately led to the formation of our planet's major animal and plant species.

Therefore, this "co-evolution" between organisms and their environment is an intrinsic feature of living systems. Life survives and sometimes thrives under what seem to be harsh conditions on Earth. For example, some microbes thrive at temperatures of 113°C. Others exist only in highly acidic environments or survive exposures to intense radiation. While all organisms are composed of nearly identical macromolecules, evolution has enabled such microbes to cope with a broad range of physical and chemical conditions. What are the features that enable some microbes to thrive under extreme conditions that are lethal to many others? An understanding of the tenacity and versatility of life on Earth, as well as an understanding of the molecular systems that some organisms utilize to survive such extremes, will provide a critical foundation that will help us to understand the molecular adaptations that define the physical and chemical limits for life on Earth.

2. Photosynthetic extremophiles

When we think of extremophiles, prokaryotes come to mind first. Thomas Brock's pioneering studies of extremophiles carried out in Yellowstone's hydrothermal environments, set the focus of life in extreme environments on prokaryotes and their metabolisms (Brock, 1978). However, eukaryotic microbial life may be found actively growing in almost any extreme condition where there is a source of energy to sustain it, with the only exception of high temperature (>70°C) and the deep subsurface biosphere (Roberts, 1999). The development of molecular technologies and their application to microbial ecology has increased our knowledge of eukaryotic diversity in many different environments (Caron et al., 2004). This is particularly relevant in extreme environments, generally more difficult to replicate in the laboratory.

Recent studies based on molecular ecology have demonstrated that eukaryotic organisms are exceedingly adaptable and not notably less so than the prokaryotes, although most habitats have not been sufficiently well explored for sound generalizations to be made. In fact, molecular analysis has also revealed novel protist genetic diversity in different extreme environments (Roberts, 1999).

Temperature is one of the main factors determining the distribution and abundance of species due to its effects on enzymatic activities (Alexandrof, 1977). Most extremophiles that survive at high temperatures (95-115°C) are microorganisms from the archaeal or bacterial domains. On the contrary, for eukaryotic microorganisms, the highest temperature reported is 62°C, and most of the metazoans are unable to grow above 50°C (Rothschild & Mancinelli, 2001). Surprisingly, photosynthetic prokaryotes, such as cyanobacteria, have never been found in hot acidic aquatic systems (Brock, 1973). Instead, these ecological niches are usually profusely colonized by species of the order Cyanidiales, red unicellular algae (Brock, 1973). Thus, species from the genera *Galdieria* and *Cyanidium* have been isolated from hot sulfur springs, showing an optimal growth temperature of 45°C and a maximum growth temperature of 57°C (Seckbach, 1994; Ciniglia et al., 2004). These extreme hot springs are usually acidic (pH 0.05-4) and frequently characterized by high concentrations of metals such as cadmium, nickel, iron or arsenic, that are highly toxic to almost all known organisms.

Additionally, phototrophic eukaryotic microorganisms have colonized environments characterized by temperatures at or below 0°C. Some algal species bloom at the snow surface during spring (Fujii et al., 2010), and complex microbial communities have been found on glaciers, probably the most widely studies environments after marine ice habitats. Aplanospores of *Chlamydomonas nivalis* are frequently found in high-altitude, persistent

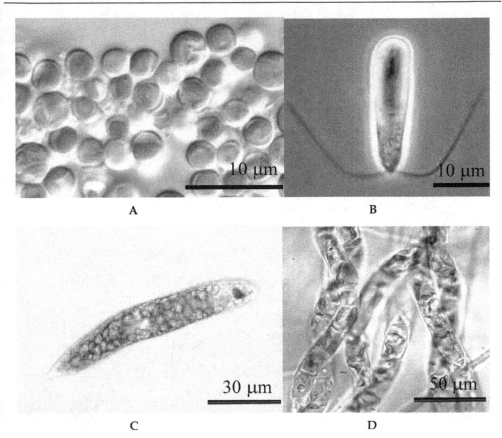

Fig. 1. Some examples of eukaryotic photosynthetic extremophiles. A.- *Cyanidium caldarium*, B.- *Dunaliella salina*, C.- *Euglena mutabilis*, D.- *Klebsormidium* sp.

snowfields where they are photosynthetically active despite cold temperatures and high levels of ultraviolet radiation (Stibal et al., 2006). Distinct microbial communities composed of psychrophilic bacteria, microalgae and protozoa colonize and grow in melt pools on the ice surface, or in brine channels in the sub-ice platelet in the Arctic even during winter, at extremely low temperatures of -20°C (Garrison & Close, 1993).

Non aquatic environments, where desiccation is common and water is a scarce resource, are also colonized by photosynthetic microorganisms. In these ecosystems, open spaces are usually covered by biological soil crusts, a highly specialized community of cyanobacteria, mosses and lichens. Without a doubt, the most colorful coatings on rocks are produced by lichens, a remarkable symbiotic relationship between microscopic algal cells and fungal filaments. Although lichens can also withstand extreme environmental conditions, they generally cannot survive as well in the dry, sun-baked deserts (Garthy, 1999). The most recurrent species of lichens found in deserts are the large colonies of the lime-green map lichen *Rhizocarpon geographicum*, the ashy-gray *Aspicilia cinerea* and the orange *Caloplaca saxicola*. They might be thousands of years old. In fact, the colorful chartreuse rock lichen *Acarospora chlorophana* may grow only a few millimeters per century.

Besides photosynthetic eukaryotes, in extreme environments, phototrophic microorganisms such as cyanobacteria are also frequently found forming thick microbial mats. Cyanobacterial mats are found in a broad range of environments, some of which can be considered extreme, such as hypersaline ponds and lakes, thermal springs, dry and hot deserts and the cold environment of polar regions (Stal, 2000; Zakhia et al., 2007). These communities often dominate total ecosystem biomass and productivity, and must contend with persistent low temperatures, repeated freeze–thaw cycles and highly variable light, nutrient and osmotic regimes (Vincent, 2000). These extreme habitats typically exhibit spatial gradients of chemical and physical factors, including extreme variations in temperature or salinity over relatively short distances that may influence local community structure (Miller & Castenholz, 2000; Nübel et al., 2001). In photosynthetic mats, daily shifts in oxygen production contribute to significant variation in depth-related chemical structure, including periodic hyperoxia near the surface and highly variable oxygen penetration (Des Marais, 2003).

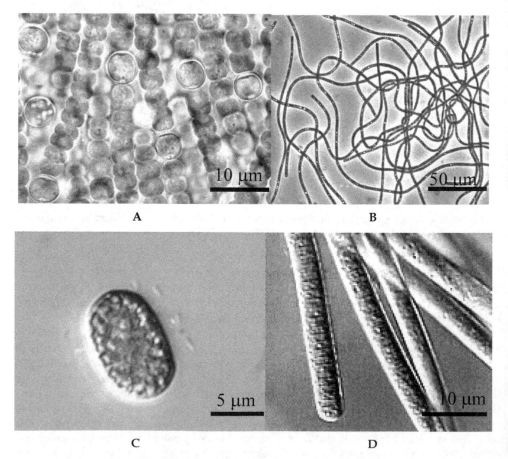

Fig. 2. Some examples of prokaryotic photosynthetic extremophiles. A.- *Pseudoanabaena* sp., B.- *Chloroflexus* sp., C.- *Synechococcus* sp., D.- *Phormidium* sp.

Filamentous, mucilage-producing Oscillatoriales are responsible for much of the biomass and three-dimensional structure of these polar mat consortia. They have been shown to tolerate a wide range of conditions and to maintain slow net growth despite the frigid ambient temperatures (Tang et al., 1997; Jungblut, 2010). In addition, other cyanobacterial taxonomic groups are also present in cold environments. Species of the genus *Phormidium*, *Pseudanabaena* or *Leptolyngbya* are also wide distributed in the polar regions (Fernández-Valiente, 2007) (Fig.2). Usually, the photosynthetic rates per unit of chlorophyll a of the microbial mats formed by these species are low compared with other cyanobacterial communities from more temperate environments (Ariosa et al., 2006).

Hot springs are other extreme environment in which photosynthetic cyanobacteria can also be found. These springs are well-isolated habitats occurring as clusters in globally distant regions, and the microorganisms that inhabit them are extremophiles adapted to conditions quite different from the ambient. Thus, in well studied North American hot springs, such cyanobacterial mats are formed by rod-shaped unicellular cyanobacteria of the genus *Synechococcus* with an upper temperature limit of 72 °C (Papke et al., 2003) as well as the green nonsulfur bacterium *Chloroflexus*. *Chloroflexus* is a thermophilic filamentous anoxygenic phototrophic bacterium, and can grow phototrophically under anaerobic conditions or chemotrophically under aerobic and dark conditions. According to 16S rRNA analysis, Chloroflexi species are the earliest branching bacteria capable of photosynthesis, and they have been long regarded as a key organism to resolve the obscurity of the origin and early evolution of photosynthesis. Chloroflexi species contains a chimeric photosystem that comprises some characters of green sulfur bacteria and purple photosynthetic bacteria, and also has some unique electron transport proteins compared to other photosynthetic bacteria (Tang et al., 2011).

3. Photosynthesis in extreme environments

3.1 Acidic environments. The Río Tinto (SW, Spain) case

Highly acidic environments are relatively scarce worldwide and are generally associated with volcanic activity and mining operation (Baffico et al., 2004). The natural oxidation and dissolution of the sulfidic minerals exposed to oxygen and water results in acid production, and the process can be greatly enhanced by microbial metabolism (Nordstrom & Southam, 1997; González-Toril et al., 2003). At the same time, low pH facilitates metal solubility in water, particularly cationic metals (such as aluminum and many heavy metals), and therefore acidic water tends to have high concentrations of heavy metals (Johnson, 1998).

Although low pH and high metals concentrations are restrictive to most aquatic life, large phototrophic biofilms and mats composed of filamentous green algae such *Zygnemopsis* or phototrophic protists such *Euglena* are often observed to thrive in extreme acidic environments (reviewed by Das et al., 2009a,b). Thus, acidic ecosystems are frequently colonized by organisms at an oligotrophic or mesotrophic level with dominant alga taxa belonging to the Crysophyceae, Chlorophyta and Bacillariophyta. These taxa are sometimes found in considerable quantities indicating a remarkable potential for primary production (Amaral-Zettler et al., 2002). Besides, presence of *Klebsormidium* sp. and *Euglena mutabilis* in a given environment is an ecological indicator of low pH and high level of metals (Valente & Gomes, 2007).

Thus, since extreme acidic environments are often the consequence of anthropogenic influences (e.g., mining activity or acid rain), most ecological studies of acidic waters have been focused on environments affected by human activity. In this regard, Río Tinto (Iberian Pyritic Belt), is one of the most unique examples of extreme acidic environments, not only because of its natural origin (Fernández-Remolar et al., 2003), but also for its peculiar microbial ecology (Amaral et al., 2002). What makes Río Tinto a unique acidic extreme environment is the unexpectedly high degree of eukaryotic diversity found in its waters and that these eukaryotic organisms are the principal contributors to the biomass in the river (Aguilera et al., 2006a; Amaral et al., 2002). Over 65% of the total biomass is due to photosynthetic protists. Members of the phylum Chlorophyta such as *Chlamydomonas*, *Chlorella*, and *Euglena*, are the most frequent species found followed by two filamentous algae belonging to the genera *Klebsormidium* and *Zygnemopsis*. The most acidic part of the river is inhabited by a eukaryotic community dominated by two species related to the genera *Dunaliella* (Chlorophyta) and *Cyanidium* (Rhodophyta), well known for their metal and acid

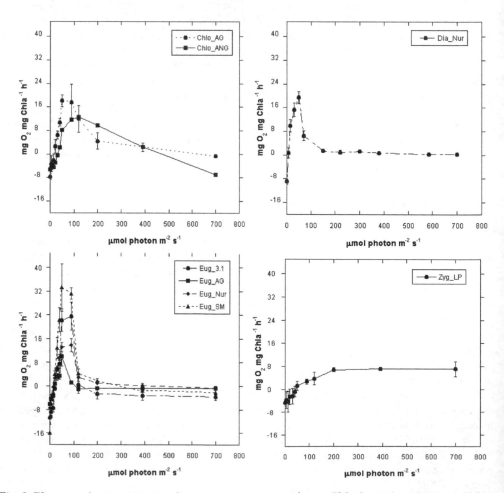

Fig. 3. Photosynthesis *versus* irradiance curves expressed on a Chl *a* basis (mg O_2 mg $^{-1}$ Chl*a* h^{-1}). Data are expressed as mean ± SD (n=3).

tolerance. Molecular ecology techniques have identified algae closely related to those characterized phenotypically, emphazising the high degree of eukaryotic diversity existing in the extreme acidic conditions of the Tinto Basin (Aguilera et al., 2006b; Aguilera et al., 2006b). Most of the phototrophic microbial communities found in the river are distributed in extensive biofilms along the riverbed, mostly formed for one dominating species (Aguilera et al., 2007).

On the other hand, photosynthesis is known to be particularly sensitive to stressful environmental conditions, such as salinity, pH or presence of toxicants. There are relatively few reports regarding photosynthesis in acidic environments in the literature, and most have focused on primary productivity measurements in acidic lakes. Thus, its been reported that minimum primary productivity is mainly due to metal stress (Niyogi et al., 2002; Hamsher et al., 2002) or soluble reactive phosphate concentration (Spijkerman et al., 2007). However, low pH itself does not reduce photosynthetic activity (Gyure et al., 1987). Light is another limiting factor for primary productivity in acidic lakes. Adaptation to low light intensities has been reported for benthic biofilms of diatoms in acidic lakes (Koschorreck & Tittle, 2002). To address this issue, the photosynthetic performance of different phototrophic biofilms isolated from Río Tinto was analyzed , in order to detect possible different photoadaptation capacities related to the environmental microhabitat conditions and the species composition of the biofilm. As far as we know this was the first attempt to determine the photosynthetic activity of low pH and heavy metal adapted phototrophic biofilms, which may give light in the understanding of the ecological importance of these organisms for the maintenance of the primary production of these extreme and unique ecosystems (Souza-Egipsy et al., 2011).

All the species analyzed showed a strong photoinhibition behaviour in their P-I curves except for the *Zygemopsis* sp. biofilm, which exhibited a photosaturated P-I curve behaviour. Generally, photoinhibition in P-I curves is commonly observed in shade adapted organisms, which cannot adequately dissipate the excessive photon flux provided by a high irradiance (Platt et al., 1980).

In our case, the biofilms exhibiting photoinhibition (*Euglena mutabilis*, *Pinnularia* sp. and *Chlorella* sp.) are usually located at the bottom of the river bed, covered by several centimetres of a highly coloured red water. However, *Zygnemopsis* sp. is a filamentous algae usually found at the water surface during the summer, when the sun irradiance is extremely high, and in this way can be considered a high light adapted species.

In addittion, the analyzed species can be considered as low light or shade adapted organisms due to their low I_c and I_k values (Fig. 4), which may be related to the fact that they develop under highly coloured waters, that affect quantitatively and qualitatively the light available for phototrophic organisms. Even at irradiances as low as 5 mmol m^{-2} s^{-1}, in the case of the diatom *Pinnularia* sp., photosynthetic activity was detected. These results are in agreement with previous data from sediments of acidic lakes, where photosynthetic ability at low light intensities (< 1.2 mE m^{-2} s^{-1}) were found in a benthic biofilm of diatoms suggesting an efficient absorption of red light, the dominant wavelength available in these iron-rich acidic waters, by these organisms (Koschorreck & Tittle, 2002). Maximum photosynthesis values (P_{max}) were also low in comparison with other environments, in which rates higher than 200 mmol O_2 mg Chla^{-1} h^{-1} are usually reached (Harting et al., 1998; Ritchie, 2008). In acidic mining lakes, planktonic primary productivity is usually low probably due to the low phytoplankton biomass (Nixdorf et al., 2003). In our case, this cannot be the reason, since phototrophic biofilms are the principal contributors of biomass in Río Tinto, representing over 65% of the total biomass (Lopez-Archilla et al, 2001). Another suggested reason for the low productivity in these extreme environments could be the lack of nutrients such as ammonium, phosphate or nitrate (Spijkerman et al., 2007).

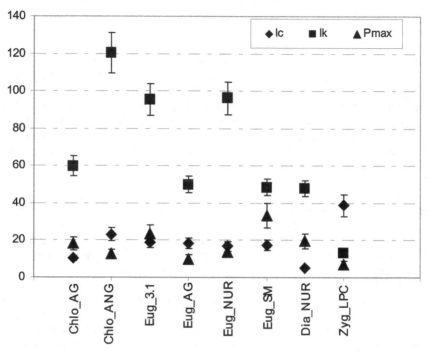

Fig. 4. Photosynthetic parameters of the different biofilms (*Chlorella*, *Euglena*, Diatom and *Zygnemopsis*) isolated from different locations at Río Tinto (AG, ANG, 3.1, NUR, SM and LPC). Compensation light intensity (I_c) and light saturation parameter (I_k) are expressed on photon basis (mmol photons m^{-2} s^{-1}). Maximum photosyntesis (P_{max}) are expressed on Chl *a* basis (mg O$_2$ mg Chla^{-1} *a* h^{-1}).

All the photosynthetic parameters analyzed showed statistical significant differences among species and sampling locations. Thus, the *Euglena* biofilms isolated from different habitats of the Río Tinto (3.1, AG, NUR, and SM) showed different photosynthetic values despite they are mainly formed by the same phototrophic species. The cells were able to maintain differences in their photosynthetic activity although all the experiments were carried out in artificial BG11 media, since the natural water from the river cannot be used in the oxymeter due to its high heavy metal concentrations. Following Falkowski & La Roche (1991), these results could be explained by photoadaptation processes instead of photoacclimation procedures. Photoadaptation refers to changes in the genotype that arise either from mutations or from changes in the distribution of alleles within a gene pool, while photoacclimation refers to phenotypic adjustments that arise in response to variations of environmental factors.

Besides nutrient limitation, presence of high concentrations of heavy metals could also explain the low values of maximum photosynthetic activity, since inhibition of photosynthesis is one of the most important cellular responses to metal stress conditions (Hanikenne, 2003; Spijkerman et al., 2007). Thus, a significant decrease in the amount of some photosynthetic proteins (RuBisCo, cytrochrome C peroxidase or photosystem I 11K protein precursor) has been reported in an isolate of *Chlamydomonas* spp. form Río Tinto when the cells were grown in natural metal rich water from the river (Cid et al., 2010). In this study, the metabolic response to metal-rich natural acidic waters from Río Tinto compared with artificial media was studied by determining proteomic differences between the two growth conditions. Although the cellular response induced by heavy metals on the growth and development of phytoplankton has been widely studied (Pinto et al., 2003; Gillet et al., 2006) most of these studies have focused on a single metal and most were carried out on freshwater algal species. However, little is currently known about heavy metal response in acidophilic organisms, despite the fact that they thrive in waters containing high levels of these toxicants.

To our knowledge, this was the first report regarding these issues using proteomic analysis of global expression patterns of cellular soluble proteins in an acidophilic strain of *Chlamydomonas* sp. Our results revealed that several stress-related proteins are induced in the cells growing in natural river water, along with a complex battery of proteins involved in photosynthesis, primary and energy metabolism or motility. When the 2-DE gels were compared, some of the most dramatic changes observed were related to proteins involved in the Calvin cycle and photosynthetic metabolism. In fact, three of the nine identified down-regulated proteins found in cells grown in the presence of metals, were described from these metabolic pathways. The amount of the ribulose-1,5-bisphosphate carboxylase/oxygenase (RuBisCo) decreases significantly when cells grow in metal rich water (Fig. 5). This decrease correlates with other proteins described from photosynthesis, such us cytrochrome C peroxidase, oxygen-evolving enhancer protein or photosystem I 11K protein precursor. These results are closely related to the presence of high levels of heavy metals present in the natural acidic waters since, inhibition of photosynthetic activity is one of the most important cellular responses to metal stress conditions (Hanikenne, 2003; Pinto et al., 2003). Similar results were found in *C. reinhardtii* in the presence of cadmium and copper (Boswell et al., 2002; Gillet et al., 2006), as well as in other photosynthetic organisms (Takamura et al., 1989; Wang et al., 2004). Although growth in extreme acidic environments is expected to require specific cellular adaptations of photosynthetic organisms, other studies have reported stress symptoms in acidophilic *Chlamydomonas* growing under acidic or metal-enriched natural water (Spijkerman et al., 2007; Langner et al., 2009).

On the contrary, phytochrome B, phosphoribulokinase and phosphoglycerate kinase were up-regulated when cells were grown in metal rich acidic water . Phytochromes are a family of light-sensing proteins required for plant developmental responses to light (Furuya, 1993; Quail, 1991). Plants perceive the intensity, direction, and quality of light and use this information to optimize photosynthesis. Phytochrome is the best characterized of the photoreceptors involved in these light dependent responses. In our case, the induction of this protein in the cells under study could be due to the intense red color of the Río Tinto water, caused by the high concentration of soluble ferric iron at the low pH of the river. This color has a marked effect on the quality and intensity of the light that reaches the cells. Experiments carried out in acidic mining lakes showed that only red light reaches the sediments of iron-rich water (Koschorreck & Tittel, 2002). The increased levels of phytochrome could be an adaptation process to these environmental conditions. The remaining induced enzymes, phosphoglycerate kinase and phosphoribulokinase are described from carbon fixation metabolism in phototrophic organisms (Merchant et al., 2007). Similar results were found for *C. reinhardtii* under cadmium exposure suggesting a limitation of the photosynthetic electron transfer that might force the cell to reorganize its whole metabolism (Gillet et al., 2006).

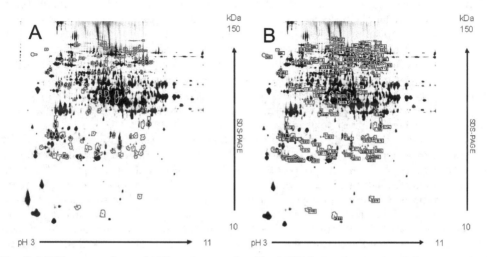

Fig. 5. 2-DE preparative gels. The spots resolved by 2-DE from preparative gels were stained with Coomassie Blue or with MALDI-MS-compatible silver reagent for peptide mass fingerprinting analysis. A)- gel obtained with cells growing under BG11/f2 artificial media at pH2. B)- gel obtained with cells growing under natural metal-rich water NW/f2 at pH2.

3.2 Cold environments. Benthic phototrophic biofilms from Antarctica

A very prevalent group of oxygenic phototrophs found in low-temperature environments are the Chromophytes, of which diatom algae in particular dominate marine and sea ice habitats. The diatoms possess a typical oxygenic photochemical apparatus; however, chlorophyll *b* is replaced by chlorophyll *c*, and fucoxanthin is a major carotenoid (Green et al., 1996). Green algae play various roles in low-temperature environments, which are often more likely to be dominated by prokaryotic photosynthetic microorganisms. Notable exceptions are found in two divergent low-temperature environments, the alpine snow ecosystem, which is dominated by psychrophillic *Chlamydomonas* and *Chloromonas* spp., and the permanent ice-covered lakes of the McMurdo Dry Valleys, which are vertically stratified layers of green algae (Morgan-Kiss et al., 2006).

However, in the polar regions, cyanobacterial mats are the most widespread and abundant organisms in freshwater environments such as lakes, ponds, streams, glaciers, and ice shelves, and they often dominate the total biomass and biological productivity due to their adaptation to the polar environment (Vincent, 2000). The ubiquitous presence of cyanobacterial mats as well as their taxonomic composition and physiological activities in streams, ponds, lakes and melt-waters of different places in continental Antarctica are well documented (Fernández-Valiente et al., 2007; Jungblut et al., 2010). The ice shelves of the polar areas support shallow pond ecosystems that are created during the summer season when pockets of the ice melt to form bodies of liquid water of various sizes. While the formation of liquid water is seasonally transitory, the ponds often melt out in the same location every year, and the microorganisms, particularly the cyanobacterial microbial mats exhibit several decades of seasonal growth. The microorganisms that colonize these extreme habitats must be capable of surviving daily and annual freezing-thawing cycles, persistent low temperatures, continuously high exposure to solar radiation during the summer, and long periods of dormancy. Despite these constraints, there exist diverse and productive consortia of microorganisms in the form of microbial phototrophic mats. These biota are representative modern-day examples of how life survived and evolved during global glaciations and extended periods of extreme cold (Vincent et al., 2000).

Nevertheless, in the area of the maritime Antarctica these aspects are less well defined. This region is characterized by a less extreme climatic regime than other Antarctic areas, with higher mean temperatures and precipitation (Camacho, 2006). These climatic conditions as well as the complex geology lead to a great variety of water bodies and to a high number of ice-free freshwater ecosystems during summer months. Taxonomical studies of microbial mats in Antarctic Peninsula (Vinocur & Pizarro, 1995) and King George Island (Vinocur & Pizarro, 2000) showed a richer species composition than in continental Antarctica, reflecting the broad range of physical and chemical conditions of the studied lakes and ponds. Polar cyanobacteria withstand the extremes of their environment through production of photoprotective screening and quenching pigments, as well as by their highly efficient light-capturing systems, nutrient storage ability and freeze–thaw tolerance (Hawes & Schwarz, 2001; Zakhia et al., 2007).

In 2011, we carried out several *in situ* studies investigating the photosynthetic characteristics of aquatic microbial mats and biofilms ecosystems from in King George island (South Shetland Islands) and to evaluate their sensitivity to potential climate change.

The microbial mats analyzed were mainly dominated by filamentous cyanobacteria of the order Oscillatoriales (*Oscillatoria* sp. and *Phormidium* sp.) and Nostocales. Microbial mats are extremely abundant in Byers Peninsula, particularly in the puddle soils of the catchment areas and at the bottom of small lakes and ponds of the central plateau, where they formed large expanses up to several hundred square metres in extent. In the coastal areas, microbial mats are usually restricted to the shore and bottom of the streams as the flat lowlands are covered by extensive carpets of mosses and in some areas by the vascular plants *Deschampsia antarctica* and *Colobanthus quitensis*. The photosynthetic rates per unit of chlorophyll a of the mats analyzed were low compared with other cyanobacterial communities of more temperate environments (Ariosa et al., 2006), but are in the range of other Antarctic microbial mats (Howard-Williams et al., 1989; Davey, 1993; Vincent et al., 1993). It is possible that the depth layers where photosynthesis reached its maximal values and the overcast weather conditions of this region are conducive to a low P_{max} and α, which were lower and higher, respectively, than in other cyanobacterial communities. Low P_{max}, low Ek and high a values are characteristics of photosynthesis in a shaded environment (Boston & Hill, 1991).

The polar areas are one of the environments most vulnerable to climate warming, increasing global temperatures has already had a significant negative impact on accelerating the fragmentation and loss of the ice shelf. For that reason, in this study, our aim was to compare how the temperature leads photosynthetic performance in these microbial mats (Fig.6). In general, we can conclude that these mats showed higher photosynthetic rates at high temperatures (ca. 20°C). Tang et al. (1997) found that many of the Arctic and Antarctic mat-forming cyanobacteria are not psychrophilic and exhibit growth temperature optima far above the temperatures found in their natural environments. Therefore it appears that adaptation to growth at low temperatures is not a requirement for successful colonization of these habitats, and other characteristics, such as UV screening and protection against photoinhibition, may have a greater selective advantage in the Arctic ice shelf environment. In addition, analysis of 16S rRNA genes from oscillatorians isolated from Antarctic and Arctic ice-shelf microbial mat communities indicates that filamentous cyanobacteria in both polar environments originated from temperate species (Nadeau et al., 2001).

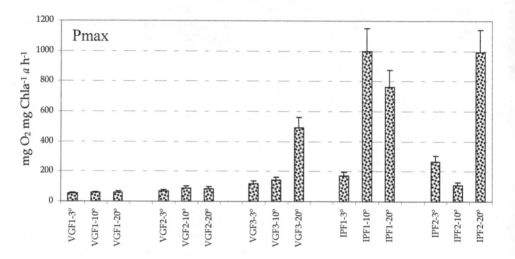

Fig. 6. Maximal photosynthetic rates of the different microbial mats isolated in King George island

4. Conclusion

In summary, although considerable work is yet to be done to understand the metabolism under extreme acidic environmental conditions, our results have shown that low-light adaptation of the phototrophic species could be an important determinant in the competitive success for colonizing these environments. Our work represents, we believe, the first attempt to use a proteomic approach to analyze the effects of acidic metal-rich natural water on one species adapted to these stress conditions. It demonstrates that naturally occurring metal rich water induces a stress response in acidophilic *Chlamydomonas* forcing algal cells to vary, not only their photosynthetic systems, but also to reorganize their metabolic pathways as an adaptive response to these environmental conditions, mainly driven by the high levels of heavy metals. Although further molecular and biochemical studies will properly elucidate the photosynthetic adaptation mechanisms to acidic waters, this study provides important information concerning these response pathways at the protein level, which are the fundamental functioning molecules in the cells.

In addition, permanently cold ecosystems make up one of the largest biospheres on the Earth. Paradoxically, the microorganisms that not only survive but thrive in these extreme habitats are still poorly understood. These two factors make low-temperature environments one of the last unexplored frontiers in our world. Research regarding the physiology and biochemistry of the primary producers, the microorganisms relying on photoautotrophic metabolism, of many of the food webs is still scarce.

Genome sequencing is the next step in understanding the adaptation of microorganisms to life in extreme environments, although a more powerful tool will be the integration of genomics with metabolic function through physiological and biochemical investigations. Therefore, our sequencing efforts should focus on extremophilic phototrophic genomes in order to achieve a better knowledge regarding the natural ecology, physiology, and biochemistry of these environments.

5. Acknowledgment

This work has been supported by the Spanish Science and Innovation Grants CGL2008-02298/BOS and CGL2009-08648-E.

6. References

Aguilera, A.; Manrubia, S.C.; Gómez, F.; Rodríguez, N. & Amils, R. (2006a). Eukaryotic community distribution and their relationship to water physicochemical parameters in an extreme acidic environment, Río Tinto (SW, Spain). *Applied and Environmental Microbiology*, Vol.72, pp.5325-5330

Aguilera, A.; Gómez, F.; Lospitao, E. & Amils, R. (2006b). A molecular approach to the characterization of the eukaryotic communities of an extreme acidic environment: methods for DNA extraction and denaturing gradient electrophoresis analysis. *Systematic and Applied Microbiology*, Vol.29, pp.593-605

Aguilera, A.; Souza-Egipsy, V.; Gómez, F. & Amils, R. (2007a) Development and structure of eukaryotic biofilms in an extreme acidic environment, Río Tinto (SW, Spain). *Microbial Ecology*, Vol. 53,pp.294-305

Alexandrof, V.Y. (1977). *Cells, molecules and temperature. Conformational flexibility of macromolecules and ecological adaptations.* Springer-Verlag, Berlin, Germany

Amaral, L.A.; Gómez, F.; Zettler, E.; Keenan, B.G.; Amils, R. & Sogin, M.L. (2002) Eukaryotic diversity in Spain's river of fire. *Nature*, Vol. 417, pp.137

Ariosa, Y.; Carrasco, D.; Quesada, A. & Fernández-Valiente, E. (2006). Incorporation of different N sources and light response curves of nitrogenase and photosynthesis by cyanobacterial blooms from rice fields. *Microbial Ecology*, Vol.51, pp.394–403

Baffico, G.D.; Díaz, M.M.; Wenzel, M.T.; Koschorreck, M.; Schimmele, M.; Neu, T.R. and Pedrozo, F. (2004) Community structure and photosynthetic activity of epilithon from a highly acidic (pH<2) mountain stream in Patagonia, Argentina. *Extremophiles*, Vol.8, pp.465-475

Boston, H.L. & Hill, W.R. (1991). Photosynthesis–light relations of stream periphyton communities. *Limnology and Oceanography*, Vol.36, pp.644–656

Boswell, C.; Sharma, N.C. & Sahi, S.V.. (2002). Copper tolerance and accumulation potential of *Chlamydomonas reinhardtii*. *Bulletin of Environmental Contamination Toxicology*, Vol.69, pp.546-553

Brock, T. (1973). Lower pH limit for the existence of blue-green algae: Evolutionary and ecological implications. *Science*, Vol.179, pp. 480-483

Brock, T. (1978). *Thermophilic Microorganisms and Life at High Temperatures*, Springer-Verlag, New York.

Camacho, A. (2006). Planktonic microbial assemblages and the potential effects of metazooplankton predation on the food web of lakes of the maritime Antarctica and Subantarctic Islands. *Reviews in Environmental Science and Biotechnology*, Vol.5, pp.167–185

Caron, D.A.; Countway, P.D. & Brown, M.V. (2004) The growing contributions of molecular biology and immunology to protistan ecology: molecular signatures as ecological tools. *Journal of Eukaryotic Microbiology*, Vol.51, pp.38-48

Cid, C.; Garcia-Descalzo, L.; Casado-Lafuente, V.; Amils, R. & Aguilera, A. (2010). Proteomic analysis of the response of an acidophilic strain of *Chlamydomonas* sp. (Chlorophyta) to natural metal-rich water. *Proteomics, Vol.*10, pp.2026-2036

Ciniglia, C.; Yoon, H.S.; Pollio, A.; Pinto, G. & Bhattacharya, D. (2004). Hidden biodiversity of the extremophilic Cyanidiales red algae. *Molecular Ecology*, Vol.13, pp.1827-1838

Das, B.K.; Roy, A.; Koschorrech, M.; Mandal, S.M.; Wendt-Potthoff, K. & Bhattacharya, J. (2009a). Occurrence and role of algae and fungi in acid mine drainage environment with special reference to metals and sulphate immobilization. *Water Research*, Vol.43, pp.883-894

Das, B.K.; Roy, A.; Singh, S. & Bhattacharya, J. (2009b) Eukaryotes in acidic mine drainage environments: potential applications in bioremediation. *Reviews in Environmental Science Biotecnology*, Vol. 8, pp.257-274

Davey, M.C. (1993). Carbon and nitrogen dynamics in a maritime Antarctic stream. *Freshwater Biology*, Vol.30, pp.319–330

De los Ríos, A.; Ascaso, C.; Wierzchos, J.; Fernández-Valiente, E. & Quesada, A. (2004). Microstructural characterization of cyanobacterial mats from the McMurdo Ice Shelf, Antarctica. *Applied and Environmental Microbiology*, Vol.70, pp.569–580

Des Marais, D.J. (2003). Biogeochemistry of hypersaline microbial mats illustrates the dynamics of modern microbial ecosystems and the early evolution of the biosphere. *Biological Bulletin*, Vol..204, pp.160–167

Falkowski, P.G. & La Roche, J. (1991) Acclimation to spectral irradiance in algae. *Journal of Phycology*, Vol. 27, pp.8-14

Fernández-Remolar, D.C.; Rodríguez, N.; Gómez, F. & Amils, R. (2003) Geological record of an acidic environment driven by the iron hydrochemistry: the Tinto river system. *Journal of Geophysical Research*, Vol.108, pp.5080-5095

Fernández-Valiente, E.; Camacho, A.; Rochera, C.; Rico, E.; Vincent, W.F. & Quesada, A. (2007). Community structure and physiological characterization of microbialmats in Byers Peninsula, Livingston Island (SouthShetland Islands, Antarctica). *FEMS Microbiology and Ecology*, Vol.59, pp. 377–385

Fujii, M.; Takano, Y.; Kojima, H.; Hoshino, T.; Tanaka, R. & Fukui, M. (2010). Microbial community structure, pigment composition, and nitrogen source of red snow in Antarctica. *Microbial Ecology*, Vol.59, No.3, pp.466-475

Furuya, M. (1993). Phytochromes: their molecular species, gene families, and functions. *Annual Review in Plant Physiology and Plant Molecular Biology*, Vol.44, pp.617-641

Garrison, D.L. & Close, A.R. (1993). Winter ecology of the sea ice biota in Weddel Sea pack ice. *Marine Ecolology Progress Series*, Vol.96, pp. 17-31

Garthy, J. (1999). Lithobionts in the eastern mediterranean. In: *Enigmatic Microorganisms and Life in Extreme Environments*, Seckbach, J. (Ed.), 257-276, Kluwer Academic Publ. London, England

González-Toril, E.; Llobet-Brossa, E.; Casamayor, E.O.; Amann, R. & Amils, R. (2003). Microbial ecology of an extreme acidic environment. The Tinto River. *Applied and Environmental Microbiology*, Vol.69, pp. 4853-4865

Gillet, S.; Decottignies, P.; Chardonnet, S. & Maréchal, P. (2006). Cadmium response and redoxin targets in *Chlamydomonas reinhardtii*: a proteomic approach. *Photosynthesis Research, Vol.89*, pp.201–211

Green, B.R. & Durnford, D.G. (1996). The chlorophyll-carotenoid proteins of oxygenic photosynthesis. *Annual Reviews in Plant Physiology and Plant Molecular Biology*, Vol.47, pp.685–714

Guyre, R.A.; Konopka, A.; Brooks, A. & Doemel, W. (1987). Algal and bacterial activities in acidic (ph3) strip mine lakes. *Applied and Environmental Microbiology*, Vol.53, pp.2069-2076

Hanikenne, M. (2003). *C. reinhardtii* as a eukaryotic photosynthetic model for studies of heavy metal homeostasis and tolerance. *New Phytologist*, Vol..159, pp.331-340

Harting, P.; Wolfstein, K.; Lippemeier, S. & Colijn, F. (1998). Photosynthetic activity of natural microbenthos populations measured by fluorescence PAM and ^{14}C-tracer: a comparison. *Marine Ecology Progress Series*, Vol.166, pp.53-62

Hawes, I. & Schwarz, A.M. (2001). Absorption and utilization of irradiance by cyanobacterial mats in two icecovered Antarctic lakes with contrasting light climates. *Journal of Phycology*, Vol.37, pp.5–15

Horikoshi, K. (1998) Barophiles: deep-sea microorganisms adapted to an extreme environment. *Current Opinion in Microbiology*, Vol.1, No.3, pp. 291-295

Howard-Williams, C.; Pridmore, R.D.; Downes, M.T. & Vincent. W.F. (1989). Microbial biomass, photosynthesis and chlorophyll a related pigments in the ponds of the McMurdo Ice Shelf, Antarctica. *Antarctic Science*, Vol.1, pp.125–131

Johnson, D.B. (1998). Biodiversity and ecology of acidophilic microorganisms. *FEMS Microbiology Ecolology*, Vol.27, pp.307-317

Jungblut, A.D.; Lovejoy, C. & Vincent, W.F. (2010). Global distribution of cyanobacterial ecotypes in the cold biosphere. *The ISME Journal*, Vol.4, pp. 191–202

Koschorrek, M. & Tittel, J. (2002). Benthic photosynthesis in acidic mining lake (pH. 2.6). *Limnololy and Oceanography*, Vol.47, No.4, pp.1197-1201

Langner, U.; Jakob, T.; Stehfest, K. & Wilhelm, C. (2009). An energy balance from absorbed protons to new biomass for *C. reinhardtii* and *C. acidophila* under neutral and extremely acidic growth conditions. *Plant Cell Environment*, Vol.32, pp.250-258

Lebedinsky, A.V.; Chernyh, N.A. & Bonch-Osmolovskaya, E.A. (2007). Phylogenetic systematics of microorganisms inhabiting thermal environments. *Biochemistry*, Vol.72, No.12, pp.1299-1312

López-Archilla, A.I.; Marín, I. & Amils, R. (2001). Microbial community composition and ecology of an acidic aquatic environment: the Tinto river, Spain. *Microbial Ecology*, Vol.41, pp.20-35

Ma, Y.; Galinski, E.A.; Grant, W.D.; Oren, A. & Ventosa, A. (2010). Halophiles 2010: life in saline environments. *Applied and Environmental Microbiology*, Vol.76, No.21, pp. 6971-6981

Merchant, S.; Prochnik, S.E.; Vallon, O. & Harris, E.H. (2007). The *Chlamydomonas* genome reveals the evolution of key animal and plant functions. *Science*, Vol.*318*, pp.245-251

Miller, S.R. & Castenholz, R.W. (2000). Evolution of thermotolerance in hot spring cyanobacteria of the genus *Synechococcus*. *Applied and Environmental Microbiology*, Vol.66, pp.4222–4229

Morgan-Kiss, R.M.; Priscu, J.C; Pocock, T.; Gudynaite-Savitch, L. & Huner, N.P (2006). Adaptation and acclimation of photosynthetic microorganisms to permanently cold environments. *Microbiology and Molecular Biology Reviews*, Vol.70, No.1, pp.222–252

Nadeau, T.L.; Milbrandt, E.C. & Castenholz, R.W. (2001). Evolutionary relationships of cultivated Antarctic oscillatorians (cyanobacteria). *Journal of Phycology*, Vol.37, pp.650–654

Niyogi, D.K.; Lewis, W.M. & McKnight, D.M. (2002). Effects of stress from mine drainage on diversity, biomass, and function of primary producers in mountain streams. *Ecosystems*, Vol.5, pp. 554-567

Nixdorf, B.; Krumbeeck, H.; Jander, J. & Beulker, C. (2003). Comparison of bacterial and phytoplankton productivity in extremely acidic mining lakes and eutrophic hard water lakes. *Acta Oecologica*, Vol.24, pp.S281-S288

Nordstrom, D.K. & Southam, G. (1997). Geomicrobiology of sulphide mineral oxidation, In: *Geomicrobiology:Iinteractions Between Microbes and Minerals*, J.F. Banfield & K.H. Nealson (Eds.), 361-390, Mineralogical Society of America, Washington DC, USA

Nübel, U.; Bateson, M.M.; Madigan, M.T.; Kuhl, M. & Ward, D.M. (2001). Diversity and distribution in hypersaline microbial mats of bacteria related to *Chloroflexus* spp. *Applied and Environmental Microbiology*, Vol.67, pp.4365–4371

Papke, R.T., Ramsing, N.B., Bateson, M.M. & Ward, D.M. (2003). Geographical isolation in hot spring cyanobacteria. *Environmental Microbiology*, Vol. 5, No.8, pp. 650–659

Pinto, E.; Sigaud-Kutner, T.; Leitão, M.; Okamoto, O. (2003). Heavy-metal induced oxidative stress in algae. *Journal of Phycology*, Vol.*39*, pp.1008-1018

Platt, T.; Gallegos, C.L. and Harrison, W.G. (1980). Photoinhibition of photosynthesis in natural assemblages of marine phytoplankton. *Journal of Marine Research*, Vol. 38, pp.687-701

Pointing, S.B.; Chan, Y.; Lacap, D.C.; Lau, M.C.; Jurgens, J.A. & Farrell, R.L. (2009).Highly specialized microbial diversity in hyper-arid polar desert. *Proceedings of the National Academy of Science USA*, Vol.106, No.47, pp.19964-19969

Quail, PH. (1991). Phytochrome: a light-activated molecular switch that regulates plant gene expression. *Annual Reviews in Genetic*, Vol.*25*, pp.389-409

Ritchie, R.J. (2008). Fitting light saturation curves measured using modulated fluorometry. *Photosynthesis Research*, Vol.96, pp.201-215

Roberts, D.M.L. (1999). Eukaryotic cells under extreme conditions, In: *Enigmatic Microorganisms and Life in Extreme Environments*, J. Seckbach (Ed.), 165-173, Kluwer Academic Publ., London, England

Rothschild, L.J. & Mancinelli, R.L. (2001). Life in extreme environments. *Nature*, Vol.409, pp. 1092-1101

Seckbach, J. (1994). Evolutionary pathways and enigmatic algae: *Cyanidium caldarium* (*Rhodophyta*) and related cells. In: *Developments in Hydrobiology*, Seckbach, J. (Ed.), 245-255, Kluwer Academic Publ., Dordrecht

Souza-Egipsy, V.; Altamirano M.; Amils R. & Aguilera, A. (2011). Photosynthetic performance of phototrophic biofilms in extreme acidic environments. *Environmental Microbiology*, DOI: 10.1111/j.1462-2920.2011.02506.x

Spijkerman, E.; Barua, D.; Gerloff-Elias, A.; Kern, J.; Gaedke, U. & Heckathorn, S.A. (2007). Stress responses and metal tolerance of *Chlamydomonas acidophila* in metal-enriched lake water and artificial medium. *Extremophiles*, Vol.11, pp.551-562

Stal, L.J. (2000). *Cyanobacterial mats and stromatolites. The Ecology of Cyanobacteria: Their Diversity in Time and Space*, B.A. Whitton & M. Potts, (Eds.), Kluwer Academic Press, Dordrecht

Stibal, M., Elster, J., Šabacká, M. & Kaštovská, K. (2006). Seasonal and diel changes in photosynthetic activity of the snow alga *Chlamydomonas nivalis* (*Chlorophyceae*) from Svalbard determined by pulse amplitude modulation fluorometry. *FEMS Microbiology and Ecology*, Vol.59, pp.265-273

Takamura, N.; Kasai, F.; Watanabe, M.M. (1989). Effects of Cu, Cd and Zn on photosynthesis of fresh water benthic algae. *Journal of Applied Phycology*, Vol.1, pp.39-52

Tang, E.P.Y.; Tremblay, R. & Vincent, W.F. (1997). Cyanobacterial dominance of polar freshwater ecosystems: are high-latitude mat-formers adapted to low temperature? *Journal of Phycology*, Vol.33, pp.171-181

Tang, K.H.; Barry, K.; Chertkov, O.; Dalin, E.; Han, C.S.; Hauser, L.J.; Honchak, B.M.; Karbach, L.E.; Land, M.L.; Lapidus, A.; Larimer, F.W.; Mikhailova, N.; Pitluck, S.; Pierson, B.K.; Blankenship, R.E. (2011). Complete genome sequence of the filamentous anoxygenic phototrophic bacterium *Chloroflexus aurantiacus*. *BMC Genomics*, Vol.12, No.1, pp.334

Taton, A.; Grubisic, S.; Balthasart, P.; Hodgson, D.A.; Laybourn-Parry, J. & Wilmotte, A. (2006). Biogeographical distributionand ecological ranges of benthic cyanobacteria inEastAntarctic lakes. *FEMS Microbiology and Ecology*, Vol.57, No.2, pp. 272-289

Valente, T.M. & Gomes, C.L. (2007). The role of two acidophilic algae as ecological indicators of acid mine drainage sites. *Journal of Iberian Geology*, Vol. 33, pp.283-294

Vincent, W.F.; Castenholz, R.W.; Downes, M.T. & Howard-Williams, C. (1993). Antarctic cyanobacteria: light, nutrients, and photosynthesis in the microbial mat environment. *Journal of Phycology*, Vol.29, pp.745-755

Vincent, W.F. (2000). Cyanobacterial dominance in the polar regions. In: *The Ecology of Cyanobacteria*, B.A. Whitton & M. Potts (Eds.), 321-340, Kluwer Academic Publishers, Dordrecht, The Netherlands

Vincent, W. F.; Gibson, J.A.; Pienitz, R.; Villeneuve, V.; Broady, P.A.; Hamilton, P.B. & Howard-Williams, C. (2000). Ice shelf microbial ecosystems in the high arctic and implications for life on snowball earth. *Naturwissenschaften*, Vol.87, pp.137-141

Vinocur, A. & Pizarro, H. (1995). Periphyton flora of some lotic and lentic environments of Hope Bay (Antarctic Peninsula). *Polar Biology*, Vol.15, pp.401-414

Vinocur, A. & Pizarro, H. (2000). Microbial mats of twenty-six lakes from Potter Peninsula, King George Island, Antarctica. *Hydrobiologia*, Vol.437, pp.171-185

Wang, S.; Chen, F.; Sommerfeld, M. & Hu, Q. (2004). Proteomic analysis of molecular response to oxidative stress by the green alga *Haematococcus pluvialis* (Chlorophyceae). *Planta*, Vol.220, pp.17-29

Zakhia, F.; Jungblut, A.D.; Taton, A.; Vincent, W.F. & Wilmotte, A. (2007). Cyanobacteria in cold environments. In: *Psychrophiles: from Biodiversity to Biotechnology*, R. Margesin, F. Schinner & J.C. Marx (Eds,), 121–135, Springer-Verlag, Berlin, Germany

13

Effect of 5-Aminolevulinic Acid (ALA) on Leaf Diurnal Photosynthetic Characteristics and Antioxidant Activity in Pear (*Pyrus Pyrifolia* Nakai)

Ming Shen, Zhi Ping Zhang and Liang Ju Wang
College of Horticulture, Nanjing Agricultural University,
Nanjing
China

1. Introduction

Photosynthesis is the basis of fruit growth and development. The higher photosynthetic efficiency of tree leaves, the more photosynthate is accumulated, which is beneficial to tree growth, root development, flower bud initialization, and the ultimate guarantee of quality and yield of fruits.

5-Aminolevulinic acid (ALA) is a key precursor of all porphyrin compounds, such as chlorophyll (Chl), heme, and phytochrome (von Wettstein et al., 1995). Exogenous application of ALA at low concentrations was found to promote growth and yield of several crops and vegetables (Hotta et al., 1997a). It also improved chlorophyll content and gas exchange capacity of melon seedlings under low light and chilling conditions (Wang et al., 2004), increased CO_2 fixation in the light, and suppressed the release of CO_2 in darkness (Hotta et al., 1997a), and promoted salt tolerance of cotton plants by manipulating the Na^+ uptake (Watanabe et al., 2000). ALA-based fertilizer also enhanced the photosynthetic rate, chlorophyll content, and stomatal conductance in spinach and date palm seedlings under salinity (Nishihara et al., 2003; Youssef and Awad, 2008). However when the plant was treated with exogenous ALA at high concentrations (such as ≥ 1000 mg/L), it was assumed that the induced chlorophyll intermediate accumulation acted as a photosensitizer for the formation of $^1O_2$, triggering photodynamic damage in ALA-treated plants (Chakrabory et al., 1992). Thus, ALA could be used as a natural bioherbicide.

ALA has been suggested to be a new natural and environmental friendly regulator, which can be widely used in agriculture (Wang et al., 2003). However, whether it can be used in woody trees such as pear has not been reported, and the mechanisms of ALA regulation on plant growth have not yet been elucidated. In the work, we found ALA promotion on pear photosynthesis might be related with the increase of antioxidant enzyme activities, and well as H_2O_2, which might act as signaling molecules involved in the regulation process.

2. Materials and methods

2.1 Plant growth and treatment

The experiment was started in the late of May 2010 in the Horticultural Experimental Station, Nanjing Agricultural University, Jiangsu Province China. Ten years old pear trees (*Pyrus pyrifolia* Nakai. 'Akemizu') were used, which were grew in the brown yellow soil with space of 4x5m. ALA solution in concentration of 0.5 mg $\cdot L^{-1}$ was sprayed to the leaves, with the distilled water as the control. The experimental trees were arranged in complete random design, with ten tree repeats for ALA treatment or control, respectively. The measurements of the gas exchange and chlorophyll fast fluorescence parameters were conducted one week after ALA spray at an interval of 2 hours from morning to dusk. And meanwhile, the leaf samples were taken and stored in liquid nitrogen for subsequent analysis.

2.2 Gas exchange parameters

The measurements of diurnal gas exchange parameters were carried out according to the method described by Wang et al. (2004) with a portable photosynthesis system CIRAS-2 (PP Systems, UK). At each time point, from 6:00 am to 18:00 pm, the net photosynthetic rate (P_n), intercellular CO_2 concentration (C_i) and stomata conductance (G_s) were measured simultaneously under the photon flux densities (PFD) from a built-in light source equal to the natural light intensities with the ambient temperatures. Each measurement was conducted at least 10 times, and the means were used to compare the ALA's effect.

2.3 Chlorophyll a fast fluorescence and JIP test

Chlorophyll fast fluorescence transient was measured by a Plant Efficiency Analyzer (Hansatech, UK), according to methods of Strasser et al. (1995) and Sun et al. (2009a). All the leaves were immediately exposed to a saturating light pulse (3000 mmol $m^{-2} s^{-1}$ PFD) for 1 s after dark adapted for 20min. Each transient obtained from the dark-adapted samples was analyzed according to the JIP-test (Srivastava et al., 1997; Li et al., 2005; Sun et al., 2009a).

2.4 Determination of Rubisco initial activity and its coding gene expression

Rubisco activity was determined spectrophotometrically by monitoring NADH oxidation at A340 (Lilley et al., 1974; Xia et al., 2009). Leaf samples were homogenized in a chilled mortar within the ice-cold extraction buffer solution, 40mM Tris–HCl (pH 7.6), which contained 10mM $MgCl_2$, 0.25mM EDTA, 5mM glutathione, 2% β-mercaptoethanol and 1.5% PVP (W/V). The homogenate was centrifuged at 4℃ for 15 min at 10,000g. The resulting supernatant was used for assay of the enzyme. The reaction mixture contained 100 mM Tris–HCl buffer solution, which contained 12 mM $MgCl_2$ and 0.4 mM EDTA (pH 7.8), 0.2 M $NaHCO_3$, 5 mM NADH, 50 mM ATP, 50 mM creatine phosphate, 1 U of glyceraldehyde 3-phosphodehydrogenase and 1 U of 3-phosphoglycerate kinase. The activity was estimated after the addition of enzyme extract and 0.2 mM ribulose-1, 5- bisphosphate (RuBP). Enzyme activity was expressed as mmol CO_2 fixed per min g FW.

To measure the gene expression, leaf RNA was extracted according to Louime et al. (2008). Firstly, a washing buffer, which contained 100 mM Tris-boric acid (pH 7.4), 0.35 mol/L sorbitol and 10% (w/v) PEG-6000, was added to remove the secondary material of samples in the free state. Then an extraction buffer, containing 0.25 M Tris-Boric acid (pH 7.4), 0.05 M EDTA, 2.5 M NaCl, 2% CTAB, 3% PVP, and 5% β-mercaptoethanol was used. The yield and quality of total RNA were measured by absorbance at 230, 260, and 280 nm (A_{260}/A_{230} and A_{260}/A_{280} ratios) using UV-spectrophotometer and by running samples on a 1.5% non-denaturing agarose gel electrophoresis. Message RNAs in total RNA solution were reversed transcribed to their complementary cDNA (I strand) by oligo-dT primer using MLV reverse transcriptase Kit (Takara Bio) according to the manufacturer's recommendations.

To amplify the cDNA produced from RNA by the RT reaction, PCR was performed according to the protocol. As a template, the RT product was used. According to published pear *Actin* (GenBank: GU830958.1) and *Rubisco small subunit* sequences (GenBank: D00572.1), two pairs of oligonucleotide primers were designed for expression analysis. Gene-specific primers for *Actin* (forward: 5′-CAATGTGCCTGCCATGTATG-3′; reverse: 5′-CCAGCAGCTTCCATTCCAAT-3′) and for *Rubisco small subunit* (forward: 5′-CTTGGAATTTGAGTTGGAGAC-3′; reverse: 5′-GTAA GCGATGAAACTGATGC-3′) were used in RT-PCR. Each pair of primers cycling parameters were *Actin*: 29 cycles, Tm 51℃ and *Rubisco small subunit*: 32 cycles, Tm 59℃. PCR products were analyzed following electrophoresis on a 1% agarose gel containing ethidium bromide.

2.5 Determination of antioxidant enzymes

One hundred milligrams of leaves were homogenized in 2 ml of 50 mM phosphate buffer (pH 7.8) which contained 0.4 % polyvinyl pyrrolidone (PVP), an inhibitor of phenolic compounds in a pre-chilled mortar and pestle on ice. The homogenate was centrifuged at 10 000×g for 20 min at 4 °C and the supernatant was collected as crude enzyme extraction (Tan et al., 2008).

Superoxide dismutase (SOD, EC 1.15.1.1) activity was measured by monitoring the inhibition of nitro blue tetrazolium (NBT) reduction at 560 nm. The reaction mixture (3 ml) contained 195 mM methionine, 1.125 mM NBT, 3 μM EDTA, and 100 μl of enzyme extract in 50 mM PBS (pH 7.8). After addition of 20 μM riboflavin, the cuvettes were exposed to a 15-W circular "white light" tube for photoreaction 10 min. Then the reaction mixture was measured as absorbance of 1 cm cuvette at 560 nm. One unit of SOD activity was defined as the amount of enzyme per fresh mass sample causing 50 % inhibition of the photochemical reduction of NBT (Beauchamp et al., 1971).

Ascorbate peroxidase (APX, EC 1.11.1.11) activity was assayed by measuring the oxidation of ascorbate at 290 nm according to Nakano and Asada (1981). Total 3 ml of reaction solution contained 50 mM PBS (pH 7.0), 1 mM H_2O_2, and 1mM ascorbate. The reaction was started by adding 100 μl of enzyme extraction. Changes of absorbance at 290 nm were then recorded within 3 min after the start of the reaction at 1 min intervals.

Catalase (CAT, EC 1.11.1.6) activity was determined according to the method of Zavaleta-Mancera et al. (2007). The total reaction mixture (3 mL) contained 50 mM PBS pH 7.0 and 100 μL of enzyme extract. The reaction was initiated by the addition of 10 mM H_2O_2. The decomposition was followed directly by the decrease in absorbance at 240 nm every 20 s for 3 min.

2.6 Determination of hydrogen peroxide (H_2O_2) and malondialdehyde (MDA)

Hydrogen peroxide content was estimated through the formation of a titanium-hydro peroxide complex (Agarwal et al., 2005). One hundred milligram of leaf sample was ground with liquid nitrogen and the fine powdered material was mixed with 2 ml cooled acetone. Then the mixture was centrifuged at 10,000×g for 10 min and the supernatant was collected as hydrogen peroxide extraction. One milliliter of the extract was added by 0.1 ml 5% titanium sulfate and 0.2 ml ammonium solution to precipitate the titanium-hydro peroxide complex. Reaction mixture was centrifuged at $5,000 \times g$ for 10 min in the centrifuge and the precipitate was washed 5 times by cooled acetone. Precipitate was dissolved in 10 ml 2 M H_2SO_4 and then recentrifuged. Supernatant was read at 415 nm against reagent blank in UV-spectrophotometer. The hydrogen peroxide content was calculated by comparing with a standard curve drawn with known hydrogen peroxide concentrations.

The MDA content was measured following the method of Heath et al. (1968) with modification and expressed as nmol per g of fresh weight. Five hundred milligram of frozen powder was added to about 5 ml of 5% trichloroacetic acid (TCA) and centrifuged at $10,000 \times g$ for 5 min. Two milliliter aliquot of supernatant was added to 2 ml of 0.67% 2-thiobarbituric acid (TBA). The mixture was incubated in boiling water for 30 min and then quickly cooled in an ice bath. After centrifugation at $10,000 \times g$ for 10 min, A_{532}, A_{600} and A_{450} of the supernatant were recorded. MDA content was estimated by the formula C (μmol/L) $=6.45(A_{532}-A_{600})-0.56A_{450}$.

2.7 Statistical analysis

All data were subjected to ANOVA test and the means were compared by the Duncan's test. Comparisons with $p<0.05$ were considered significant difference. Pearson correlation analysis between parameters was performed to test for relationships between variables by SPSS. 13 software.

3. Results

3.1 Diurnal variation of air temperature and light intensity in the orchard

On the testing day, the weather was fine when the temperature and photon flux density (*PFD*) exhibited a single peak curve (Fig.1). In the morning (6:00 am), the air temperature

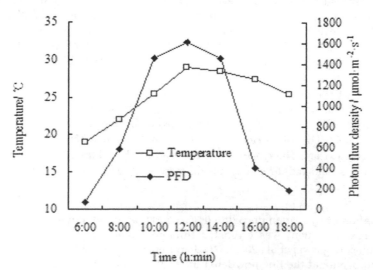

Fig. 1. Diurnal variations of air temperature and light intensity in pear orchard

was about 19℃,which linearly increased to the maximum temperature about 29℃ at noon
(12:00), then decreased to 25.5℃ at dusk (6:00 pm). In the aspect of light intensity, the PFD
was 65.6 μmol·m⁻²·s⁻¹ in the morning, and the maximum was 1612 μmol·m⁻²·s⁻¹ at noon.

3.2 Effect of ALA on gas exchange characteristics in pear leaves

The measurement of diurnal variations of leaf gas exchange characteristics showed that the
net photosynthetic rate (P_n) possessed a twin-peaks curve (Fig. 2A) and ALA treatment
significantly increased P_n of pear leaves compared with the control, especially at noon time.
P_n/C_i, representing the instantaneous carboxylation efficiency, exhibited a single peak curve
in diurnal variation (Fig. 2B), where ALA treatment generally promoted P_n/C_i of pear
leaves, especially at noontide. Changes in stomatal conductance was similar with P_n, and
ALA treatment promoted stomtal open in most of day time (Fig. 2C). However, there was
no difference in the intercellular CO_2 concentrations of pear leaves between control and
treatment (Fig. 2D), suggesting that 150 μmol/mol CO_2 in the experiment did not limit
photosynthesis in pear leaves.

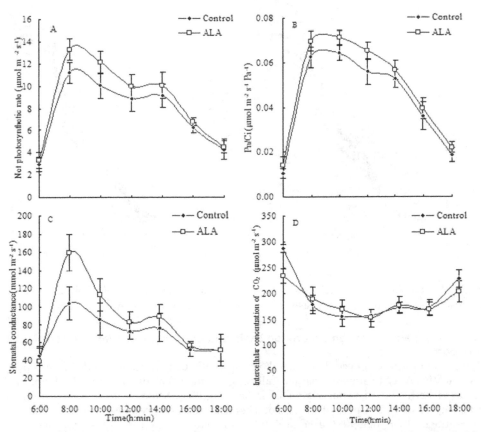

Fig. 2. Effect of ALA on diurnal variations of gas exchange parameters of pear leaves. A: Net
photosynthetic rate; B: Instantaneous carboxylation efficiency; C: Stomatal conductance; D:
Intercellular CO_2 concentration

3.3 Effects of ALA treatment on chlorophyll fast fluorescence characteristics
3.3.1 Fast induction curves of chlorophyll fluorescence

Fig. 3 displays the fast fluorescence transients measured from 6:00 am to 18:00 pm in the control and ALA-treated pear leaves. There were many differences between OJIP curves at different time, however, the most important difference was found from the P-step, which was the highest at 6:00 am, then decreased to the lowest at 12:00 at noon, and recovered to higher levels in the afternoon. The valley of P level at noon might be a characteristic of photosynthetic midday nap, or photoinhibition under high light condition. From Fig. 3, it can be seen that the P was generally higher in ALA-treated leaves than that in the control, especially at noontide, which may suggest that ALA treatment was favorable to leaf photosynthesis against photoinhibition under high light stress.

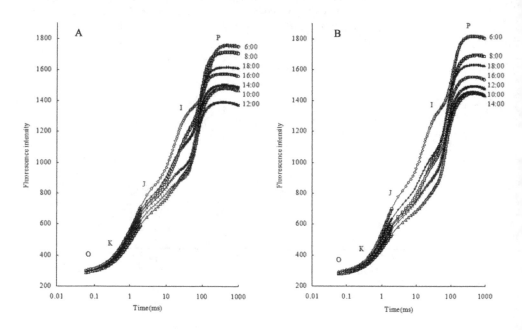

Fig. 3. Effect of ALA on diurnal variations of fast induction curves of chlorophyll a fluorescence (OJIP curve) of pear leaves. A: control, B: ALA treatment

3.3.2 Performance index on absorption basis and performance index of electron transport

The result of JIP-test showed that PI_{ABS}, the photosynthetic performance index on absorption basis, in the ALA treatment was generally higher than that of the control, although the trends of both diurnal variations were similar (Fig. 4A). Moreover, P_{ET}, the performance index of electron transport of PSII photochemical reaction was also higher in the ALA pretreatment than that of the control (Fig. 4B), suggesting that ALA treatment could improve photochemical electron transport and photosynthesis. The diurnal means of PI_{ABS} and P_{ET} were about 38% and 26% higher in the ALA treatment than that of the control, respectively.

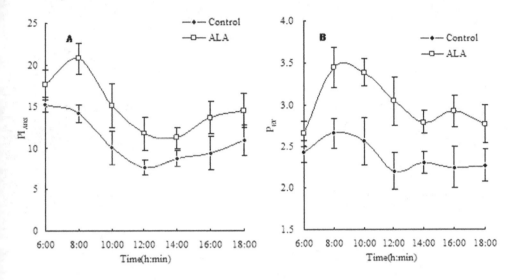

Fig. 4. Effect of ALA on diurnal variations of performance index on absorption basis (A) and
performance due to electron transport probability (B) of pear leaves

3.3.3 Flux ratios of PSII photochemical reaction

Fig. 5 is the results of photosynthesis of pear leaves from the perspective of flux ratios of
PSII photochemical reaction. There were differences in diurnal variation curve of φP_o, Ψ_o,
φE_o and φD_o.

Fig.5A shows that ALA treatment significantly increased the maximal photochemical
efficiency of PSII ($\varphi P_o \equiv F_v/F_m$) of pear leaves compared with the control. The highest of φP_o
in diurnal variation curves was found at 8:00, and dropped to the lowest at 12:00, and then
recovered in the afternoon, suggesting that the high light at noon time depressed the
maximal photochemical efficiency of PSII, and ALA treatment was prone to alleviate the
photoinhibtion.

In Fig. 5B, Ψ_o, a parameter of the PSII acceptor-side which means the possibility of a trapped
exciton moves an electron into the electron transport chain beyond Q_A^-, was generally
higher in ALA-treated leaves than that of the control, and the diurnal means in the former
was 6.5% higher than that of the latter.

In Fig. 5C, φE_o, another parameter of quantum yield for electron transport exhibited obvious
diurnal variation in the day, which was lowest at 12:00 of noon time. ALA treatment
significantly improved φE_o of pear leaves, and the diurnal means of the former was 7%
higher than that of the latter.

Conversely, a single peak curve was found in the aspect of energy dissipation through heat
(φD_o), which reached the highest at 12:00, and ALA treatment depressed φD_o, suggesting
that ALA decreased non-photochemical energy dissipation in the pear leaves (Fig.5D).

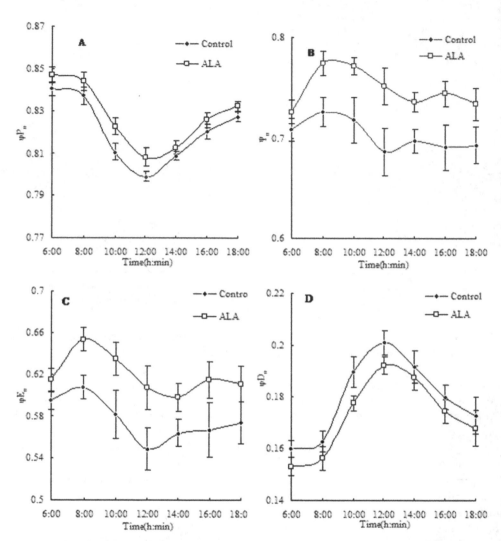

Fig. 5. Effect of ALA on diurnal variations of flux ratios of PSII photochemical reaction of pear leaves. A: Maximum quantum yield for primary photochemistry (φP_o), B: Probability that a trapped exciton moves an electron into the electron transport chain beyond Q_A^- (Ψ_o), C: Quantum yield for electron transport (φE_o), D: Quantum yield for heat dissipation (φD_o).

3.3.4 Activity of donor side and acceptor side of PSII reaction

Amplitude of the K step (W_k) as a parameter of the PSII donor-side, expresses the inactivity of the oxygen evolving complex (OEC). The smaller of the W_k, the stronger of the OEC activity is. As in the Fig. 6A, the diurnal variation of W_k showed a plateau from 10:00 am to 16:00 pm in the control, which was obviously higher than that of ALA treatment. This suggests that ALA treatment alleviated OEC inactivity at high light environment.

M_o, an approximate slope at the origin of the fluorescence rise, represents the maximum rate Q_A reduction. From Fig.6A, M_0 of pear leaves rose gradually in the morning and kept high level in the afternoon. However, ALA treatment significantly reduced M_0, which was about 80% of the control in the diurnal mean, suggesting that ALA could decrease Q_A reduction rate in pear leaves.

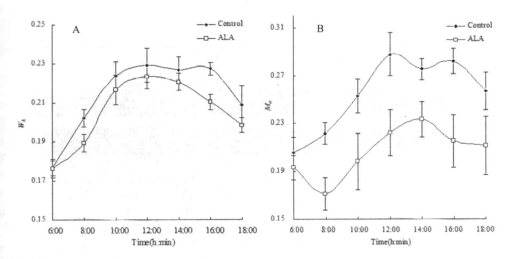

Fig. 6. Effect of ALA on diurnal variations of donor side and acceptor side parameter of PSII reaction of pear leaves. A: Amplitude of the K step (W_k); B: An approximation of the slope at the origin of the fluorescence rise (M_o)

3.4 Rubisco initial activity and RT-PCR analysis

The diurnal variation of the Rubisco initial activity of pear leaves showed a bimodal curve, where the first maximum occurred at 8:00 am, and the second at 16:00 pm. A significant valley occurred at noontide (Fig. 7A). In most cases, ALA treatment stimulated the activity, compared with the control. From the result of RT-PCR of the coding gene (Fig.7B and C), it can be seen that expression of *Rubisco small subunit* gene in pear leaves also revealed a bimodal curve, which was similar with the change of the enzyme activity in Fig. 7A. The relative expression was significantly higher in ALA-treated leaves than that of the control, especially at 8:00, which was more than 2 times. Therefore, ALA treatment improved the expression of *Rubisco small subunit* gene at transcript level.

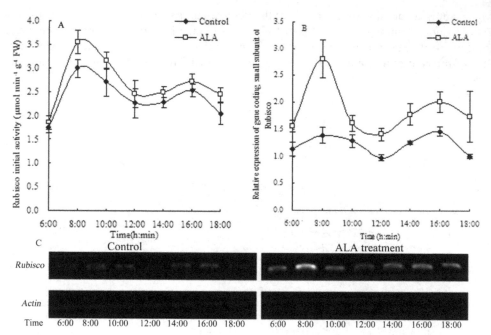

Fig. 7. Effect of ALA on diurnal variations of ribulose-1, 5-bisphosphate carboxylase oxygenase (Rubisco) initial activity (A) and relative expression of *Rubisco small subunit gene* (B) of pear leaves. C is electrophorogram of the gene expression detected by RT-PCR.

3.5 Effect of ALA treatment on the H_2O_2 and MDA content

There is a difference in the H_2O_2 and MDA content in pear leaves between ALA treatment and control (Fig. 8). The H_2O_2 content in ALA treated-leaves maintained at a relatively

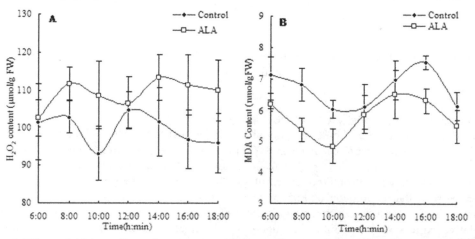

Fig. 8. Effect of ALA on diurnal variations of H_2O_2 and MDA content of pear leaves. A: H_2O_2 content; B: MDA content

stable higher level in one day than that of the control. The diurnal mean of the former was
about 10% higher that of the latter, suggesting that ALA treatment could increase the H_2O_2
content of pear leaves (Fig. 8A).

As for the MDA content of pear leaves, it revealed a wave daily variation (Fig.8B). The
lowest in the diurnal variation curve was at about 10:00 am and the highest at 16:00 pm.
ALA pretreatment decreased the content of MDA, and the significant differences were
found in the morning and afternoon. At noon time, the MDA content in the ALA-treated
leaves was also slightly higher than that of the control, however, the difference was not
significant at $P=0.05$ level.

3.6 Activities of antioxidant enzymes

The diurnal variation of SOD, APX and CAT activities in response to ALA treatment are
shown in Fig. 9. In SOD, the diurnal variation was a two-peak curve, where the first big
peak was recorded at 8:00 am and the second small one at 16:00 pm (Fig. 9A). Compared
with the control, SOD activities in ALA-treated leaves were generally increased, especially
in the early morning and evenfall.

Different with SOD, no peak could be found in the diurnal variation of APX activities of
pear leaves (Fig. 9B). Instead, it was lowest at noontide, but kept higher levels in the
morning or afternoon. ALA treatment significantly stimulated the enzyme activity in all
day time, and the diurnal mean in the ALA-treated was 37% higher than that of the
control.

The diurnal variation of CAT activities in pear leaves was similar with a sine curve, which
exhibited a peak at 8:00 am and valley at noontide, and then recovered to earlier levels (Fig.
9C). At any time, the activities in ALA-treated leaves were generally higher than that of the
control, suggesting that ALA improved the CAT activity in pear leaves.

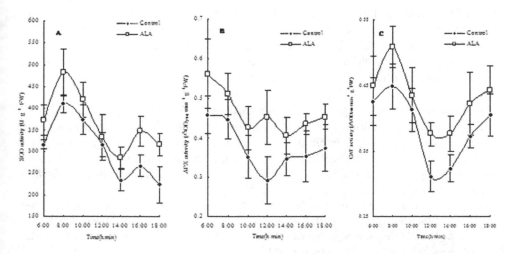

Fig. 9. Effect of ALA on diurnal variations of antioxidant enzymes activities of pear leaf. A:
SOD; B: APX; C: CAT

3.7 Analysis of correlation between environmental factors, photosynthetic parameters and antioxidant activities

The results of Pearson correlation analysis by SPSS 13.0 showed that there were a lot of high correlations between environmental factors, gas exchange, chlorophyll fast fluorescence, antioxidant activity and Rubisco initial activity in pear leaves, significant at either the 0.05 level or the 0.01 level (Table 1).

Firstly, the light intensity (PFD) was positively correlated with air temperature, P_n, P_n/C_i, W_k and φD_0, but negatively correlated with C_i, φP_0 and CAT activity, which means that high light intensity led to increase of temperature, net photosynthetic rate and instantaneous carboxylation efficiency, and meanwhile, OEC was possibly inactivated and the absorbed energy dissipation through heat was increased. Additionally, the intercellular CO_2 concentration was prone to decrease, and the maximal photochemical efficiency and CAT activity were also inhibited under high light stress.

Secondly, air temperature was positively correlated with W_k, M_0, and φD_0, but negatively with C_i, PI_{ABS}, φP_0, APX and CAT activity. This may mean that high temperature led to inactivate both donor and acceptor sides of PSII reaction center in pear leaves. Therefore, photosynthetic performance index on absorption basis and the maximal photochemical efficiency decreased under high temperature. Additionally, the activities of two H_2O_2 eliminating enzymes APX and CAT decreased as temperature increased.

Among the gas exchange characteristics, P_n was closely related with P_n/C_i and G_s, but negatively with C_i, which means that under the experimental condition, C_i was beyond a limited factor for photosynthesis. P_n was also positively with P_{ET}, SOD and Rubisco initial activity. Therefore, the electron transport activity of PSII reaction center, $O_2{}^{.-}$ scavenging enzyme SOD activity and the enzyme for CO_2 fixation were key for net photosynthetic rate.

Among the fluorescence indexes, PI_{ABS} was positively correlated with P_{ET}, Ψ_0, φP_0 and φE_0. Meanwhile, PI_{ABS} was also highly correlated with the activities of SOD, APX and CAT. Additionally, it was correlated with the relative expression of *Rubisco small unit* coded gene ($P=0.005$).

The initial activity of Rubisco was significantly correlated with P_n, P_n/C_i, P_{ET}, Ψ_0, φE_0, expression of the coding gene, and SOD activity, which means that $O_2{}^{.-}$ scavenging enzyme activity and photochemical electron transfer activity were closely related with the CO_2 fixation enzyme activity. Furthermore, the expression of gene coding *Rubisco small unit* was highly correlated with PI_{ABS}, P_{ET}, Ψ_0, φE_0, H_2O_2 content and three antioxidant enzyme activity.

H_2O_2, a famous reaction oxygen species, was correlated with P_{ET}, Ψ_0, φE_0, the activities of SOD, APX and CAT, the Rubisco initial activity and the relative expression of *Rubisco small unit* coded gene, which might imply the ROS was a active factor for photosynthesis of pear leaves under the experimental condition. MDA, a lipid peroxidation product, was the only parameter, negatively correlated with P_{ET}, Ψ_0, φE_0, and SOD activity, which might be an adverse factor for photosynthesis.

Among the antioxidant enzymes, APX and CAT activities were negatively correlated with PFD and (or) temperature, although the coefficient between APX and PFD was -0.518, slight higher than the 0.05 level, which means the enzymes was affected by environmental factors. Nevertheless, SOD activity was not affected by PFD or temperature. SOD was significantly correlated with P_n, PI_{ABS}, P_{ET}, φ_0, φE_0, APX, CAT, Rubisco initial activity, and the relative expression of *Rubisco small unit* coded gene, but negatively with M_0 and MDA content,

Effect of 5-Aminolevulinic Acid (ALA) on Leaf Diurnal Photosynthetic Characteristics and
Antioxidant Activity in Pear (Pyrus Pyrifolia Nakai)

283

	Temperature	P_n	P_n/C_i	G_s	C_i	PI_{ABS}	P_{ET}	W_k	Ψ_o	M_o	φP_o	φD_o	φE_o	Rubisco	Gene expression	H_2O_2	MDA	SOD	APX	CAT
PFD	0.660*	0.692*	0.791**	0.425	-0.743*	-0.507	0.130	0.754**	0.107	0.359	-0.832**	0.831**	-0.221	0.267	-0.158	0.110	-0.228	0.008	-0.518	-0.563*
Temperature		0.306	0.424	0.016	-0.759*	-0.745*	-0.131	0.921**	-0.152	0.635*	-0.882**	0.885**	-0.451	0.122	-0.143	0.134	-0.013	-0.468	-0.706*	-0.722*
P_n			0.979**	0.918**	-0.707**	0.043	0.558*	0.430	0.522	-0.096	-0.312	0.310	0.315	0.835**	0.401	0.320	-0.378	0.540*	-0.130	0.056
P_n/C_i				0.831**	-0.804**	-0.10	0.490	0.556*	0.457	0.018	-0.462	0.460	0.205	0.772**	0.289	0.258	-0.358	0.470	-0.233	-0.077
G_s					-0.423	0.359	0.677**	0.105	0.633*	-0.352	0.013	-0.014	0.528	0.870	0.593	0.379	-0.475	0.683	0.112	0.354
C_i						0.440	-0.199	-0.824**	-0.173	-0.364	0.695**	-0.694**	0.110	-0.570*	-0.130	-0.150	0.213	-0.124	0.460	0.367
PI_{ABS}							0.699**	-0.826**	0.707**	-0.952**	0.864**	-0.864**	0.901**	0.336	0.698*	0.436	-0.426	0.730**	0.910**	0.915**
P_{ET}								-0.217	0.995**	-0.804**	0.299	-0.297	0.924**	0.708**	0.768**	0.675**	-0.725**	0.758**	0.594*	0.640*
W_k									-0.247	0.747**	-0.921**	0.922**	-0.544*	0.191	-0.288	-0.133	0.071	-0.383	-0.824**	-0.738**
Ψ_o										-0.828**	0.320	-0.318	0.936**	0.673**	0.760**	0.685**	-0.701**	0.750**	0.632*	0.654*
M_o											-0.745**	0.744**	-0.952**	-0.331	-0.663**	-0.542*	0.517	-0.722**	-0.910**	-0.867**
φP_o												-1.00**	0.633*	0.050	0.427	0.113	-0.062	0.443	0.821**	0.857**
φD_o													-0.632	-0.050	-0.426	-0.112	0.062	-0.447	-0.821**	-0.857**
φE_o														0.575	0.785**	0.604*	-0.599*	0.782**	0.820**	0.854**
Rubisco															0.688**	0.396	-0.473	0.692**	0.091	0.438
Gene expression																0.671**	-0.399	0.612*	0.556*	0.646*
H_2O_2																	-0.441	0.352	0.395	0.238
MDA																		-0.548*	-0.281	-0.418
SOD																			0.560*	0.726**
APX																				0.793**

* Correlation is significant at the 0.05 level (2-tailed).
** Correlation is significant at the 0.01 level (2-tailed).

Table 1. Correlations between environmental factors, gas exchange, chlorophyll fast fluorescence, antioxidant activity and Rubisco initial activity in pear leaves

suggesting that SOD was not only important in prevention of lipid peroxidation, but also in prevention PSII reaction center close, and therefore promotion of photochemical electron transfer and photosynthetic dark reaction.

Additionally, the correlations of APX and CAT activities were similar with the SOD in the most parameters. This means that three antioxidant enzymes synergized in eliminating reaction oxygen species to prevent peroxidation of lipid in plant cells. However, the correlations of MDA with APX and CAT were not significant ($P>0.05$), implying the role of APX and CAT activity was not enough to impact lipid peroxidation, i.e, the H_2O_2 level was not adverse in the experimental condition.

4. Discussion

Previous reports have demonstrated that ALA treatment can improve the net photosynthetic rate in spinach (Nishihara et al., 2003), melon (Wang et al., 2004), pakchoi (Wang et al., 2004b), radish (Hotta et al., 1997b; Wang et al., 2005b), strawberry (Liu et al., 2006) and watermelon (Sun et al., 2009a), under normal or stress conditions. The result of this work confirmed that exogenous ALA at a concentration of 0.5 mg/L could increase the net photosynthetic rate of pear leaves (Fig. 2A), which were related with the increase of P_n/Ci (Fig. 2B) and Gs (Fig.2C). This means that ALA treatment did not only promote stomatal opening, but also affect the non-stomatal factors related with photosynthesis (Farquhar and Sharkey, 1982). In other studies, exogenous ALA promotion on stomatal opening has been reported in melon (Wang et al., 2004a) or watermelon (Kang et al., 2006). A transgenic tobacco, which could over-produce endogenous ALA, also possessed higher stomatal conductivity (Zhang et al., 2010). Therefore, ALA inducing stomatal conductance might be a universal phenomena. However, the mechanism need further to be elucidated.

The shape of the OJIP transient has been found to be sensitive to stress such as excess light, temperature and drought (Appenroth et al., 2001; Thach et al., 2007). In our data (Fig. 3), the P-step in pear leaves was significantly lower at noon than that in the morning or at dusk, suggesting the environmental factors at noon impaired chlorophyll fast fluorescence yield. Wang et al. (2005a) suggested that the optimal temperature for photosynthesis of pear leaves was about 27°C. In this work, the highest temperature was 29°C (Fig. 1), which was near to the theoretic optimal temperature, and might not be the key inhibitory factor for pear photosynthesis. Instead, the typical midday nap characteristic of pear leaves was possible to be result of the high light intensity at noontide (Fig. 2, Fig.3). ALA treatment increased the fluorescence yield, especially at noontide, suggesting that ALA could promote resistance of pear leaves against high light stress. Sun et al. (2008) also found that ALA treatment could alleviate photoinhibition of watermelon seedlings switched to high light from shaded condition. Our results here were similar with the previous observation. Because ALA treatment could also improve leaf photosynthesis of plants grown under low light condition (Wang et al., 2004; Sun and Wang, 2007), it can be deduced that ALA might stabilize photosynthetic capacity against light stresses. In fact, this effect of ALA is important because plants are not always grown under optimal light intensity condition, and low light or high light often affects plant photosynthesis. ALA improvement on photosynthesis under light stresses can increase photosynthate accumulation in many crops.

The mechanisms of ALA improvement on photosynthesis by non-stomata have been mentioned in many aspects. Firstly, ALA increased the chlorophyll content, since it is the key biosynthetic precursor of all tetrapyrrole compounds (von Wettstein et al., 1995; Jahn and Heinz, 2009), which has been suggested to contribute to increase of photosynthesis (Tanaka et al., 1993). However, in the most cases, the chlorophyll content was not a limiting factor for leaf photosynthesis.

ALA has been suggested to increase the activity of OEC at the donor side of PSII reaction center under stress condition. Sun et al. (2009b) found that W_k, which represented the inhibition of OEC activity, was lower in ALA-treated leaves than that in the control of watermelon seedlings under chilling stress. Zhang et al. (2010) observed that W_k in the transgenic tobacco with capacity to over- produce endogenous ALA was also lower than that of the wild type. In the work, W_k in ALA-treated pear leaves was always lower than that of the control (Fig. 6A). Thus, ALA promotion in OEC activity might be a general effect. ALA might improve the photochemical efficiency of PSII reaction center. Whether the dark-adapted or light-adapted maximal photochemical efficiency, it has been reported that ALA had significant effect (Sun et al., 2009a; Wang et al., 2010). Recently, the similar effect was confirmed in the transgenic tobacco (Zhang et al., 2010). In this work, we observed that φP_0 of pear leaves treated by ALA was higher than that of the control (Fig. 5A), suggesting the maximal photochemical efficiency was improved.

ALA might also improve the activity of acceptor side of PSII reaction center. Two important fluorescence parameters M_0 and ψ_0, where the M_0 represents the proximate rate of Q_A completely being reduced, and ψ_0 was the probability of a trapped exciton moves an electron into the electron transport chain beyond Q_A^-, often responded to ALA treatment. In most situations, ALA decreased M_0 but stimulated ψ_0, which was beneficial to electron transfer through Q_A^- electron acceptor of PSII reaction center (Strasser et al., 1995; Li et al., 2005). The results in the work approved the previous observations that Q_A was retardant to be completely reduced (Fig. 6B) and the electron was easily transferred to the downstream electron acceptors beyond Q_A^- in the chain after ALA treatment (Fig. 5B).

Liu et al. (2010) suggested that ALA treatment alleviated the decrease of Rubisco activity of cucumber under suboptimal temperature and light intensity stress. In this work, we found the diurnal variation of Rubisco initial activity in pear leaves (Fig. 7A), which was improved by ALA treatment and highly correlated with P_n (r=0.835, $P<10^{-5}$). It was the first time to observe that ALA could up-regulate transcription of gene coding Rubisco small unit in pear leaves (Fig. 7). The level of transcript in ALA-treated leaves at 8:00 am was more than 2 times as high as that of the control, which means that ALA treatment did not only affect light reaction of photosynthesis, but also dark reaction, even the expression of the key enzyme. The effect of ALA has been not mentioned before. However, the mechanism of ALA regulation on gene transcription needs to be elucidated further.

In aspect of antioxidant enzymes, it has been suggested that ALA treatment stimulated SOD activity around PSI reaction center, which can scavenge ROS aroused from photosynthetic electron transport in electron transfer chain to improve photochemical electron transfer rate (Sun et al., 2009a, b). Diethyldithiocarbamate (DDC), an inhibitor of Cu-Zn-SOD, could eliminate ALA' effect on photochemical efficiency (Liu et al., 2006; Sun et al., 2009a), which suggested the important role of SOD on ALA promotion. In this work, ALA treatment also induced SOD activity in pear leaves (Fig. 9A), which was positively correlated with many photosynthesis and chlorophyll fluorescence parameters but negatively with M_0 and the MDA content (Table 1), suggesting that it might play an important role on the acceptor side activity of PSII reaction center and preventing lipid peroxidation of photosynthetic apparatus during daytime. On the other hand, Jung et al. (2008) found higher levels of SOD activity in transgenic rice than in the wild type. The same was true in transgenic tobacco and *Arabidopsis* (Wang et al., unpublished). Thus, enhancement of SOD activity was accompanied with increase of endogenous or exogenous ALA levels. However, the mechanism of SOD activity induced by ALA has no been known.

That ALA induced the increase of enzymes eliminating H_2O_2, such as APX, CAT and POD, has been suggested (Nishihara et al., 2003; Liu et al., 2006). Since ALA is the essential biosynthetic precursor of tetrapyrrole compounds including heme, and the latter is a necessary component for the activity of all three enzymes (Tsiftsoglou et al., 2006), it is reasonable to deduce that ALA induces heme accumulation, which is beneficial for H_2O_2 eliminating enzyme activity. In this work, it was also observed that ALA induced increase of activities of APX and CAT in pear leaves (Fig. 9). However, ALA treatment also increased the content of H_2O_2 in pear leaves, which might be at a safe level, because MDA content with ALA treatment was significantly lower than the control (Fig. 8). In Table 1, H_2O_2 level was correlated with P_{ET}, Ψ_0, φE_0, the activities of SOD, APX, CAT and Rubisco initial activity, as well as transcript of gene coding *Rubisco small unit*. This means that H_2O_2 might be an active signal molecule rather than an adverse ROS, involved in regulation of antioxidant enzyme activity and physiological or molecular processes. H_2O_2 has been suggested as a cellular signal, and has wide-ranging effects in many biological processes (Finkel and Holbrook, 2000). It can also regulate gene expression in plants (Neill et al., 2002; Apel and Hirt, 2004). However, whether ALA promotion on plant photosynthesis is dependent on H_2O_2 signal need further study.

5. Acknowledgement

The work was financially supported by the Priority Academic Program Development of Jiangsu Higher Education Institutions (PAPD).

6. References

Agarwal S, Sairam RK, Srivastava GC, Tyagi A, Meena RC. Role of ABA, salicylic acid, calcium and hydrogen peroxide on antioxidant enzymes induction in wheat seedlings. Plant Sci, 2005, 169:559-570

Apel K, Hirt H. Reactive oxygen species: metabolism, oxidative stress, and signal transduction. Ann Rev Plant Biol, 2004, 55: 373-399

Appenroth KJ, Stockel J, Srivastava A, Strasser RJ. Multiple effects of chromate on the photosynthetic apparatus of *Sprirodela polyrhiza* as probed by OJIP chlorophyll a fluorescence measurements. Environ Pollu, 2001, 115:49-64

Beauchamp CO, Fridovich I. Superoxide dismutase: Improved assays and an assay applicable to acrylamide gels. Anal Biochem, 1971, 44:276-287

Chakrabory N, Tripathy BC. Involvement of singlet oxygen in 5-aminolevulinic acid induced photodynamic damage of cucumber chloroplast. Plant Physiol, 1992, 98: 7-11

Farquhar GD, Sharkey TD. Stomatal conductance and photosynthesis. Ann Rev Plant Physiol, 1982, 33: 317- 345

Finkel T, Holbrook NJ. Oxidants, oxidative stress and biology of ageing. Nature, 2000, 408:239-247

Heath RL, Packer L. Photoperoxidation in isolated chloroplasts: I. Kinetics and stoichiometry of fatty acid peroxidation. Arch Biochem Biophys 1968, 125: 189-198

Hotta Y, Tanaka T, Takaoka H, Takeuchi Y, Konnai M. New physiological effects of 5-aminolevulinic acid in plants: the increase of photosynthesis, chlorophyll content, and plant growth. Biosci Biotech Biochem, 1997b, 61:2025-2028

Hotta Y, Tanaka T, Takaoka H, Takeuchi Y, Konnai M. Promotive effects of 5-aminolevulinic acid on the yield of several crops. Plant Growth Regul, 1997a, 22:109-114

Jahn D, Heinz DW. Biosynthesis of 5-aminolevulinic acid. In: Tetrapyrroles: Birth, Lift and Death, ed by Warren M, Smith AG. Landes Bioscience and Springer Science+Business Media. 2009, 29-42

Jung S, Back K, Yang K, Kuk YI, Chon SU. Defence response produced during photodynamic damage in transgenic rice overexpressing 5-aminolevulinic acid synthase. Photosynthetica, 2008, 46: 3-9

Kang L, Cheng Y, Wang LJ. Effects of 5-aminoluveulinic acid (ALA) on the photosyntehsis and anti-oxidative enzyme activity of leaves of greenhouse watermelon in summer and winter. Acta Bot Boreal-Occident Sin, 2006, 26: 2297-2301

Li PM, Gao HY, Strasser RJ. Application of the fast chlorophyll fluorescence induction dynamics analysis in photosynthesis study. J Plant Physiol Mol Biol, 2005, 31: 559-566

Lilley RM, Walker DA. An improved spectrophotometric assay for ribulose-bisphosphate carboxylase. Biochim Biophys Acta, 1974, 358: 226-229

Liu WQ, Kang L, Wang LJ. Effect of 5-aminolevulinic acid (ALA) on photosynthesis and its relationship with antioxidant enzymes of strawberry leaves. Acta Bot Boreal Occident Sin, 2006, 26:57-62

Liu YM, Ai XZ, Yu XC. Effects of ALA on photosynthesis of cucumber seedlings under suboptimal temperature and light intensity. Acta Hort Sin, 2010, 37: 65-71

Louime C, Vasanthaiah HKN, Jittayasothorn Y, Jiang L, Basha SM, Thipyapong P, Boonkerd N. A simple and efficient protocol for high quality RNA extraction and cloning of chalcone synthase partial cds from muscadine grape cultivars (Vitis Rotundifolia Michx.). Euro J Sci Res, 2008, 22:232-240

Nakano Y, Asada K. Hydrogen peroxide is scavenged by ascorbate specific peroxidase in spinach chloroplasts. Photochem Photobiol, 1981, 37: 679-690

Neill S, Desikan R, Hancock J. Hydrogen peroxide signaling. Curr Opin Plant Biol, 2002, 5: 388-395

Nishihara E, Kondo K, Parvez MM, Takahashi K, Watanabe K, Tanaka K. Role of 5-aminolevulinic acid (ALA) on active oxygen-scavenging system in NaCl treated spinach (Spinacia oleracea). Plant Physiol, 2003, 160: 1085- 1091

Srivastava A, Guissé B, Greppin H, Strasser RJ. Regulation of antenna structure and electron transport in PSII of Pisum sativum under elevated temperature probed by the fast polyphasic chlorophyll a fluorescence transient: OKJIP. Biochim Biophys Acta, 1997, 1320: 95-106

Strasser RJ, Srivastava A, Govindjee. Polyphasic chlorophyll-alpha fluorescence transient in plants and cyanobacteria. Photochem Photobiol, 1995, 61: 32-42

Sun YP, Wang LJ. Effects of 5-aminolevulinic acid (ALA) on chlorophyll fluorecence dynamics of watermelon seedlings under shade condition. Acta Hort Sin, 2007, 34: 901-908

Sun YP, Wei ZY, Zhang ZP, Wang LJ. Protection of 5-aminolevulinic acid (ALA) on high light photoinhibition of watermelon grown under shade condition. Acta Bot Boreal-occident Sin, 2008, 28: 1384-1390

Sun YP, Zhang ZP, Wang LJ. Promotion of 5-aminolevulinic acid treatment on leaf photosynthesis is related with increase of antioxidant enzyme activity in watermelon seedlings grown under shade condition. Photosynthetica, 2009a, 47(3):347-354

Sun YP, Zhang ZP, Xu CX, Shen CM, Gao C, Wang LJ. Effect of ALA on fast chlorophyll fluorescence induction dynamics of watermelon leaves under chilling stress. Acta Hort Sin, 2009b, 36: 671-678

Tan W, Liu J, Dai T, Jing Q, Cao W, Jiang D. Alterations in photosynthesis and antioxidant enzyme activity in winter wheat subjected to post-anthesis water-logging. Photosynthetica, 2008, 46:21-27

Tanaka Y, Tanaka A, Tsuji H. Effects of 5-aminolevulinic acid on the accumulation of chlorophyll b and apoproteins of the light-harvesting chlorophyll a/b-protein complex of photosystem II. Plant Cell Physiol, 1993, 34: 465–472

Thach LB, Shapcott A, Schmidt S, Critchley C. The OJIP fast fluorescence rise characterizes *Graptophyllum* species and their stress responses. Photosynth Res, 2007, 94:423-436

Tsiftsoglou AS, Tsamadou AI, Papadopoulou LC. Heme as key regulator of major mammalian cellular functions: Molecular, cellular, and pharmacological aspects. Pharmacol Therapeut, 2006, 111: 327-345.

Von Wettstein DV, Gough S, Kannangara CG. Chlorophyll biosynthesis. Plant Cell, 1995, 7: 1039-1057

Wang LJ, Jiang WB, Huang BJ. Promotion of 5-aminolevulinic acid on photosynthesis of melon (*Cucumis melo*) seedling under low light and chilling stress conditions. Physiol Plant, 2004a, 121:258-264

Wang LJ, Jiang WB, Gao GL, Han HZ, Kuang YL, Liang SQ. Studies on leaf photosynthesis of young pear trees with various cultivars. Acta Hort Sin, 2005a, 32: 571-577

Wang LJ, Jiang WB, Zhang Z, Yao QH, Matsui H, Ohara H. Biosynthesis and physiological activities of 5-aminolevulinic acid (ALA) and its potential application in agriculture. Plant Physiol Commun, 2003, 39:185-192

Wang LJ, Liu WQ, Sun GR, Wang JB, Jiang WB, Liu H, Li ZQ, Zhuang M. Effect of 5-aminolevulinic acid on photosynthesis and chlorophyll fluorescence of radish seedlings. Acta Bot Boreal-Occident Sin, 2005b, 25: 488- 496

Wang LJ, Shi W, Liu H, Liu WQ, Jiang WB, Hou X L. Effects of exogenous 5-aminolevulinic acid treatment on leaf photosynthesis of pak-choi. J Nanjing Agri Univer, 2004b, 47:1084-1091

Wang LJ, Sun YP, Zhang ZP, Kang L. Effects of 5-aminolevulinic acid (ALA) on photosynthesis and chlorophyll fluorescence of watermelon seedlings grown under low light and low temperature conditions. Acta Hort, 2010, 856: 159-166

Watanabe K, Tanaka T, Hotta Y, Kuramochi H, Takeuchi Y. Improving salt tolerance of cotton seedlings with 5-aminolevulinic acid. Plant Growth Regul, 2000, 32:99-103

Xia XJ, Huang LF, Zhou YH, Mao WH, Shi K, Wu JX, Asami A, Chen ZX, Yu JQ. Brassinosteroids promote photosynthesis and growth by enhancing activation of Rubisco and expression of photosynthetic genes in *Cucumis sativus*. Planta, 2009, 230:1185-1196

Youssef T, Awad MA. Mechanisms of enhancing photosynthetic gas exchange in date palm seedlings (*Phoenix dactylifera* L.) under salinity stress by a 5-aminolevulinic acid-based fertilizer. J Plant Growth Regul, 2008, 27:1-9

Zavaleta-Mancera HA, López-Delgado H, Loza-Tavera H, Mora-Herrera M, Trevilla-García C, Vargas-Suárez M, Ougham H. Cytokinin promotes catalase and ascorbate peroxidase activities and preserves the chloroplast integrity during dark-senescence. J Plant Physiol, 2007, 164: 1572-1582

Zhang ZP, Yao QH, Wang LJ. Expression of yeast *Hem1* controlled by *Arabidopsis HemA1* promoter enhances leaf photosynthesis in transgenic tobacco. Mol Biol Rep, 2011, 38: 4369-4379

Permissions

The contributors of this book come from diverse backgrounds, making this book a truly international effort. This book will bring forth new frontiers with its revolutionizing research information and detailed analysis of the nascent developments around the world.

We would like to thank Mohammad Mahdi Najafpour, for lending his expertise to make the book truly unique. He has played a crucial role in the development of this book. Without his invaluable contribution this book wouldn't have been possible. He has made vital efforts to compile up to date information on the varied aspects of this subject to make this book a valuable addition to the collection of many professionals and students.

This book was conceptualized with the vision of imparting up-to-date information and advanced data in this field. To ensure the same, a matchless editorial board was set up. Every individual on the board went through rigorous rounds of assessment to prove their worth. After which they invested a large part of their time researching and compiling the most relevant data for our readers. Conferences and sessions were held from time to time between the editorial board and the contributing authors to present the data in the most comprehensible form. The editorial team has worked tirelessly to provide valuable and valid information to help people across the globe.

Every chapter published in this book has been scrutinized by our experts. Their significance has been extensively debated. The topics covered herein carry significant findings which will fuel the growth of the discipline. They may even be implemented as practical applications or may be referred to as a beginning point for another development. Chapters in this book were first published by InTech; hereby published with permission under the Creative Commons Attribution License or equivalent.

The editorial board has been involved in producing this book since its inception. They have spent rigorous hours researching and exploring the diverse topics which have resulted in the successful publishing of this book. They have passed on their knowledge of decades through this book. To expedite this challenging task, the publisher supported the team at every step. A small team of assistant editors was also appointed to further simplify the editing procedure and attain best results for the readers.

Our editorial team has been hand-picked from every corner of the world. Their multi-ethnicity adds dynamic inputs to the discussions which result in innovative outcomes. These outcomes are then further discussed with the researchers and contributors who give their valuable feedback and opinion regarding the same. The feedback is then collaborated with the researches and they are edited in a comprehensive manner to aid the understanding of the subject.

Apart from the editorial board, the designing team has also invested a significant amount of their time in understanding the subject and creating the most relevant covers. They scrutinized every image to scout for the most suitable representation of the subject and create an appropriate cover for the book.

The publishing team has been involved in this book since its early stages. They were actively engaged in every process, be it collecting the data, connecting with the contributors or procuring relevant information. The team has been an ardent support to the editorial, designing and production team. Their endless efforts to recruit the best for this project, has resulted in the accomplishment of this book. They are a veteran in the field of academics and their pool of knowledge is as vast as their experience in printing. Their expertise and guidance has proved useful at every step. Their uncompromising quality standards have made this book an exceptional effort. Their encouragement from time to time has been an inspiration for everyone.

The publisher and the editorial board hope that this book will prove to be a valuable piece of knowledge for researchers, students, practitioners and scholars across the globe.

List of Contributors

Mohammad Mahdi Najafpour and Sara Nayeri
Chemistry Department, Institute for Advanced Studies in Basic Sciences, (IASBS), Zanjan, Iran

Bahar Ipek and Deniz Uner
Middle East Technical University, Turkey

Brandon D. Burch and Cindy Putnam-Evans
East Carolina University/Biology Department, USA

Terry M. Bricker
Louisiana State University/Department of Biological Sciences, USA

Cristian Gambarotti, Lucio Melone and Carlo Punta
Department of Chemistry, Materials and Chemical Engineering "Giulio Natta", Politecnico di Milano, Italy

Mohammad Mahdi Najafpour
Chemistry Department, Institute for Advanced Studies in Basic Sciences (IASBS), Zanjan, Iran

David Iluz, Irit Alexandrovich and Zvy Dubinsky
The Mina and Everard Goodman Faculty of Life Sciences, Bar-Ilan University, Ramat-Gan, Israel

David Iluz
The Department of Geography and Environment, Bar-Ilan University, Ramat-Gan, Israel

Takayuki Kajikawa and Shigeo Katsumura
Kwansei Gakuin University, Japan

María José Quiles, Helena Ibáñez and Romualdo Muñoz
Department of Plant Biology, Faculty of Biology, University of Murcia, Murcia, Spain

John Paul Délano Frier and Axel Tiessen
Unidad de Biotecnología e Ingeniería Genética de Plantas, Cinvestav - Irapuato, Guanajuato, México

Carla Vanessa Sánchez Hernández
Centro Universitario de Ciencias Biológicas y Agropecuarias, Universidad de Guadalajara, Zapopan, Jalisco, México

Mohamed Mohamed Ibrahim
Botany and Microbiology Department, Faculty of Science, Alexandria University, Egypt

Yulia Pinchasov-Grinblat and Zvy Dubinsky
The Mina & Everard Goodman Faculty of Life Sciences, Bar-Ilan University, Ramat-Gan, Israel

Angeles Aguilera and Ricardo Amils
Centro de Astrobiología (INTA-CSIC), Spain

Virginia Souza-Egipsy
Centro de Investigaciones Agrarias (CSIC), Spain

Ricardo Amils
Centro de Biología Molecular (UAM-CSIC), Spain

Ming Shen, Zhi Ping Zhang and Liang Ju Wang
College of Horticulture, Nanjing Agricultural University, Nanjing, China

Printed in the USA
CPSIA information can be obtained
at www.ICGtesting.com
JSHW011502221024
72173JS00005B/1169